UNDERSTANDING

SCIENTIFIC

REASONING

RONALD N. GIERE

*Professor of Philosophy of Science
Department of History and Philosophy
of Science, Indiana University*

HOLT, RINEHART and WINSTON

*New York   Chicago   San Francisco   Dallas
Montreal   Toronto   London   Sydney*

To the Memory of ALLEN JAMES HARDER
1942–1977

Material in Exercise 1.1 reprinted with permission. Copyright © 1977, The Courier-Journal.

Material in Exercise 6.4 reprinted with permission of W. H. Freeman and Company, Publishers, from the *Scientific American* article, "The Confirmation of Continental Drift" by Patrick M. Hurley, April 1968.

Figures 7.2 and 7.3 have been reprinted, by permission, from THE LIMITS TO GROWTH: A report for THE CLUB OF ROME's Project on the Predicament of Mankind, by Donnella H. Meadows, Dennis L. Meadows, Jorgen Randers, William H. Behrens III. A Potomac Associates book published by Universe Books, New York, 1972. Graphics by Potomac Associates.

Material in Exercise 8.5 reprinted with permission from *Chariots of the Gods* by E. von Daniken, Bantam edition. Copyright © 1968, Econ-Verlag GmbH. Translation copyright © 1969 by Michael Heron and Souvenir Press.

Material in Exercise 8.6 reprinted with permission of the World News Corporation from "Cliff Robertson tells of moment he saw UFO" in *The Star*, July 12, 1977, p. 11.

Material in Exercises 12.1, 12.2, and 12.3 reprinted with permission from *Vitamin C and the Common Cold* by Linus Pauling. Copyright © 1970 by W. H. Freeman and Company, Bantam edition.

Table 12.2 and Figures 12.11 and 12.12 have been reprinted, by permission, from *The Journal of the American Medical Association*, 116 (1958): 1159, 1295. Copyright 1958, American Medical Association.

Figures 12.5, 12.6, 12.7, 12.8, 12.9, and 14.14 have been reprinted by permission of the American Cancer Society.

Library of Congress Cataloging in Publication Data

Giere, Ronald N
  Understanding scientific reasoning.

  Bibliography: p. 364
  Includes index.
  1. Science—Philosophy. 2. Reasoning. I. Title.
Q175.G49   501'.8   78-31503
ISBN 0-03-044631-7

PREFACE

I have written this text to help beginning students to learn to evaluate and utilize scientific information. It can be used in any beginning course devoted in whole or in part to "scientific method." This includes standard courses such as logic (or practical reasoning), probability and induction (or inductive logic), and the philosophy of science. It is particularly appropriate for a variety of new courses in "applied philosophy," that is, courses entitled "Philosophy and X," where X might be medicine, engineering, technology, the environment, law, business, or even "the consumer." This text could also be used outside philosophy departments by teachers who are not professional philosophers—teachers in science, medicine, and engineering, or even business and criminal justice. Those who teach science writing to journalism students could certainly benefit from reading this text.

I began teaching "scientific reasoning" in the fall of 1969 as a direct response to students' demand for "relevance." It seemed to me that many of the issues of that time had a substantial scientific component and that one thing I could do was teach students how to deal intelligently with the scientific issues. The issues and examples changed, but the class grew from roughly 30 the first year to nearly 300 a semester in 1976–1977. These classes were composed of a majority of freshmen and sophomores, but included some juniors and seniors as well. The students' interests ranged literally from accounting to zoology, with large numbers from both business and biomedical sciences, such as medicine, nursing, and speech therapy. In 1977–1978 I used an earlier version of this text with two classes of roughly 250 students each. That experience is reflected in the present book.

I considered subtitling the book "A Consumer's Guide to Scientific Method," but rejected this idea as not being sufficiently serious. Studying this text should indeed help students become more critical consumers, but it emphasizes something more fundamental than the consumption of products, namely, the consumption of *scientific information*. As a society we are becoming more and more aware of the role of science and technology in our lives. One result of this increased awareness is that we are now bombarded with scientific information of all sorts. Almost every issue of any large newspaper features a report of some new scientific study, for example, on the dangers of air pollution or the benefits of jogging. People are expected to understand and make use of this information, and perhaps even to vote on related issues, as Californians voted on the use of nuclear power. It seems to be presumed that most people can use the information available to make intelligent decisions even though they are not specialists in nuclear engineering, environmental medicine, or any other science. Yet one need only read "Letters to the Editor" or the pronouncements of some U.S. Senators to realize that this presumption is often mistaken. Many people are willing to dismiss solid evidence that smoking and saccharin cause cancer and to treat reports of extraterrestrial visitors on a par with reports concerning the depletion of the ozone layer. The ability to evaluate and utilize scientific information is thus a skill that needs to be learned—and to be taught.

Can college freshmen and sophomores really learn to be intelligent consumers of scientific information? They can. This text exhibits one way of teaching such skills. I have employed concepts from logic, the philosophy of science, statistics, and decision theory to develop a unified framework that can accommodate a wide range of cases. The framework is developed in the context of specific examples which in turn provide models for analyzing new cases—a good selection of which appear as exercises. The secret of the approach is not to expect students to learn how to be scientists, but merely how to read and interpret *reports* of scientific findings, especially reports that appear in the popular magazines, newspapers, and even supermarket tabloids. The framework reveals that there are only a few basic patterns of argument that can justify scientific hypotheses. One does not need to be a specialist to learn to spot the patterns in diverse cases and to act accordingly.

The text itself has four parts. Part I provides motivation and introduces the basic logical concepts needed to begin talking about reasoning: statement, truth and falsity, argument, and justification. The main emphasis is on conditional statements and conditional arguments because they play so great a role in scientific reasoning.

Part II teaches one how to identify theories (and "models"), and to recognize statements that make use of theories. It also shows how nonspecialists can judge for themselves whether or not reported data justify the hypotheses. The techniques developed are applied both to standard sciences (for example, physics and biology), and to popular sciences and theories (for example, astrology and claims of clairvoyance or extraterrestrial visitation). One chapter is devoted to examining the arguments of the book *The Limits to Growth*.

Part III deals with statistical correlation and simple causal hypotheses, for example, that smoking causes cancer. It explains why a "statistically significant difference" may or may not be important. It shows how to identify different types of experimental designs from popular or semitechnical reports. And it explains why different types of designs provide different amounts of justification for causal claims.

Part IV shows how to judge the relevance of scientific findings, particularly statistical findings, to personal and public decisions, such as whether to give up smoking or to ban a certain drug.

Courses in scientific method are currently being taught in a variety of departments and programs. By and large, however, semester- (or quarter-) long courses in scientific method ceased being a general feature of the philosophy curriculum in the 1950s. Although interest in scientific method seems to be reviving, it is not to be expected that large numbers of such courses could be created, or recreated, overnight. If a book like *Understanding Scientific Reasoning* is to be immediately useful, it will have to be adapted to existing courses. I have not myself faced this problem, but I can offer some suggestions.

Logic   Most elementary logic texts conclude with a section on "induction" or "scientific method." That this material is often skipped is at least partly due to the lack of excitement and relevance in traditional treatments of the method of hypothesis, Mill's methods, and the types of probability. *Understanding Scientific Reasoning* provides a contemporary treatment of this material. It will complement what students meet in other elementary courses, for example, in biology, psychology, and sociology. Chapters 5 and 6 cover theories and theoretical hypotheses while Chapters 9, 10, 11, and 12 cover probability and simple causal hypotheses. This material can be taught in five to eight weeks, that is, half a quarter or one third to one half a semester. Those more concerned with applications might wish to substitute the material on decision making, Part IV, for that on theories.

It is important to realize that *Understanding Scientific Reasoning* can be used for the inductive part of a logic course regardless of whether the deductive part tends toward "formal" or "informal" techniques. Scientific reasoning is naturally less formal than formal deduction but more structured than informal deductive reasoning. In either case, therefore, the transition from the deductive part to the inductive part will not be severe.

Probability and Induction (Inductive Logic)   Most courses on probability, induction, and inductive logic are taught at an advanced level and deal primarily with philosophical issues such as the nature of probability and the justification of induction. These courses are the inductive analog of courses in the philosophy of logic and mathematics. *Understanding Scientific Reasoning* is most appropriate for the inductive analog of an elementary logic course that emphasizes reasoning skills over metatheoretical debate. In teaching such a course one need not choose between a course that is highly structured but irrelevant and one that is relevant but frustratingly unstructured. *Understanding Scientific Reasoning* makes possible a course that has sufficient content and structure to be readily taught, even to large numbers, and still be highly relevant to the concerns of most students. As the exercises in this text illustrate (see particularly Exercises 6, 11, and 12.), it is easy to find illustrations for lectures and questions for exams in one's morning paper. And projects for individuals or small groups of students lie waiting on the racks of paperbacks in any bookstore.

Philosophy of Science   Few courses on the philosophy of science are taught at the elementary level. *Understanding Scientific Reasoning* makes possible such a course. Many topics one would naturally want to discuss are covered. Chapter 5, for example, is really about the nature of theories. It employs a "semantic" rather than the more usual "syntactic" analysis of theories. Chapter 6 deals with the confirmation and falsification of theoretical hypotheses. It emphasizes the role of successful prediction in establishing the (approximate) truth of theoretical hypotheses. Chapter 9 develops the distinction between correlation and causation. The analysis of simple causal hypotheses is explicitly counterfactual. Chapter 11 explores the nature of the statistical evidence needed to establish the existence of a correlation. Chapter 12, which stresses randomized experimental designs, does the same for causal hypotheses. And Part IV emphasizes the role of values in decision making. These topics provide the basis for more philosophical discussion than one could fit into a single semester. Indeed, to make time for more discussion it might be necessary to skip some of the more applied sections of the text, such as Chapter 7 ("The Limits to Growth"), Chapter 8 ("Fallacies of Theory Testing"), and Part IV ("Values and Decisions").

My hope is that others will discover many additional uses for *Understanding Scientific Reasoning.* Developing this material has been a rewarding , though time-consuming and often difficult, experience. Perhaps others can now enjoy some of the rewards with less effort and difficulty.

December 1978

Ronald N. Giere
Indiana University

Acknowledgments

My first debt is to Michael Scriven who took off for the coast in 1966 but left behind a course entitled "Introduction to Scientific Reasoning." Knowing that such a course could be taught was crucial to my own efforts in this direction.

My second debt is to Allen Harder who was first Scriven's student, then mine. Allen went on to teach a version of "Scientific Reasoning" at Iowa State. We took some preliminary steps toward joint authorship of a text before he was forced to leave teaching by an eventually fatal brainstem tumor. His remarkable good nature never faltered.

Next I must thank the hundreds of students whose comments and questions over the years helped to shape both the course and this text. Without their support and encouragement, this book would not have been written.

Many thanks are also due the numerous graduate students who, in addition to grading papers and meeting discussion sections, helped me to improve the material. Among those now teaching the philosophy of science are Alberto Cortes, Thomas Rogers, and Linda Wessels.

A number of people read and commented on the first version of this text. Special thanks are due Jon Ringen and Linda Sartorelli who actually taught from the text and provided me with numerous suggestions. Fred Suppe's generosity with his time, energy, and encouragement was greatly appreciated. I also benefited from extensive comments by Peter Asquith, James H. Fetzer, and Roger Rosenkrantz. My colleague, Roger Buck, has supplied numerous suggestions for corrections in the final version.

The completed manuscript owes much to the expertise of Kim Rumple, who organized, typed and drew diagrams, often under tight deadlines; Lewis Gray, who read the whole final draft as it was being written; and Kevin Korb, who prepared the index.

Finally, I should like to thank my editor, Dave Boynton, whose willingness to try something a little different and whose gentle encouragement have been much appreciated.

CONTENTS

PART TWO Reasoning About Theories

7

The Limits to Growth *116*

8

Fallacies of Theory Testing *137*

PART THREE Causes, Correlations, and Statistical Reasoning

12

Testing Causal Hypotheses *247*

PART FOUR Values and Decisions

13

The Structure of Decisions *289*

14

Decision Strategies *305*

Part One is short and mainly preliminary. Chapter 1 begins with some examples illustrating the importance of learning the kinds of things you will be studying in Parts Two, Three, and Four. The second half of Chapter 1 explains why the basic concepts we shall use are not limited to any particular science, or even to science in general. They apply to all reasoning, both in science and in everyday life.

Chapter 2 focuses on statements, which are the basic units for communicating knowledge and for reasoning. Here you will learn about the concept of truth and its relations with other important concepts such as belief, knowledge, and certainty.

Chapter 3 shows how reasoning can be analyzed in terms of arguments, which are sets of statements. Justifying a belief or a claim to knowledge is then shown to depend on the use of good arguments for your claims. We shall distinguish two types of argument: deductive and inductive. Understanding scientific reasoning requires the use of both types, but especially inductive arguments.

Chapter 4 contains most of what you will need to know about deductive reasoning. We shall focus on a class of deductive arguments known as conditional arguments and also look briefly at several other useful types of deductive argument. By the end of Chapter 4 you will have all the tools you need to begin the study of full-fledged scientific reasoning.

PART ONE

Basic Concepts of Scientific Reasoning

1

Why Study Scientific Reasoning?

This chapter begins with a preliminary look at several examples that illustrate the kinds of things you should learn from studying this text. These examples are also intended to convince you that learning about scientific reasoning and decision making is worth your time and effort. In particular, they should show you how knowing something about scientific reasoning can be useful, if not essential, in your everyday life and work. The second part of the chapter explains the general approach we shall follow and gives some tips on the best ways for studying the text and making use of the exercises.

1.1 WHY STUDY SCIENTIFIC REASONING?

Is There Really an Energy Crisis?

In the early 1970s a little book entitled *The Limits to Growth* generated heated public discussion in the public media—television, newspapers, and magazines. The reason for excitement was clear because, as the headlines put it, the authors were predicting "The End of Civilization As We Know It" within roughly the next 100 years. Of course the argument of the book was more subtle than indicated by the headlines. It was, briefly, that attempting to maintain recent rates of growth would produce a world population so large that the resources to support it would simply run out before the year 2100. The result would be social chaos, including the death of billions of people from starvation and disease. But the authors did not simply predict that this catastrophe would occur. Indeed, they argued that we could prevent it *if* we acted soon enough to limit population and industrial growth.

Some critics of the book, especially economists, called it pure "garbage." They argued that existing social and economic mechanisms would prevent any such catastrophe without the imposition of deliberate policies to limit growth.

What is the average person who reads these debates to do? Most of us are

2

not specialists in economics or any of the other relevant fields. Yet the issue is too important simply to ignore. One's views on this problem are relevant to the most important personal decisions. How many children should I have? Indeed, should I have any at all? Should I take a job in a large city and commute to a house in the suburbs? Perhaps a job in a small town in an agricultural area would be a better bet. And increasingly we are forced to decide between candidates for public office partly on the basis of their positions on energy legislation and related issues.

One cannot avoid the problem simply by appealing to the authority of those who are supposed to know. There are authorities on both sides. More- over, both sides claim to base their conclusions on scientific "models" that are supported by "empirical facts." What does it mean to have a model of the world system? How can such models tell us what might happen? Why should we believe the conclusions of one model rather than another? What does it mean to claim that a model is supported by "the facts"? What counts as a "fact" anyway? We have recently experienced periodic shortages in various energy sources: gasoline, natural gas, and electricity. Do these "facts" support the position of *The Limits to Growth?* Or are these occurrences merely the workings of the natural mechanisms that the critics say will prevent any catastrophe? Or maybe these occasional shortages have no systematic explanation at all. Maybe they are just the accidental result of strikes by unions or price fixing by large oil companies.

We shall discuss *The Limits to Growth* in greater detail in Part II ("Rea- soning about Theories"), and techniques for understanding and dealing with this type of problem will be explained both in Part II and also in Part IV ("Values and Decisions"). At the moment let us look at two similar examples that illustrate different points.

Will the Universe Expand Forever?

One of the most interesting scientific findings of the twentieth century was the discovery that the universe is "expanding"; that is, the galaxies that make up the universe (of which our Milky Way is only one) are all moving away from each other. Most astronomers now think that the present phase of expan- sion began with all the matter in the universe exploding in one "big bang." The remaining question is whether the present expansion will continue forever, or whether it will stop and all the galaxies will then come back together for another "big bang" and another expansion. Increasingly, scientific results appear in the popular press before they are published in "official" scientific journals. Thus it is not surprising to find an article in the *New York Times* (March 12, 1978) with the headline "Scientists Expect New Clue to Origin of Universe." It appears that whether the universe will continue to expand or undergo a never-ending series of expansions and contractions depends on whether the number of atoms in the universe is greater or less than one for every 88 gallons of space. If it is greater, gravitational attraction will be strong enough to pull everything back together again. If it is less, the attraction will not be strong enough, and the expansion will go on forever. The reason this article appeared when it did was that new measurements of the strength of extragalactic Xrays by earth-orbiting satellites are about to give us the answer.

Now this type of scientific discovery has few immediate practical implications for anyone. By the time the next contraction is due to occur, our sun will have long since died out or exploded. But such findings may excite your curiosity, and they are thought to be relevant to other beliefs you may have— for example, the belief in the creation of the universe by a divine being. Moreover, since these findings are being written up for the general public, one ought to be able to give some evaluation of the reported results without being an astrophysicist. How can one identify the theories in question? How can one distinguish the theories from the facts? How can one tell which facts support which theories? And how strong is the support? Should we regard one of the theories as "proven"? And would either theory be relevant to a belief in divine creation? These again are the kinds of questions Part II teaches you to answer.

Are We Being Visited by Intelligent Aliens?

The popularity of books and movies like *The Bermuda Triangle* and *Close Encounters* indicates considerable interest in the possibility that the earth has been or is being visited by intelligent beings from other worlds. While many scientists think it highly probable that there is intelligent life elsewhere in the universe, most would scoff at the idea of actual visitations by such creatures. Yet to the nonscientist, such an idea may not seem any more farfetched than the claim that the universe is expanding or contracting.

Those who believe that at least some unidentified flying objects (UFOs) do contain intelligent aliens are able to cite lots of "facts" to support this belief. Why is it that scientists do not pay more attention to this "evidence"? What makes the belief in an expanding universe scientifically acceptable and the belief in extraterrestrial visitation unacceptable? A good account of scientific reasoning ought to shed some light on this question. Our account will do just that.

Does Cigarette Smoking Cause Cancer?

Now let us preview a few examples from Part III ("Causes, Correlations, and Statistical Reasoning"). It has been roughly 20 years since the surgeon general of the United States issued an official warning that cigarette smoking can be dangerous to your health. Since that time, cigarette advertising has been banned from television and warning labels have been placed on cigarette packs and printed advertisements. Most recently, regulations have been proposed banning smoking on commercial airplanes, and the Department of Health,

Education and Welfare has announced another campaign to warn teenagers of the dangers of smoking. Yet the Tobacco Institute, the research arm of the tobacco industry, maintains that the connection between smoking and lung cancer or any other disease is "merely statistical." No "causal" connection has been proved. Moreover, millions of people continue to smoke, and the federal government itself continues to subsidize both the growing and the exporting of tobacco.

If you are not a statistician or a medical researcher, what are you to conclude? Whom are you to believe? Is it true that all the data are merely statistical? What is the difference between a statistical correlation and a genuine causal connection? Could statistical data "prove" the existence of a causal connection? If not, why not? What kind of evidence could establish the existence of a causal connection? You should be able to answer these sorts of questions after studying Part III.

To continue, suppose the surgeon general is correct: Smoking does cause lung cancer. Does it follow that you should not smoke or that you should give it up if you already do? How do you determine the risks? Can you balance off the risks against the pleasure of smoking or the discomfort of trying to give it up? Are the concerns of public health officials necessarily relevant to individuals who may be concerned only about *their own* chances of contracting smoking-related diseases? Techniques for answering these sorts of questions are developed in Part IV.

What about Saccharin?

Similar types of cases appear in the popular press almost weekly. For example, only recently the Food and Drug Administration proposed a ban on the use of saccharin because of experiments that supposedly "proved" that saccharin causes cancer in mice. The public outcry from producers and users of saccharin was so great that Congress passed a temporary ban on the ban. Critics contended that experiments on mice cannot show that saccharin causes cancer in humans. Some critics even claimed that the experiments do not prove that it causes cancer in mice. And many people argue that one ought to be able to choose for oneself whether to use saccharin. The health establishment countered with new studies showing a connection between use of artificial sweeteners and bladder cancer in men, but, surprisingly, not in women. And the debate goes on.

From the standpoint of the nonspecialist who simply reads the local newspaper, what should you conclude? And what should you do about it? Can you be sure, from the published information, that saccharin does cause cancer in mice? Is this relevant to humans? What kind of a "connection" has been found between saccharin and cancer in humans? Does it "prove" causation? Why did this connection only hold for men and not for women? The principles and techniques presented in Part III should make it possible for you to sort out these issues—without being a specialist in any of the relevant fields.

Shouldn't Everyone Be Free To Choose?

Finally, the question of "freedom of choice" that arose in the saccharin controversy deserves special mention. This same idea has become a slogan among people who want the government to legalize the use of the drug laetrile in the treatment of cancer. It has also recently been advocated by the cosmetics industry in response to reports that some hair dyes cause skin cancer. The products should not be banned, they say. Rather, the packages should contain summaries of the known facts so that people can judge for themselves whether to use the product or not. This argument is bound to come up in similar cases in the future.

Now the idea that individual citizens should have the "freedom" to make their own choices has considerable appeal in a society that values individual freedom. But this idea presupposes that the majority of citizens have the ability to weigh the facts and make their decisions "rationally" in light of these facts. Otherwise the freedom to choose is merely the freedom to be exploited. And you have much more to lose than just money.

Why Study Scientific Reasoning and Decision Making?

The first answer is that it will make you better able to understand and utilize scientific information in both your personal life and in your work. In an increasingly scientific and technological society, these abilities can literally be a matter of life and death for you, your family, and, if you should achieve a position of power in business or government, for many others as well.

I have stressed the practical value of studying scientific reasoning and decision making because this is most readily appreciated by most people. But two other related motives are worth mentioning. One is simply the ability to understand and appreciate scientific findings as they are reported in the popular media. This can be a valuable ability whether or not the findings have any practical implications. Science is an increasingly important part of modern culture. Having some ability to understand and evaluate the latest important findings makes you a more literate and cultured person. So does just being curious about the world around you.

Finally, learning something about scientific reasoning provides some insight not only into particular scientific findings but also into the general nature of science as a human activity. It is an activity that engages an increasing fraction of our population, requires an increasing fraction of our resources, and impinges on an increasing number of our other activities. The better we all understand it, the better off we all will be.

1.2 HOW TO STUDY SCIENTIFIC REASONING

Assuming that you now want to learn about scientific reasoning, the next question is how best to do this. Here one must consider both the general strategy and some particular tactics for learning how to reason "scientifically."

A General Strategy for Studying Scientific Reasoning

Most people tend to associate the phrase "scientific reasoning" with the kind of reasoning that scientists use in the process of making new scientific discoveries. The paradigm examples of scientific reasoning would then be the reasoning that led to the great discoveries in science: Newton's discovery of the law of gravitation, Mendel's discovery of the laws of inheritance, the discovery of the "double helix" by Watson and Crick, and so on. If this is what one means by "scientific reasoning," however, it is difficult to see how anyone could learn how to do it without a great deal of training in science and perhaps a touch of genius as well. Those who are not specialists in a scientific subject could not hope to reason scientifically about that subject. Even specialists in one subject would find it difficult to reason scientifically about subjects other than in their own specialty. Our understanding of what constitutes "scientific reasoning" must be somewhat different.

Not only do scientists *make* discoveries, they must also *communicate* those discoveries to others and attempt to convince others that they are correct. Now the process of convincing others of what one has discovered is very different from the process of discovering. To see just how different one need only compare the published research reports of famous scientists with their memoirs or autobiographies. The autobiographies may tell of all the ideas that failed to pan out and all the experiments that came to nothing. They may tell of flashes of inspiration and even sleepless nights and days of frustration. None of this is in the published report. Here one finds only the final conclusions and the experimental results that support those conclusions. This is what other scientists use in judging whether the claimed discovery is indeed a genuine discovery and not just another idea gone wrong. And this published report is what those who write about science for the general public use as the basis for the articles that we all read.

For our purposes, then, "scientific reasoning" does not include all the reasoning that may go on in the process of making a scientific discovery. It includes only the reasoning that is present in the finished research report. To understand this reasoning one need not be able to do what the scientists did in arriving at their conclusions. Many of the other scientists who read a report could not do this. And in the case of highly significant new findings, few other scientists could have done it themselves. But most scientists can understand published reports. And so, with a little help, can almost anyone else.

By and large, only specialists read original research reports. Most people read popular versions or summaries written either by scientists or by people with at least some training in science writing. It is the reasoning present in this sort of report that we shall be most concerned with understanding.

Now that we have a better idea of what we shall mean by "scientific reasoning" we can go on to ask how one should begin learning about such reasoning. Since scientists write research reports, presumably one could learn the reasoning of a research report by learning to be a scientist. But this is not necessary for our purposes. Nor is it even the best way. One could not hope to learn more than one or two scientific subjects, and although there is a general pattern to all scientific reasoning, different subject matters lead to great differences in emphasis. So studying physics would not prepare you to deal with biomedical issues, and studying sociology would not prepare you to deal with new findings in physics. We want to be able to deal with all of these subjects, and more.

We shall take a more general view of the problem. A research report is a type of communication, and communication takes place in a language. Moreover, scientific reasoning is a type of reasoning. So we shall begin our study of scientific reasoning by looking at language and the use of language in reasoning of all kinds. This strategy has the virtue that it prepares us to deal with a report on any subject whatsoever. And, fortunately, the amount one needs to know about language and about reasoning in general is not great. Indeed, all you will need to know is contained in the following three chapters, which make up the rest of Part I.

Now that we are clear about the general strategy to be followed, we can turn to the tactics of how the subject should be studied.

Tactics for Studying Scientific Reasoning

Like a sport or physical activity, scientific reasoning is more a skill than a subject matter. There are few facts to learn—no names, dates, places, or parts of animals. The object here is not to acquire new information, but to learn how to appraise and utilize information. So you should not approach this course as you would a subject like genetics or English history. You should approach it more like you would tennis or the piano. Thus, the tactic for studying scientific reasoning can be summed up in one word: *practice.*

In order to provide you with the necessary practice, each chapter of this text is followed by a set of exercises. Many of these are slightly simplified versions of actual reports of scientific results taken from commonly available sources. So you will be practicing with examples very much like those you will meet in real life. But as in the case of learning to play tennis or the piano, practice is of little value if it is not done correctly.

The correct way to approach the exercises is this: First study the relevant sections of the text and related lecture notes until you think you understand the general principles. Then try working the exercises without looking back at the text or your notes. Try as many as you can. If all proceeds smoothly, turn to the answers in the back of the book to check those problems for which the answers are given. This will give you immediate feedback. If you have made no mistakes, fine. You are done with that chapter. If you have made some mistakes, be sure you see how you went wrong. If not, ask your instructor. You may be missing something important.

If you get stuck while doing a problem, spend 5 or 10 minutes thinking about it, but not longer. Go on to the next problem. Maybe it will give you a clue to doing the previous one. If after a while there are still some questions you could not figure out, go back to the text and your notes. Do not go to the back of the book just yet. See if you can discover what you are doing wrong from the text and your notes. If after an honest effort, that is, another 5 or 10 minutes of hard thinking, you still do not see what is wrong, check whether the answer is given and see whether it helps you. If not, be sure to ask your instructor. The material within a given part of the text tends to be cumulative. If you miss something important at the beginning of a part, you will likely be confused all the way through.

Keep in mind that the object of doing the exercises is not simply to get the right answer down on a sheet of paper. The object is to develop the skill that will enable you to do the exercises without having the text at hand. You will most likely not have access to the text during quizzes and examinations. More important, you will not have it when you need to evaluate scientific reasoning in day-to-day situations. And that is where the real payoff from developing reasoning skills should be.

Another implication of the fact that reasoning is a skill and not just information is that you must practice regularly. Do the exercises to each chapter as they are assigned. Do not let things go and then try to cram the night before a test. No one would dream of staying up all night before a tennis match or a piano recital. Staying up all night studying for a test of reasoning skills is almost as ludicrous. And it will not work.

Finally, like any skill, becoming very good at scientific reasoning requires both practice and talent. But becoming tolerably good requires mainly practice and only a little talent. And, for most people, tolerably good is good enough. So work at developing your skills little by little. By the time you have to face a quiz, you will just be doing what comes naturally. And you will be able to do it naturally in real life too.

CHAPTER EXERCISES

1.1 This is not really an exercise in the sense that it gives you practice at doing something you are supposed to have learned. Its purpose is just to allow you to see for yourself how well you can handle everyday sorts of "scientific" information. What follows is an excerpt from an article that appeared on the front page of the *Louisville Courier-Journal* (July 29, 1977). Read the excerpt. Then, based on the information given, indicate your attitude toward the statements that follow by writing in the letter, a, b, c, d, or e in the space provided.

  (a) Strongly inclined to agree.
  (b) Somewhat inclined to agree.
  (c) Cannot say. Insufficient information to form any opinion.
  (d) Somewhat inclined to disagree.
  (e) Strongly inclined to disagree.

When you have finished, turn to the answers at the back of the book for some comments on what your answers would be if you apply the principles developed in this text.

Full Moon Causes Some People To Go off Their Beams with Sexual Offenses

Not only do people like to spoon by the light of the silvery moon, they are also more inclined to commit rape, sexual assault, and indecent exposure when the moon is full.

The evidence of moon madness can be found in police statistics. A recent study by Dr. Ronald Holmes, a Jefferson Community College sociologist, found that reports of a range of sexual offenses increased significantly during a full moon.

Dr. Holmes analyzed the timing of the 1274 sexual crimes reported to the Louisville and Jefferson County Police Departments during 1974 and 1975. He found that 404 of those reported crimes (31.6% of the total) took place in the full moon cycle—82 more cases (6.8% more) than the next most frequent of the moon's four phases.

Dr. Holmes cautioned, however, that the study does not point to the strength of moonbeams as the primary cause of a sex crime. "There's no one explanation for any kind of behavior," he said. "A person who is inclined to behave in a certain way may be more inclined to behave that way when his brain is under added pressure from certain acids normally produced by the body." When the moon is full, the "bodily tides"—levels of body water—are at their highest and more of those pressure-creating acids are shoved through the brain, he said.

Dr. Holmes rejected the tenets of astrology or lunar-based religions. "It hasn't got anything to do with the position of the planets or any religion," he said. However, Nolan Meyers, an astrologist with whom Holmes consulted on the project, said he thinks that the signs of the zodiac are important in analyzing the effect of the moon on sexual violence. The effect of those full moons during a person's "sun sign" will be especially powerful, he said.

A.   In 1974–1975 there were more reports of sexual crimes in Louisville–Jefferson County area during full moons than during any of the other three phases of the moon.   _____

B.   In 1974–1975 there were "significantly" more reports of sexual crimes in the Louisville–Jefferson County area during full moons than during any of the other three phases of the moon.   _____

C. In 1974–1975 there were more *actual* sexual crimes in the Louisville–Jefferson County area during the full moon than during the other three phases. _____

D. In general there are more sexual crimes during the full moon than in other phases; that is, there will likely be more such crimes during full moons than during other phases this coming year in other cities. _____

E. The full moon *causes* some people to go off their beams with sexual offenses. _____

F. There are more sexual crimes during the full moon because of the stronger "bodily tides" that send pressure-creating acids to the brain. _____

G. The position of the planets has nothing to do with the higher percentages of sexual crimes during full moons. _____

H. The signs of the zodiac are important in analyzing the effect of the moon on sexual violence. _____

I. A person inclined to commit sexual crimes will be more inclined to do so during full moons associated with his sun sign. _____

J. In 1974–1975 one's chances of being the victim of a sexual crime in the Louisville–Jefferson County area were roughly 7 percent greater during full moons than during other times of the month. _____

K. Women should probably be a little more careful about going out alone during full moons than during other times of the month. _____

L. Women should avoid going out alone during full moons except in cases of extreme emergency. _____

1.2 Look through a few recent issues of any available newspapers or weekly newsmagazines until you find a report of some new scientific finding. Read it carefully. Ask yourself whether you really understand what is being claimed, and why. Can you see any relevance to decisions you might make? If you cannot follow the article, it might be your fault. But it might also be the fault of the person who wrote the news story. Often information that is available and would be helpful in understanding some new finding does not get reported because reporters sometimes do not themselves know which information is really relevant. If you understand scientific reasoning, you should be able yourself to spot gaps in such reports. As a special project, just for yourself, if not for an assignment, you might start keeping a file of reports that you come across—if only in your school newspaper. As the course progresses, you should find yourself better and better able to evaluate the reports in your collection. When you have finished the text you should be able to deal effectively with most of them.

2

Statements

For the purpose of studying scientific reasoning, the basic units of language are not words but whole statements. Moreover, there are only a few characteristics of statements that are relevant. This chapter, then, is devoted to explaining what statements are and to describing those few characteristics of statements relevant to our purposes. Chapter 3 will explain how reasoning may be analyzed in terms of statements.

2.1 STATEMENTS: TRUTH AND FALSITY

A sentence, as the grammar books tell us, is a meaningful set of words that expresses an assertion, question, command, request, or the like. Although scientists use sentences in all these ways, scientific reasoning is primarily concerned with those sentences that express assertions. And although it is sometimes useful to distinguish sentences from what they express, we shall use the term *statement* to refer either to a sentence that expresses an assertion or to the assertion that is expressed. For example, if I say, "The text of this chapter is ten pages long," I have used the sentence between the quotation marks to make a statement, namely, the statement that the text to this chapter is ten pages long.

The above explanation still leaves considerable ambiguity as to which sentences are statements and which are something else. Most of this ambiguity is removed by the following very important condition, which we may regard as part of the definition of statements:

Every statement is either true or false.

This condition clearly distinguishes statements ("The door is closed") from commands ("Close the door!"), from requests ("Please close the door"), from questions ("Is the door closed?"), and so on. Only of a statement does it make sense to ask, "Is it true?" Commands, requests, questions, and so forth may be many things, but not true or false.

Having made the ability to be true or false part of the definition of statements, we need to know something about truth and falsity. "What is truth?" is obviously a perplexing philosophical question, and the practical-minded person may well wish to leave such matters to poets and philosophers. But the concepts of truth and falsity are so important to science and scientific reasoning that we have to come to grips with them.

In proper philosophical fashion, we begin by noting that the question "What is truth?" may be understood in at least two importantly different ways. One way is to understand it as asking, "Which statements are the true ones?" This is not the question we are concerned with here. Indeed, one may regard all of scientific reasoning as a tool that scientists and others use to help them decide whether a given statement is true or false. So Parts II and III of this text are concerned with methods for answering the question "Which statements are the true ones?" or, less generally, "Is such-and-such particular statement true or false?"

Our concern at the moment is with the question "What does it *mean* for a statement to be true or false?" or "What are you saying about a statement when you say that it is true or that it is false?" As you might have guessed if you did not already know, this question is still a matter of debate among philosophers. To enter the debate at this point would keep us from ever getting very far with the project of learning useful techniques for reasoning about scientific questions. So, rather than discuss the various "theories of truth" now available, I shall simply adopt the account that seems best to me and seems best for developing a simple, useful approach to scientific reasoning. I think this account also best describes the views of most practicing scientists, but this too might be disputed. In any case, those who wish to pursue the matter further should consult the readings suggested at the end of the book.

For the purposes of this text we shall adopt a simple *correspondence theory of truth*. To say that a statement is true is to say that it *corresponds* with the way things really are. Thus, for example, my statement "The text to this chapter is ten pages long" is true if the text is ten pages long, and false if it is not. Far from being wrong, this account of what it means for a statement to be true can be made to seem utterly trivial. To take a more interesting example,

> The statement "Smoking causes lung cancer" is true just in case smoking does cause lung cancer.

If this seems trivial, so much the better. Rather than spending further time on the "trivial" question "What does it mean for a statement to be true?" we can move on toward the significant question "How does one decide whether a given statement is true or not?" That is what scientific reasoning is all about.

The distinctions we have made so far can be neatly summarized as follows:

> A *statement* is a sentence that asserts the way things might or might not be.
> A *true statement* is a sentence that asserts the way things *in fact are*.
> A *false statement* is a statement that *is not true*.

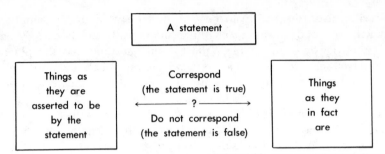

FIGURE 2.1

To repeat the earlier example, my statement that the text to this chapter is ten pages long describes, or asserts, something that might or might not correspond with the way things really are. This statement *is true* just in case it does correspond with how things are; that is, just in case the text to this chapter is in fact ten pages long. Otherwise the statement is false. These relationships are pictured in Figure 2.1.

It is a consequence of our account of truth that the truth or falsity of a statement is "objective" in the sense that it is independent of the specific hopes, fears, or desires of any particular person or group of people. Of course, it takes people to create the language in which statements are formulated, but once the language is fixed, the truth or falsity of any particular statement is solely a matter of whether or not things are as the statement asserts them to be.

2.2   DEFINITIONS, TAUTOLOGIES, AND CONTRADICTIONS

There are several subcategories of statements that turn up in discussions of science and scientific reasoning. This is a convenient place to examine their special characteristics.

Definitions

Definitions play an important role in the study of language, reasoning, and science. Yet there is considerable controversy over the nature of definitions themselves. It is even debated whether definitions are statements or some other kind of expression. For our purposes it is not necessary to discover the "right" account of definitions. We only need a simple account that will permit us to understand the special role of definitions in science.

As we will see in Part II, the kind of definitions that play an especially important role in science are rather like abbreviations: They say that one expression is to stand for another expression. For example, let me create a new definition (at least I think it is new).

Something is a Canadian maple tree just in case it is a maple tree growing in Canada.

In this definition, I have introduced the expression "Canadian maple tree" to stand for the expression "maple tree growing in Canada." The definition tells you that things are to be called "Canadian maple trees" just in case they are maple trees growing in Canada.

If we regard the definition itself as a statement, then it must be regarded as a *true* statement. This follows immediately from the correspondence theory of truth. The definition itself cannot fail to correspond to the way things are. For example, the tallest tree in my front yard is a Canadian maple just in case it is a maple tree growing in Canada. That is, it would be a Canadian maple tree if it were a maple tree growing in Canada. It could not be otherwise.

If you find it strange to think of definitions as true statements, you may reformulate them so that they are not statements at all. The above definition, for example, could be reformulated as follows:

Let us call something a "Canadian maple tree" just in case it is a maple tree growing in Canada.

This is not a statement, but a recommendation, stipulation, or even a command. It is a recommendation that we use the expression "Canadian maple tree" to refer to maple trees growing in Canada. The suggestion might not be very helpful, but it cannot be false. It does not make sense to call a recommendation false.

Definitions have little value in themselves. They are only important because they can be used to make other statements. For example, I could use the above definition to make the statement:

The tallest tree in my front yard is a Canadian maple.

This is a statement in that it asserts something that might be true or false. As it happens, it is false, because my yard is not in Canada. Even if the statement were true, however, it would not be very interesting. So my definition is not very useful. The scientific definitions we will consider will be useful because they can be used to make very interesting statements about the world.

Tautologies

Definitions are not the only sorts of statements that are by their very nature always true. Consider the following statement:

Today is either Monday or some other day of the week.

This statement also cannot fail to correspond with the way things are. No matter what day of the week it happens to be, this statement will be true. Every day is either Monday or some other day of the week.

Although it is often given a narrower, more technical meaning, let us use the word *tautology* to refer to any statement that is so constructed that it must come out to be true no matter what the facts might be. Such statements are sometimes also called "logical truths" because their truth is guaranteed by the logical structure of the statement itself.

To be told that today is either Monday or some other day is really to be told nothing at all about which day today is. This may be taken as a general feature of tautologies. They do not really assert anything at all. But the fact that there can be such statements tells us something interesting about truth. The mere fact that a statement is true may be of little significance because its truth may be due simply to its not really saying anything at all. It may just be an "empty truth."

Contradictions

Knowing that there are statements that have to be true, it may have occurred to you that there should also be statements that have to be false. Indeed, why not just take any tautology and assert the opposite. This will yield a statement that must be false, provided that by "opposite" you mean what is known as the *negation* of the tautology. Let us make this notion explicit before we proceed. In general,

> A statement is the *negation* of a given statement just in case it is false if the given statement is true and true if the given statement is false.

Applying this definition to the tautology "Today is either Monday or some other day," we get:

> It is false that today is either Monday or some other day.

Now this statement says that today is not Monday and that it is not some other day either, which is to say that it is no day at all. But that must be false since today has to be some day or other. So this statement cannot correspond with anything. Such statements are called *contradictions*. They are also often called "self-contradictions" since they implicitly contain within themselves both an assertion and its denial. An example of an *explicit* self-contradiction is the statement "Today is Tuesday and today is not Tuesday."

If tautologies are true because they say nothing at all, then one may think of contradictions as being false because they try to say too much. They say so much that it is impossible that they should ever correspond with things as they are.

Contingent Statements

Now that we have recognized definitions, tautologies, and contradictions as special subcategories of statements, it will be useful to have a term to refer to statements that are not in any of these subcategories. Let us call such state-

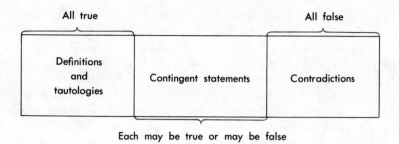

Each may be true or may be false

FIGURE 2.2

ments *contingent* statements. Other common terms are "factual" and "empirical." In order to define contingent statements positively and not merely in terms of what they are not, let us say:

> A statement is a *contingent statement* just in case it is a statement that both might be true and also might be false.

As you would expect, the important statements in science, the ones whose truth or falsity might be of some real concern, are all contingent statements. Noncontingent statements, however, do play a role in science and in scientific reasoning, so it helps to be clear about the difference from the beginning. To help you remember these distinctions, all the various types of statements are pictured in Figure 2.2.

It must be noted that not all philosophers would agree that it is possible to draw so sharp a line between contingent and noncontingent statements. If this distinction is rejected, one ends up with a somewhat different view of science and of scientific reasoning. Here again I have adopted the view that seems to me to provide the best framework for developing useful techniques for evaluating scientific reasoning. References to other approaches are given at the end of the book.

2.3 BELIEF, KNOWLEDGE, AND CERTAINTY

Belief

Scientific reasoning has often been characterized as the means by which one attempts to arrive at true, or at least "rationally justified," beliefs. There is nothing wrong with this as a general characterization, but we shall not talk much about beliefs or believing in this text. We should, however, make some attempt to relate the concept of belief to other concepts we will be using.

Sometimes the word "belief" is used to mean roughly what we mean by "statement." Thus people often ask whether such-and-such belief is true. Simply being aware of this usage should eliminate any possibility of confusion.

The primary use of the word "belief" is to refer to an attitude of some person toward a particular statement. Thus one might say, for example, that Linus Pauling believes that vitamin C prevents colds. There are several ways of attempting to explain further what this means. One is simply to say that Pauling regards the statement "Vitamin C prevents colds" as being true. Another is to say that Pauling has a high degree of confidence in the truth of this statement. This suggests the idea that belief may come in degrees; that is, one may believe something more or less strongly. Either way of understanding statements about beliefs will be fine for our purposes. The important question about beliefs for us will be whether they are "justified." We will examine the concept of justification in the next chapter and methods of justification in the rest of the text.

It is important to realize that there is no direct connection between belief and truth. Whether it is true that vitamin C prevents colds or not, you may or may not believe. And what goes for you goes for everyone else, both individually and collectively. If there was indeed a time when everyone believed that the earth was flat, that would not mean that the earth was indeed flat at that time. Similarly, there are many statements that are in fact true even though no one at all believes that they are true. There are many true statements that no one has even thought of.

Occasionally someone will say that something is "true for me" even though it may not be "true for others." According to the framework we are developing here, such talk cannot be taken literally. Any given statement is either true or false, not both. It could not literally be true relative to one person and false relative to another. The only sense we can make of such claims is that this person *believes* that the statement in question is true—perhaps even believes it very strongly. But this, as we know only too well, is quite compatible with there being other people who believe equally strongly that the statement is false.

The belief that a given statement is true is not "objective" in the same way as the truth or falsity of the statement itself. Within fairly wide limits, people have the power to believe what they want to believe. But one's power to make statements true is very limited, particularly when it comes to general scientific claims—for example, that smoking causes lung cancer. The interesting question to be taken up in Chapter 3 is whether the *justification* of beliefs can be objective in the way that truth is. For easy reference, the relationships involved in the concept of belief are pictured in Figure 2.3.

Knowledge

Scientific method has also been characterized as the means of obtaining *scientific knowledge*. This suggests the classic philosophical question: What is knowledge? This question may be cut down to manageable size by reformulating it as: What does it mean to say that a person *knows* that a given statement is true? For example, what would it mean to say that Linus Pauling knows that vitamin C prevents colds?

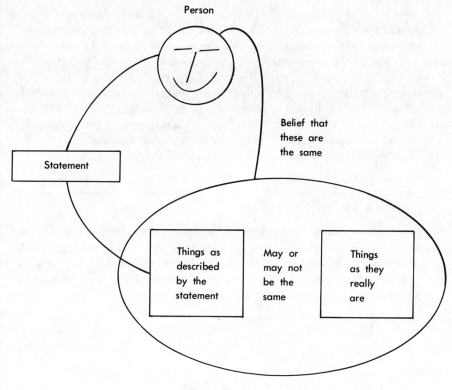

Person

Belief that
these are
the same

Statement

Things as
described
by the
statement

May or
may not
be the
same

Things
as they
really
are

FIGURE 2.3

It seems rather clear that at least part of knowing is believing. If Pauling were to say that he did not believe vitamin C prevents colds, most of us would immediately assume that he could not then know it either. How could he know that this claim is true if he does not even believe it?

But there is obviously more to knowing than just believing. We all believe lots of things we would never claim to know. What makes the difference? One difference has something to do with having "good reasons" or some type of "justification" for holding to the belief in question. Imagine that Pauling were to say that he knows vitamin C can cure *cancer*. But suppose that when asked for his evidence, he denies having any. Suppose he is not able to offer a single fact in favor of his claim. Most of us would then deny that he could *know* what he claims, even though he might, of course, still believe it.

That knowledge requires "reasons" or "justification" is also illustrated by the following example. Suppose I am talking to a new acquaintance and just get a hunch that this person is a Leo. I have no reasons whatsoever for thinking that this is true, but I just cannot help believing it. Finally I say, "I'll bet you're a Leo." It turns out I am right and the person responds, "How did you know?" If I am honest I will say, "I didn't. I just had a hunch." Believing something to be true on the basis of a hunch is not knowing—even if the hunch happens to be right.

There is at least one more ingredient to knowing. To be known to be true, a statement must *be* true. You can't know what ain't so. To illustrate, suppose I were to claim that there is a magnolia tree in my front yard. And suppose that I have very good reasons for making this claim because there is a tree there that looks just like a magnolia tree. If it turns out that this tree is an entirely different species that happens to look like a magnolia when grown fairly far north for magnolias, then I am just wrong in claiming to know I have a magnolia. If it is false that it is a magnolia tree, then I cannot possibly know that it is, however strong my belief and however good my reasons.

Let me summarize our discussion of knowledge so far as follows:

> If some person knows that a given statement is true, then (a) the person believes that the statement is true, (b) the person is justified in so believing, and (c) the statement is in fact true.

The recent philosophical literature is full of examples showing that there must be still more to knowing than these three components. These, however, will be enough for our purposes. They are all pictured in Figure 2.4.

Of these three components, the second, justification, is the most important and the most characteristic of knowledge. As for belief, the beliefs one has

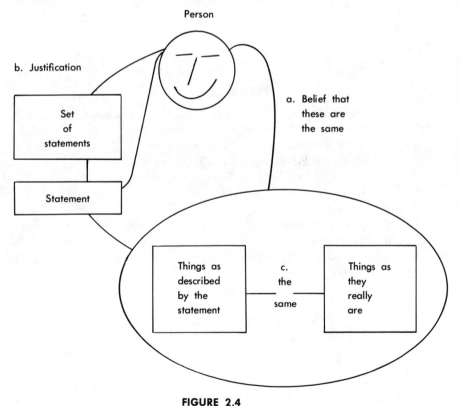

FIGURE 2.4

at any given time are what they are. The question to ask is not what they are, but whether they are justified. And for those beliefs concerning scientific matters, to ask whether they are justified is just to ask whether they are supported by good scientific reasoning. Similarly for truth. As we shall see in more detail later, there is nothing one can do to ensure that any particular scientific hypothesis is true. The best one can do is to make sure that those hypotheses one believes to be true are justified by sound scientific reasoning, and justified in a way that provides some reliable correspondence between those hypotheses that are so justified and those that are in fact true.

So all roads lead us to the concept of justification. But before turning to that concept, there is one other to consider, one that is closely related to both belief and knowledge.

Certainty

Of all the concepts considered in this chapter, certainty is the least precise and least well understood. In everyday speech it seems to have two different uses, one associated with belief and one with knowledge. The first use is illustrated by claims like "I am certain that such and such." Here it seems that what is claimed is a very strong belief or a very high degree of belief—perhaps the strongest possible belief.

The second use is illustrated by expressions like "I know for certain that such and such." One might understand this simply as a claim to knowledge with the belief component being a very strong belief. But such expressions might also be taken to mean that the *justification* offered for the belief is especially strong, so not only is the belief strong but its being strong is justified.

So long as all that is being claimed is that there is a very strong justification for the belief, we need have no worries. It is to be expected that some types of justification are stronger than others, and we shall see good examples of this in Part III of this text. But often it seems that more is being claimed, namely, that to "know for certain" is to have such a strong guarantee that the believed statement is true that it has got to be true. There is no chance whatsoever that it is false. As we will see in detail in Parts II and III, this notion must be rejected. Scientific claims are expressed in contingent statements of considerable generality, and no method of investigation can guarantee the truth of any such statement beyond all possibility of mistake. No matter how great the current evidence in its favor, any scientific conclusion might in fact be false and later face equally strong evidence in favor of its negation.

It is one of the major theses of twentieth-century thought that science cannot provide an absolute guarantee of the truth of its conclusions. Many thinkers seem to have concluded that science is therefore worthless and that we must look elsewhere for "real" knowledge. But one might also conclude that if science cannot guarantee its conclusions, then nothing can. There is no "else-where" to look. The fact that there is no absolutely perfect justification does not mean that some justifications are not better than others. One should learn to live without guarantees and to distinguish carefully between better and worse justifications. That is what we shall be learning here.

CHAPTER EXERCISES

2.1 Some of the following sentences express statements and others do not. Identify the ones that are not statements by writing "no" in the space provided. The statements can then be further identified as either definitions, tautologies, contradictions, or contingent statements. Indicate which. Note that you are not being asked whether the statements are true or false.

A. The surgeon general has determined that smoking is dangerous to your health. _____

B. Please pass the salt. _____

C. Hair dyes may be dangerous, but then again they may not be. _____

D. A statement is a contingent statement just in case it both may be true and may be false. _____

E. She won but she lost. _____

F. They won the battle but lost the war. _____

G. Drinking water causes lung cancer. _____

H. I'll be there when I get there. _____

I. I know I can't give up smoking, but I will. _____

J. Do you use sugar in your coffee? _____

K. A sentence expresses a statement just in case it expresses something that could be true or false. _____

L. This problem (2.1) requires a dozen separate answers. _____

2.2 In each of the following problems there is one answer that is clearly better than the others. Indicate the answer you think best by circling (a), (b), or (c).

A. The negation of a contradiction is:
   (a) A definition.
   (b) A true contingent statement.
   (c) A tautology.

B. The negation of a false contingent statement is:
   (a) A contradiction.
   (b) A true contingent statement.
   (c) A tautology.

C. Suppose there were a statement that everyone in the world believed to be true. If there were such a statement:
    (a) It would have to be true.
    (b) It would have to be false.
    (c) It could be either true or false.

D. Imagine a statement that is in fact false. That statement:
    (a) Could not be believed to be true by anyone.
    (b) Could not be known to be true by anyone.
    (c) Might be known to be true by someone.

E. Imagine a statement that no one has any reasons to believe is true but nevertheless is in fact true. This statement:
    (a) May be believed to be false by someone.
    (b) May be known to be true by someone.
    (c) May be known to be false by someone.

F. Imagine there is a statement that you do not know to be true.
    (a) You must not believe it to be true either.
    (b) You must know it to be false.
    (c) You may not know it to be false either.

2.3 The following questions are more difficult than the preceding, but thinking about them will help you get a better grasp of the concepts introduced in this chapter.

A. Give an example of a statement that you believe to be true, but would not claim to know to be true.

B. Give an example of a statement that you neither know to be true nor know to be false.

C. There must be statements that you believe to be true but are in fact false. Can you give an example of such a statement? If not, why not?

D. Can you give an example of a statement that is not a definition or a tautology but of whose truth you are absolutely certain? Explain why you are so certain this statement is true.

3

Justification and Arguments

In this chapter we shall see how reasoning and the justification of scientific hypotheses can be understood in terms of particular sets of statements called "arguments." Having explored the relationship between statements and arguments, we can then probe deeper into the concept of justifying claims to know that a statement, or hypothesis, is true. We shall then examine two general types of arguments: deductive and inductive. It will turn out that although deductive reasoning is ideal for justification, it is not sufficient for science. Scientific reasoning must be analyzed in terms of inductive arguments, arguments that provide an adequate but weaker form of justification. Justification in science, we shall learn, is based on the probability of truth, not just truth itself.

3.1 STATEMENTS AND ARGUMENTS

The conclusions of scientific investigation are expressed in contingent statements, such as "The universe will continue expanding forever" or "Smoking causes lung cancer." It is convenient to have a word to refer to scientific statements that are the object of current research or are being regarded as such for the purpose of a particular discussion. The term "hypothesis" is fairly well established for this purpose, and we shall adopt it here.

Ideally one would like to say simply that the purpose of scientific inquiry is to determine whether given hypotheses are true or false. But that would be an oversimplification and could be very misleading. If indeed there is no foolproof way of determining, with certainty, the truth or falsity of hypotheses like "Smoking causes lung cancer" we must be more careful. Let us say, then, that the purpose of scientific inquiry is to provide a "good justification" for believing that such a statement is true or that it is false—where it is understood that even the best justification may leave one believing a false statement to be true, or vice versa.

Now we must face the difficult question: What is justification? Or, What

24

does it mean to justify claiming that a given hypothesis is true? What goes on in the process of justifying a scientific conclusion?

If to examine the justification for a scientific claim one had to follow everything that went on in a long process of investigation, it would be impossible for a nonspecialist ever to make an independent judgment about the justifiability of any scientific claim. But this is not necessary. For us, scientific reasoning is not the reasoning of the laboratory but of the finished research report. Most of what actually went on in the investigation never appears in the report. What we get in a report is the statement of a hypothesis—for example, that smoking causes lung cancer—and some information about the results of observations or experiments—for example, the relative numbers of smokers and nonsmokers who got lung cancer. This information is contained in other statements. So to determine whether the stated hypothesis is justified or not, all we need to know is something about these other statements *and* about their relationship to the hypothesis.

It is now time to introduce some new concepts, although most of the words we shall use to designate these concepts will be familiar words used in a special way. First, consider the whole set of statements that includes the hypothesis in question together with all the other statements involved in the justification of the hypothesis. This set of statements constitutes what we shall call an *argument*. The hypothesis is the *conclusion* of the argument. All the other statements taken together are called the *premises* of the argument. These concepts are summarized in the following definition:

> A set of statements is an argument just in case it is divided into two parts, the premises and the intended conclusion.

According to this definition, an argument may have any number of statements as premises and as conclusions, but typically an argument will have several premises and a single conclusion. The schematic form of a simple argument with two premises and a conclusion is shown in Figure 3.1.

You should be aware that we are using the word "argument" in a special sense. In everyday language, this word may refer to a conclusion alone, to premises alone, or to a whole dispute, as in the common expression "They're having an argument." Unless there are clear indications to the contrary, whenever the word "argument" appears in this text, it has the special meaning given above. The study of scientific reasoning, like the study of science or any

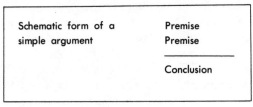

FIGURE 3.1

other subject, does require that one learn some new concepts and, correspondingly, some new vocabulary.

In general the study of reasoning, or logic, is the study of arguments. When approached from this general perspective, one finds that there are many different types of arguments to consider. Most interesting scientific arguments, however, are just variations on a single type of argument. Once one grasps the basic theme, learning the variations is relatively easy. But we still have a few more basic concepts to learn before we take up real scientific arguments in Part II.

3.2   JUSTIFICATION

Although we do not yet know much about the details of arguments or justification, the relationship between arguments and the concept of justification is easy to state. One justifies a hypothesis by exhibiting it as the conclusion of an appropriate argument. Speaking somewhat more objectively, we may also say that the argument itself justifies its conclusion, or that the premises of the argument justify the conclusion. The interesting question, of course, is what counts as an "appropriate" argument. We shall begin working on that shortly. For the moment there is more to be said about justification.

There are two general conditions that must be satisfied if an argument is to provide justification for its conclusion. These are most easily seen by considering how an argument can *fail* in justifying its conclusion.

Failure of Justification: Unjustified Premises

Let us now look at an example of an argument that fails to justify its conclusion. Although this argument has a "scientific" subject matter, it is not really a "scientific argument." But we are not yet ready to deal with "real" scientific arguments, and looking at a few examples of other kinds of simple arguments will help fix in mind the concepts we are now studying: statement, argument, and justification.

There have recently been numerous discussions, even official public hearings, concerning the possibility of harmful effects due to low doses of various chemicals released into the environment, such as PCBs. Imagine an exasperated representative of an industry that uses these chemicals saying: "Look, either small doses of this stuff are deadly or it's harmless. But these doses are obviously not deadly—not a single case of death has been attributed to PCBs. So they're harmless. Why all the fuss?"

First we must exhibit this reasoning as an argument. This is not always easy to do with the arguments one meets in everyday life. But let us try. When analyzing any reasoning, scientific or not, a good strategy is to look for the conclusion first. That is the statement that the reasoner is trying to justify. In this case the conclusion is: PCBs are harmless. Omitting rhetorical flourishes, the first premise seems to be: Either small doses of PCB are deadly or they are harmless. The second premise is: They are not deadly. The statement that no case of death has yet been attributed to PCBs is given to justify the second premise. So there is really a subargument within the larger argument, but let us ignore this. On my analysis, then, the argument may be set out in standard argument form as follows:

> First Premise: Either small doses of PCB are deadly or they are harmless.
> Second Premise: They are not deadly.
> Conclusion: Small doses of PCB are harmless.

This argument illustrates a common rhetorical ploy. Set up an alternative that is so extreme as to be easily rejected, and then reach the conclusion you want. You should not be taken in by this ploy. Why not? Because there is no reason for agreeing that the alternatives offered in the first premise are the only alternatives available. Indeed, the most plausible alternative has been left out, namely, that low level doses of PCB are harmful, though not necessarily deadly. So while one may be quite willing to agree that small doses are not deadly, one need not agree with the conclusion that they are therefore harmless.

So one condition that must be met if an argument is to provide a justification for believing its conclusion is that the premises of the argument must themselves be justified premises. Once you think about it, this requirement should seem intuitively obvious. Within the framework of an argument, it is the premises that provide the reasons for believing the conclusion. But if you have no reason for believing the premises, how can they provide you with any reasons for believing the conclusion? Indeed, if one were allowed to use unjustified premises to justify conclusions, any statement whatsoever could be justified. Just pick for premises any statements that have the right sort of connection with the conclusion. This is always possible so long as you do not have to worry whether these statements are themselves justified. As a matter of fact, if you could use unjustified premises to justify a conclusion, why not just use the conclusion itself as the premise to your argument? So any statement could be used to justify itself. This is clearly an absurd notion of justification. So we need the requirement that to use an argument to justify a conclusion, the premises must themselves be justified.

Now let us look more closely at the "connection" between premises and conclusion. This is the focus of the second requirement for justification by the use of arguments.

Insufficient Connection between Premises and Conclusion

The second way an argument can fail to provide justification for its conclusion is by lacking a sufficient connection between its premises and its conclusion. Consider the following little argument whose premises are undoubtedly true, but which fails to provide any justification for the conclusion because there is not sufficient connection between the premises and the conclusion.

> Grass is green.
> Trees are green.
> Thus, my house is green.

There is no question here of the premises being justified. These are both statements most of us would count among those statements we know to be true. Yet you would not take this argument as providing any reason to believe that I live in a green house. It is not that the premises are unjustified. It is just that the premises fail to make contact with the conclusion in such a way that they could, so to speak, transmit any of their justification to the conclusion.

For those who would like a more serious example of this type of failure of justification, consider again the PCB argument, but with the additional alternative added to the first premise. The amended argument looks like this:

> First Premise: Either small doses of PCB are deadly, or they are harmful but
>   not deadly, or they are harmless.
> Second Premise: They are not deadly.
> Conclusion: They are harmless.

In this version of the argument, the premises are pretty well justified. The first covers all the relevant possibilities and the second is supported by the fact that there have been no PCB-related deaths. But the argument clearly fails to provide an adequate justification for its conclusion. The argument jumps from the falsity of one extreme alternative to the other extreme alternative, completely ignoring the intermediate possibility. But how can one justify believing PCBs are harmless when nothing whatsoever has been said about the possibility that they might be harmful in a nonfatal way? Here again there is not sufficient connection between these premises and the conclusion to transmit much justification to the conclusion. So the second requirement for an argument to provide justification for its conclusion is that it exhibit an appropriate connection between its premises and its conclusion.

The immediate questions are: "What constitutes an appropriate connection?" and "How can you tell when the connection is appropriate?" We shall take up these questions in the next two sections of this chapter. Only a few more points remain to wind up our general account of justification.

Unjustified Statements

Suppose that you have analyzed an argument and have determined that it *fails* to justify its conclusion; either it has unjustified premises or it lacks an appropriate connection between its premises and the conclusion. In this situation it is tempting to think that you are now justified in believing that the conclusion is false. But that would be a mistake. To justify believing that the conclusion is false would require producing *another* argument, one that has as its conclusion the *negation* of the original conclusion. Just showing that the original argument failed does not automatically provide you with an argument to the contrary. Moreover, showing that one argument fails to justify the conclusion does not preclude there being some *other* argument that does justify it. That is, there may be other justified premises that do have an appropriate connection with the statement in question.

Justification is different from truth. If you consider a hypothesis and its negation, then you know one of these two statements is true and the other false, though you may not know which is the true one. Both cannot be true. Neither can they both be false. But both can be unjustified. That is, you may not be justified in believing either the hypothesis or its negation. Indeed, this is usually the case at the start of any scientific investigation. The object of the investigation is to justify one or the other.

Logic and Justification

Logic has traditionally been thought of as the study of those arguments that do transmit justification from premises to conclusions. One does not ask a logician where the argument came from, and one does not ask whether the premises are in fact justified. All a logician will tell you is whether the given argument is of a type that would produce a justified conclusion from justified premises. If you then want to know whether the premises are justified, you will have to ask someone else—for example, a scientist.

The point of view of this text is somewhat broader. It is assumed here that what most people really want to know is whether or not some given information justifies their believing a particular hypothesis. This requires that one consider both the argument *and* the plausibility of the given premises. There are, of course, severe limits on how much a nonspecialist can do. Going back to the original report is usually much too difficult, and repeating the experiment or observations is out of the question. But one can learn which kind of premises are most difficult to justify, and thus most questionable. And one can learn which sources of information are likely to be most reliable. We shall be doing this when we consider actual scientific arguments.

Does the Process of Justification Ever End?

Though it seems obviously correct, the fact that justified conclusions require justified premises may be troubling—especially to those with a "philosophical" turn of mind. How would you justify the premises of any given argument? By constructing further arguments that have these statements as conclusions. But then these further arguments have premises that need to be justified. How are these to be justified? By constructing further arguments.

And so on. It looks as though the process might never end, in which case we would never be able to justify anything.

This problem has been considered by philosophers at least since the time of Aristotle (roughly 350 B.C.). For many centuries the accepted view had been that there must be some statements that do not need to be justified by further statements. That these statements are justified, or even true, is "self-evident." So in the end, all arguments lead back to some "self-evident" premises. The problem is that finding any such statements proved so difficult that many philosophers concluded that there are none. But what then becomes of justification—and of knowledge?

This is another problem that we dare not pursue lest we never get to analyzing any real scientific arguments. I can only refer you to the suggested readings at the end of the book. Our approach will be a very practical one. In practice one seldom has to go back very far before reaching some statements whose truth no ordinary person would dispute. The appearance of a bright comet is obvious to everyone who can see. And any person of normal intelligence can count sweet peas to determine how many are smooth and how many have wrinkled skins. And so on. Philosophical skeptics may continue to ask how we can be certain that we are not all just dreaming. And others may continue to wonder how we manage to justify anything at all. But the practical task of learning to evaluate everyday reports of scientific findings need not wait for the answers to these questions.

3.3 DEDUCTIVE REASONING

In this section we shall study a special type of reasoning that is based on a special type of argument in which the connection between premises and conclusion is as strong as it could possibly be. These arguments are called "deductive arguments." The reason for studying deductive arguments is that some deductive arguments play a crucial role in scientific reasoning. Scientific arguments as a whole, however, are not deductive, and we shall see why in the very next section.

Truth and Deductive Validity

We shall begin with an example that you should recognize as being similar to the PCB example of the previous section. The present example, however, has the advantage that its structure can be pictured in a way that makes the special nature of the connection between premises and conclusion quite apparent. The reasoning represented by this argument is very simple, so simple that one would not ordinarily bother actually writing down the argument. But for the purpose of seeing what deductive arguments are, a simple argument is best.

Suppose that you are talking to several friends and one asks how far it is from New York City to Topeka, Kansas. This person is thinking of making that drive with another friend over the coming vacation. No one in the immediate group knows for sure. One person, however, has driven from New York to St. Louis and remembers that it was roughly 1000 miles—and Topeka is farther away than that. Another person once made a trip to go skiing in Colorado and remembers that it was roughly 2000 miles to Denver—and Topeka is closer than that. So, you conclude, Topeka is between 1000 and 2000 miles from New York.

The most important premise in the above argument is not even stated in the conversation. It is, roughly, that Topeka is somewhere between New York and California. The stated premises give information about where it is not. The whole argument, including the unstated premise, may be set out as follows:

First Premise: The distance between New York City and Topeka is either less than 1000 miles, or between 1000 and 2000 miles, or more than 2000 miles.

Second Premise: It is not less than 1000 miles.

Third Premise: It is not more than 2000 miles.

Conclusion: The distance between New York City and Topeka is between 1000 and 2000 miles.

It is not hard to see that there is something very special about the above argument. Just ask yourself this question: If I assume that the premises are true, is there any way that the conclusion could not be true? The answer is, "No way!" Topeka is somewhere between New York and California. If it is not east of St. Louis and not west of Denver, it has got to be somewhere in between. These special relationships are easily pictured. The large rectangle in Figure 3.2 represents the United States. It is divided into three parts. The first premise of the argument says that Topeka is in one of these three parts. The section on the right represents the area east of St. Louis. The second premise says that Topeka is not in this section. The section on the left represents the area west of Denver. The third premise says that Topeka is not in this section. The crosshatching covering the left- and right-hand parts represents the information given in the second and third premises. What is left is the conclusion. Topeka must be in the middle section, which corresponds to something between 1000 and 2000 miles from New York City. There is no where else it could be.

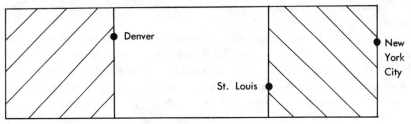

FIGURE 3.2

Any argument in which the truth of the premises makes it impossible that the conclusion could be false is called a *deductively valid argument*. Let us put this into an official definition.

An argument is a *deductively valid argument* just in case the truth of the premises makes it impossible that the conclusion is false.

The property of being deductively valid can be described in a variety of ways. One may say that the conclusion *follows from* (or logically follows from) the premises, or that the premises *logically imply* the conclusion. Similarly, one can say that the conclusion may be *deduced* (or validly deduced) from the premises or that the conclusion *follows deductively* from the premises. It is also sometimes said that the truth of the premises makes it *necessary* that the conclusion be true as well. These are all just different ways of saying the same thing. The argument is deductively valid.

The word "valid" like the word "argument" is commonly used in a variety of ways to refer to many different things. Often it seems to be applied to statements, meaning that the statement is true, or justified, or in some sense good, as in the expression, "That's a valid reason." In this text we will apply the word "valid" only to *arguments,* and then only with the meaning "deductively valid" as given in the above definition. In learning to think clearly it is helpful to keep the meanings of key terms clear and unambiguous.

Arguments that are *not* deductively valid are called "deductively invalid," or just "invalid." The Topeka argument would be invalid if one left out any of the three stated premises. For example, suppose we left out the third premise. Then the remaining two premises do not rule out the possibility that Topeka is west of Denver and thus more than 2000 miles from New York. So the conclusion that Topeka is between 1000 and 2000 miles from New York no longer follows necessarily. Even if the first two premises are true, the conclusion could be false.

In order to help to fix the meaning of deductive validity clearly in mind, consider the four different possible combinations of truth and falsity for the premises and conclusion of *any* argument, valid or invalid. The premises are either all true or not all true (that is, at least one is false). Likewise, the conclusion is either true or false. This makes four possibilities, which can be represented in a simple two by two diagram as in Figure 3.3. The restrictions that characterize *valid* arguments are written in the boxes representing the four original possibilities.

Figure 3.3 shows that validity is a more abstract concept than you might have first thought. Validity cannot be described simply in terms of the truth or falsity of the premises and conclusion. Rather, it is the *possibility, impossibility,* or *necessity* of the various combinations that define validity. A deductively valid argument is defined as an argument for which the *possibility* of a false conclusion with true premises is *ruled out*. This combination is *impossible*. This is the same as saying that if the premises are true, then it is *necessary* that the

The Truth or Falsity of the
Conclusion Is:

		Truth	Falsity
If the Premises of a Valid Argument Are:	All True	Necessary	Impossible
	Not All True	Possible	Possible

FIGURE 3.3

conclusion be true. However, if the premises are *not* all true, then a valid argument might have either a true conclusion or a false conclusion. Either combination is *possible*.

We have just examined an example of a valid argument with true premises and a true conclusion. As an example in which false premises lead validly to a true conclusion, consider the following argument:

> The earth is larger than the sun.
> The sun is larger than the moon.
> Thus, the earth is larger than the moon.

Here the first premise is clearly false, but together with the true second premise it leads validly to a true conclusion. And the argument is clearly valid. That is, if both premises *were* true, the conclusion *would have to be* true. It just so happens that the first premise is false. So a valid argument might just happen to get you from false premises to a true conclusion. Validity allows this possibility.

Similarly, an argument may lead from false premises to a false conclusion and still be perfectly valid. For example:

> The moon is larger than the earth.
> The earth is larger than the sun.
> Thus, the moon is larger than the sun.

Here both premises and the conclusion are false. But the argument is valid! If the premises were true, the conclusion would have to be. One can reason validly from falsehoods to falsehoods.

For *invalid* arguments, of course, anything goes. In particular, an invalid argument might still have true premises and a true conclusion. The argument in the preceding section concluding that I live in a green house is an argument of this sort. Any invalid argument might just accidentally get you from true premises to a true conclusion.

What Makes a Valid Argument Valid?

At this point you may be wondering how it can be that there are arguments for which the falsity of the conclusion given the truth of the premises is impossible—literally inconceivable! To try to answer this question would take us off into the philosophy of language, logic, and mathematics. The usual answer appeals to the *concepts* used in the argument. It is the concepts employed that determine the limits of possibility and impossibility. The validity of the Topeka argument, for example, is determined by spatial concepts, distance, betweenness, and so on. It is the structure of these spatial relationships that makes this argument valid. Similarly, the validity of the sun/earth/moon examples depends on the geometrical relationships of spheres of different sizes—that is, the relationship of "being larger than." Other valid arguments depend on other conceptual relationships. We shall look at a few other types of valid arguments in Chapter 4.

Deduction and Justification

As we have seen, justifying a conclusion requires both justified premises and an "appropriate" connection between the premises and the conclusion. It should now be obvious that a deductively valid argument provides not only an "appropriate" connection but the best possible connection. A valid deductive argument is a *perfect* transmitter of truth itself. Or, to put it another way, deductive arguments are perfect "truth preservers." If you begin with true premises and employ only valid arguments in drawing conclusions, you are guaranteed that your conclusions are true as well. Truth is preserved as you move from premises to conclusions.

The relationship between deduction and justification is thus obvious. If you are justified in believing the premises of a valid argument, then you cannot be any less justified in believing the conclusion. Any reason for believing the premises to be true is automatically a reason for believing the conclusion to be true. Being perfect transmitters of truth makes valid arguments perfect transmitters of justification as well.

3.4 INDUCTIVE REASONING

Unfortunately there are statements, including most interesting scientific hypotheses, that cannot be justified by deductive reasoning alone. In the first part of this section we shall see why the justification of typical scientific hypotheses requires a form of nondeductive reasoning that goes by the name "inductive reasoning." Here we consider only a few general characteristics of inductive arguments, particularly those that distinguish inductive arguments from deductive arguments. We shall get into the details of inductive reasoning in Part II.

Why Deductive Reasoning Is Not Enough

The great virtue of deductive arguments, as we have just seen, is that they are perfect transmitters not only of justification but of truth as well. But this virtue comes at a high price. It means that deductive reasoning alone is not sufficient for the justification of typical scientific hypotheses. Let us see why this is so.

Contingent statements give information. We need not bother trying to give a precise definition of "information" in order to make intuitive use of this concept. Thinking in terms of information content, let us look again at the Topeka argument discussed in the previous section. Only this time let us go through the argument more carefully, noting the information content of each premise in turn.

The first premise gives us the information that Topeka is in one of three regions—roughly east, middle, or west. This premise is represented in Figure 3.4 by the rectangle with three open regions.

The second premise adds the information that Topeka is not in the eastern region. This additional information is represented by shading in the right-hand region of the rectangle. The third premise adds the further information that Topeka is not in the western region. This additional information is represented by shading in the left-hand region of the rectangle.

Now that the information from all of the premises has been added, the conclusion is automatically represented. Topeka must be in the middle region since it is the only place left unshaded. Thus, *no new information* needs to be added in order to get the conclusion. The premises give it all.

This is a general feature of all deductively valid arguments. Their conclusions contain no more information than that already given by their premises.

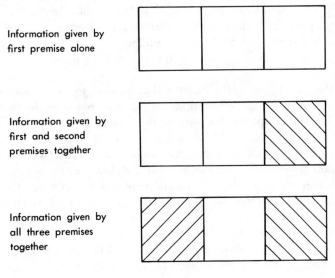

Information given by
first premise alone

Information given by
first and second
premises together

Information given by
all three premises
together

FIGURE 3.4

Indeed, this feature provides one explanation of how deductive arguments manage to be perfect transmitters of truth. The conclusion merely restates part (or all) of the information given in the premises. If what the premises say is correct, then of course what the conclusion says will be correct too, since what it says is just part of what was said by the premises.

The reason why scientific reasoning cannot be represented adequately by valid deductive arguments is that typical scientific conclusions contain *more* information than any premises that one might use to justify them. This very general feature of scientific reasoning is illustrated by the claims made in *The Limits to Growth*. The scientific claim in question is that the world cannot continue the current rates of growth in population and industrialization without suffering catastrophic consequences by the end of the twenty-first century. This is a claim about what might happen to the whole world physical-social-economic system within roughly the next 100 years.

Now if one is to offer reasons for thinking that this claim is true, these can only refer to certain features of the world system as it exists now and in the past. These are the only facts that could themselves be taken as providing the justified premises needed to justify any such conclusion. So the general form of any argument for the claim would be something like this:

Premises: Statements describing the world system at the present and in the past.

Conclusion: A statement about what will happen during the next century.

Now obviously there is information in that conclusion which is not already in the premises; that is, the premises themselves say nothing about the future. We can be sure that whatever the detailed structure of this argument, it cannot be deductively valid. A deductively valid argument must have all the information in the premises. Thus, if we insist that justification requires that the conclusion follow deductively from true premises, the indicated premises must fail to justify the given conclusion even if we could be quite sure that all these premises are true.

At this point one might suggest that we add to the given premises a general statement to the effect that the patterns observed in the past will continue into the future. Such a premise could be structured in such a way that the resulting argument would be deductively valid. But such an augmented argument would still fail to justify the conclusion because we can no longer regard all the premises themselves as being justified. Any doubts one has about the conclusion apply equally well to that new general premise. The scientific question at issue is just whether or not certain patterns of the past will continue into the future.

Finally, if we were to try to use the undisputed facts about the past and present as premises in a new argument to justify this new general premise, we would be back in our original situation. This general statement must contain information about the future that is not in our premises, so our argument based on past and present data could not be deductively valid.

In sum, any attempt to justify the original scientific conclusion using only deductive reasoning fails, either because our argument is not deductively valid or because we have a premise that is itself no more justified than the conclusion at issue.

The only way out of this dilemma is to develop forms of argument that can provide some kind of justification for conclusions even though the conclusion does not follow deductively from the premises.

Two General Features of Inductive Arguments

We have already discovered two general characteristics of inductive arguments. First, they do not "transmit" or "preserve" truth. That is, it will be possible for a good inductive argument to have a false conclusion even though all its premises are true. If this were not the case, inductive arguments would be deductive, which they are not.

This general feature of inductive arguments is one reason why scientific reasoning cannot yield certainty. Even if one could be certain of one's premises, the best inductive scientific argument would not guarantee the truth of its conclusion. Thus the very nature of scientific reasoning introduces a real and unavoidable possibility of error into the scientific process.

The second general characteristic of inductive arguments is that they are "knowledge expanding"; that is, their conclusions contain *more* information than all their premises combined. It is because of this feature of inductive arguments that science can be a source of new knowledge.

These two general features of inductive arguments are related, as we have seen. It is only by giving up the sure transmission of truth that inductive arguments can be knowledge expanding. Thus the philosophers' dream of a form of argument that would be both truth preserving and knowledge expanding is an impossible dream. You must choose one or the other. You cannot have both.

Probability and Inductive Reasoning

To be told that inductive arguments are knowledge expanding and not truth preserving does not tell you much about the details of inductive arguments. These will have to wait until Parts II and III. However, it is possible to mention one further general characteristic in a preliminary way. Indeed, we may incorporate this characteristic in a general definition of a good inductive argument.

> An argument is a *good inductive argument* just in case the truth of its premises guarantee an appropriately high probability for the truth of its conclusion.

Thus, assuming true premises, the difference between a *good inductive argu-*

ment and a *valid deductive argument* is that the deductive argument guarantees the *truth* of its conclusion while the inductive argument guarantees only an *appropriately high probability* for the truth of its conclusion.

The above definition incorporates the often expressed idea that the conclusions of science are "only probable." But we still have a lot to learn before we can be sure we understand what this means. Moreover we have also to learn just what kinds of premises yield high probabilities and which high probabilities are appropriate.

Do not let this talk of probability worry you. To understand and evaluate scientific reasoning you do not have to become a specialist in probability theory any more than you need become a specialist in any particular science. You do not even have to learn how to make calculations of probabilities. You need only learn in a general way how such calculations are made and what the results of such calculations mean for various kinds of hypotheses. We shall do this first in a very intuitive way in Part II and then more explicitly in Part III. In any case, you will be given all you need to know as we proceed.

Probability and Justification

You need not know anything in particular about probability to appreciate one final point about justification and arguments. Suppose you have an argument in which the premises tell you that the conclusion has a high probability of being true. Then, if you are justified in believing the premises, you are justified in believing that the truth of the conclusion is highly probable. Only one further step is needed to be justified in believing the conclusion to be true. It must be granted that a high probability justifies belief. But not to grant this is really to insist that justification demands a deductive argument. And that, as we have seen, would eliminate the possibility that science could be knowledge expanding. So it must be granted that good inductive arguments do transmit justification. With that reassurance, we can proceed with confidence.

3.5   IS JUSTIFICATION OBJECTIVE?

In Chapter 2 we saw that belief, by itself, need not be very "objective" in the sense that people can and do tend to believe what they want to. Truth, on the other hand, is objective in the sense that once a hypothesis is formulated, whether it is in fact true or not is usually outside anyone's influence. Knowledge has both components. It includes belief, but it also requires that to be known to be true, a statement must be true. The third component of knowledge was that the corresponding belief be *justified*. So the question immediately arises whether justification is subjective, like belief, or objective, like knowledge.

We now know that justification has two parts: justified premises and a good argument from the premises to the hypothesis in question. And we have seen that having a good argument or not is also objective. This certainly holds for deductive arguments. Whether given premises logically imply a given hypothesis is a matter of the possibilities and necessities governed by related concepts. These are not under the control of any particular person, although languages in general are a human product. The same holds for inductive arguments, as we shall see more clearly in Parts II and III. Whether given premises make it highly probable that the conclusion reached is true is, like truth itself, determined by what the premises assert, not by what anyone believes.

The remaining question concerns the requirement that the premises themselves be justified. As we have seen, the attempt to justify all premises by further arguments leads to an endless regress of argument after argument. If it is a "subjective" matter which further premises one is willing to settle on, then the whole process of justification may cease to be "objective."

This is another place where it helps to focus on scientific reasoning rather than just reasoning in general. In science it is an established norm that responsible criticism be answered responsibly. So it is not left to individual scientists to pick their premises as they wish. If some premise is questioned, it must be justified by further argument. On the other hand, if no premises are questioned by other knowledgeable scientists, one is free to regard one's conclusions as sufficiently justified—at least until someone else raises a serious objection.

Thus it appears that the "objectivity" of justification in science is grounded in the "rules" followed by the scientific community. This is not, to be sure, as secure a grounding as that provided by the very concepts of our language. But it at least takes justification out of the hands of individuals who are subject to individual biases.

Unfortunately there is still the possibility that the whole community of scientists might not be responsible in questioning premises that could be mistaken. In this case the scientific process itself would have broken down. How such a breakdown might be dealt with by people outside the scientific community is much too difficult a problem to consider here. Fortunately such breakdowns are rare, and contemporary scientific communities tend to be responsive to criticism even from nonspecialists. This provides still another reason why it is important for nonspecialists to develop some skill at scientific reasoning.

CHAPTER EXERCISES

3.1 For each of the following questions there is one answer that is clearly better than the others. Pick the answer you think best.

A. For the purposes of studying reasoning, an argument is:
(a) A hypothesis that is justified by justified premises.
(b) A set of justified premises that justify a hypothesis.
(c) Any set of statements.
(d) A set of statements including some designated as premises and others designated as conclusions.
(e) A set of statements including justified premises and a conclusion justified by
those premises.

B. Imagine an argument that you know to be deductively valid, but also to have an
unjustified premise. As far as this argument is concerned, you should regard the
conclusion as being:
(a) True.
(b) False.
(c) Justified.
(d) Unjustified.

C. Imagine an argument that you know to have justified premises, but also know to
have no appropriate relationship between its premises and its conclusion. As far as
this argument is concerned, you should regard the conclusion as being:
(a) True.
(b) False.
(c) Justified.
(d) Unjustified.

D. Imagine an argument that you know to be deductively valid and to have justified
premises. As far as this argument is concerned, you should regard the conclusion as
being:
(a) Valid.
(b) False.
(c) Justified.
(d) Unjustified.

E. Imagine an argument that you know to be a good inductive argument and to have
justified premises. As far as this argument is concerned, you should regard the
conclusion as being:
(a) Valid.
(b) False.
(c) Justified.
(d) Unjustified.

3.2 The following questions all concern the relationship between the truth or
falsity of statements and the deductive validity or invalidity of an argument. In each case
pick the answer you think best.

A. Imagine an argument you know to have true premises and a true conclusion. Given
no other information, you should regard the argument as being:
(a) Valid.
(b) Invalid.
(c) Either. You cannot tell which.

B. Imagine an argument you know to have one false premise but a true conclusion. Given no other information, you should regard the argument as being:
   (a) Valid.
   (b) Invalid.
   (c) Either. You cannot tell which.

C. Imagine an argument you know to have true premises but a false conclusion. Given no other information, you should regard the argument as being:
   (a) Valid.
   (b) Invalid.
   (c) Either. You cannot tell which.

D. Imagine an argument you know to have false premises and a false conclusion. Given no other information, you should regard the argument as being:
   (a) Valid.
   (b) Invalid.
   (c) Either. You cannot tell which.

E. Imagine an argument you know to be valid but to have a false conclusion. Given no other information, you should regard the premises as being:
   (a) All true.
   (b) Not all true.
   (c) Either. You cannot tell which.

F. Imagine an argument you know to be valid and to have true premises. Given no other information, you should regard the conclusion as being:
   (a) True.
   (b) False.
   (c) Either. You cannot tell which.

3.3 The following are for those interested in philosophical questions about the nature of justification.

A. Take a simple contingent statement you regard as being well justified—for example, "The sun will rise tomorrow." Consider how you might actually justify this statement using only deductive arguments and premises that are better justified than the statement in question—for example, "The sun rose this morning." Can you see how someone might argue quite convincingly that it is impossible to justify even so simple a statement as that the sun will rise tomorrow? Can you make a convincing case that this is so?

B. Now suppose you are allowed to use inductive arguments as well. Is it still possible to make a convincing case for its being impossible "really" to justify believing that the sun will rise tomorrow? If so, do it. If not, explain why not.

4

Conditional Arguments

Although one cannot understand scientific reasoning solely in terms of deductive arguments, several forms of deductive argument do play a major role in scientific reasoning. This is particularly true of what are called "conditional arguments." Conditional arguments take their name and their structure from a particular type of statement—"conditional statements," of course. So we shall begin by studying conditional statements and then move on to conditional arguments. We shall see that conditional arguments come in four forms, two of which are valid and two invalid. It will be important to be able to recognize and distinguish the valid from the invalid. Finally, there are two other forms of deductive argument that turn up occasionally in scientific contexts. After identifying these arguments we shall move on to actual cases of scientific reasoning.

4.1 CONDITIONAL STATEMENTS

The following are examples of conditional statements. The first is obviously true; the truth of the second and third is still being debated.

If today is Monday, then tomorrow is Tuesday.
If you take vitamin C regularly, your resistance to colds will increase.
If the world's population continues to grow at the present rate, there will be mass famines before the year 2100.

As you can see, conditional statements are a type of *compound statement*. That is, they are formed by compounding two separate statements. In the first of the above examples, the two components are the statements "Today is Monday" and "Tomorrow is Tuesday." The whole compound statement asserts that the second statement is true *if* the first is true. That is, the whole conditional statement asserts that the statement "Tomorrow is Tuesday" is true *if* the statement "Today is Monday" is true.

Conditional statements are so important that even their components have special names. The statement that follows the "if" is called the *antecedent* of the conditional (literally, "that which comes before"). The statement that follows the "then" is called the *consequent* of the conditional (literally, "that which follows after"). So a conditional statement asserts that its consequent is true if its antecedent is true.

Note in particular that a conditional statement does not say that its antecedent *is* true. Nor does it say that its consequent *is* true. It only says that *if* the antecedent is true, then the consequent is true too.

The same relationship between antecedent and consequent can be expressed by saying that the antecedent will be *false* if the consequent is *false*. Thus the vitamin C example may be written as follows:

If your resistance to colds does not increase, you did not take vitamin C regularly.

In logic texts this statement is known as the "contrapositive" of the previous statement, but there is no need for you to remember the name so long as you realize that these two conditional statements express the same relationship between the component statements.

4.2 SUFFICIENT CONDITIONS AND NECESSARY CONDITIONS

Let us begin with a simple conditional statement we all know to be true. For example, pounding on metal makes it heat up. So, referring to a particular piece of wire, we could say:

If the wire is pounded, it will heat up.

What this means is that if the statement "The wire is pounded" is true, then the statement "It heats up" is true as well.

Sufficient Conditions

Another way of expressing the same conditional relationship is to say:

The truth of the statement "The wire is pounded" *is sufficient for* the truth of the statement "It heats up."

These are two ways of saying the same thing.

The complicated statement above could be stated more simply:

Pounding the wire is sufficient to heat it.

This is a perfectly good way of rephrasing the statement, but it tends to suggest more than just the conditional relationship. It suggests that the heating *is caused by* the pounding. The conditional relationship requires only that the *truth* of the consequent is related to the *truth* of the antecedent. It makes no commitment as to *why* the two statements are so related—for example, that the one circumstance causes the other. Still, statements expressing a causal relationship are one type of conditional statement. Indeed, one of the reasons conditional statements are so important for scientific reasoning is that causal statements are typically conditional statements. But not all conditional statements are also causal. For example, today being Monday does not *cause* tomorrow to be Tuesday. We just designate the days of the week in that sequence.

With this qualification in mind, asserting a conditional statement is the same as saying that the antecedent is a *sufficient condition* for the consequent. Pounding the wire is a sufficient condition for its heating up.

Necessary Conditions

Another way of expressing the same conditional relationship is to say:

The truth of the statement "It heats up" *is necessary for* the truth of the statement "The wire is pounded."

This is a third way of saying the same thing.

Keeping in mind the qualification that not all conditional statements express a causal relationship, asserting a conditional statement is the same as saying that the consequent is *a necessary condition* for the antecedent.

At this point you should be warned that the word "necessary" in the expression "a necessary condition" does not have the same meaning as it has when describing the conclusion of a valid argument as "following necessarily" from the premises. There is a way of connecting the two meanings, as you will see in Section 4.5, but they are not the same. It is just an unfortunate fact of the language that has come to be used in these contexts that the same word is used in different, though related, ways.

Causal conditional statements tend to emphasize sufficient conditions. This is probably because most causes are sufficient for some effect, but not necessary. There are almost always other ways of producing any effect. For conditional statements that emphasize necessary conditions, one has to look to another category of conditional statement—for example, statements that express requirements for some goal.

Let us suppose that your school requires a certain number of credit hours for a degree—for example, 120 for the B.A. degree. Accumulating 120 credits, then, is *a necessary condition* for being awarded a B.A. Most likely, it is not also a sufficient condition. Most schools have further requirements for a degree, such as a certain number of hours each in natural sciences, social sciences, and so on. All these are necessary conditions for a degree. None by itself is sufficient. Only if you put them all together do you get a sufficient condition. That is, completing *all* requirements is sufficient for the degree.

Suppose that 120 hours are necessary for a B.A. This relationship may be expressed in the standard "If . . . , then . . ." form as:

If you receive a B.A., then you have completed 120 credits.

This sounds a little strange until you think about it. To say that 120 credits is necessary for a B.A. is the same as saying that getting a B.A. is sufficient for having 120 credits. And that is what the above conditional statement says. If the statement "You receive a B.A." is true, then the statement "You have completed 120 credits" is also true. You could not receive a B.A. degree unless you had that many credits.

In English the way to emphasize a necessary condition is to use the phrase "only if" to introduce the intended necessary condition. So the above conditional statement is equivalent to the following:

You will receive a B.A. *only if* you complete 120 credits.

This is often the way conditions are stated in official documents, such as college handbooks.

In general, a statement of the form "If . . . , then . . ." is equivalent to one of the form ". . . only if . . . ." Thus, the word "if" by itself means something very different than when it is preceded by the word "only." By itself, "if" serves to introduce a *sufficient* condition, that is, the antecedent of a conditional statement. Perhaps the easiest way to avoid confusion is to think of "only if" as a single word that standardly introduces a *necessary* condition in a conditional relationship.

Necessary and Sufficient Conditions

In most conditional relationships, the sufficient condition is not also necessary and the necessary condition is not also sufficient. Pounding a wire is sufficient for heating it, but surely not necessary. One can heat a wire in many other ways—for example, by putting it in a flame or running an electric current through it. Similarly, the wire heating up is necessary but certainly not sufficient for its being pounded. Again, completing 120 credits is a necessary condition for receiving a B.A. but not sufficient. And receiving the degree is sufficient but not necessary for accumulating 120 credits.

Sometimes, however, two conditions are both necessary and sufficient for each other. Today's being Monday, for example, is both necessary and sufficient for tomorrow's being Tuesday. The conditional statement is true with either component statement being the antecedent. When conditions are both necessary and sufficient for each other, we use the expression *"if and only if"* as in:

Today is Monday *if and only if* tomorrow is Tuesday.

In logic texts such statements are called "biconditionals."

Our main use for "if and only if" will be to replace the phrase "just in case" when stating definitions. For example,

A tree is a Canadian maple *if and only if* it is a maple tree growing in Canada.

Definitions give necessary and sufficient conditions for the use of the expressions they define.

4.3   CONDITIONAL ARGUMENTS

Probably the most important and most used types of arguments in all reasoning, including scientific reasoning, are conditional arguments. Now that we know about deductive validity and conditional statements, we know all that is needed to understand these basic types of arguments. The four simplest forms, which are all that we shall look at here, each begin with the first premise being a conditional statement in standard form: "If such and such, then so and so." The second premise is then either the antecedent of the conditional statement or its negation, or the consequent of the conditional or its negation. The conclusion is always that component of the conditional statement which does not appear in the second premise (or its negation). So each of these four arguments has just two premises and a single conclusion. The names attached to these arguments are taken from the role of the second premise. You should learn the names of the four forms because this will help you remember both what the arguments look like and which two of the four types are the valid ones.

In introducing conditional arguments, we shall proceed, as usual, by using examples. But in the discussion of arguments, both deductive and inductive, it is the underlying structure—or "form"—of the argument, not the particular subject matter, that counts. It is the structure, not the particular subject matter, that determines whether the argument is deductively valid or invalid or whether or not it is a good inductive argument.

The best way actually to see the structure of an argument is to substitute simple abbreviations, or place holders, for those parts of the argument that are not relevant to its "logical" structure. In the case of conditional arguments, it turns out that the specific subject matter of the component statements is irrelevant to the validity or invalidity of the argument. All that matters is that there are two component statements and that they be placed correctly in the overall argument. So let us use the letters P, Q, and so on to stand for individual statements. A conditional statement, then, may be abbreviated simply as "If P, then Q," where P stands for the antecedent statement and Q the consequent. Now of course you cannot tell whether the abbreviated conditional statement is true or false without knowing what statements the abbreviations stand for. But remember that deductive validity does not depend on the actual truth or falsity of the premises or conclusions. It depends only on the *impossibility* of there being a false conclusion with true premises. In the case of conditional arguments, impossibility (or possibility) can be determined solely from the abstract structure, which leaves out the subject matter of the component statements.

One warning: People who are unfamiliar with science, mathematics or logic sometimes object to the use of symbols, any symbols. Whenever they see symbols used, they just "turn off" intellectually and emotionally. Part of the reason for this response, I suspect, is that in early encounters with symbols—in high school algebra, for example—these people never saw, and perhaps never were told, the point of it all. Not knowing why this was of any value, they never did catch on, and consequently fared poorly in such encounters. If nothing like this ever happened to you, fine. But if this sounds even slightly like something you have experienced, stop right now and think about it. Do not reinforce the old response by turning off now. Be reassured that we are not going to do anything at all difficult or esoteric. Not having learned much in high school algebra will not make any difference. You really can just think of the letters like *P* as abbreviations for some as yet unspecified statement.

The use of symbols in this context serves a very important purpose. The purpose is to enable you to see that there is a similarity of structure in the reasoning of various sciences all the way from physics to sociology. To see the similarity you must be able to abstract from the particular subject matter. The symbols are to help you make this leap of abstraction. They do so by explicitly eliminating everything but the essential structure of the reasoning. Once you grasp the essential structure, finding it in different subjects is easy.

End of pep talk. We now begin with the easiest of the four forms of conditional argument.

Affirming the Antecedent

Suppose that in due time you come to the conclusion that smoking does indeed reduce life expectancy. If you then apply this knowledge to some particular person, Mr./Ms. Jones, your reasoning, which you would be unlikely to make explicit, could be represented by the following argument:

> If Jones smokes, Jones will have a reduced life expectancy.
> Jones smokes.
> Thus, Jones will have a reduced life expectancy.

This form of argument is called "affirming the antecedent" because the *second* premise asserts—that is, affirms—the statement that is the antecedent of the first premise.

You should assure yourself that this is indeed a deductively valid argument. That is, if the premises are true, the conclusion cannot be false. The first premise says that Jones' smoking is a sufficient condition for Jones to have a reduced life expectancy. The second premise says that the sufficient condition does hold. It is true that Jones smokes. So it must be true that Jones will have a reduced life expectancy. This could be false only if Jones doesn't smoke or if smoking is not a sufficient condition for a reduced life expectancy—that is, only if one of the premises is false. But that is just another way of saying that the argument is valid. It is not possible that the conclusion be false if the premises are true.

It is fairly easy to discern the abstract structure of this argument. Let the letter *P* abbreviate the antecedent of the conditional first premise: "Jones smokes." Let *Q* abbreviate the consequent, "Jones will have a reduced life expectancy." The abstract structure then is:

If *P*, then *Q*.
P.
Thus, *Q*.

You should be able to see that this structure must represent a valid argument, no matter what statements *P* and *Q* stand for. The first premise says that *Q* will be true if *P* is true—whatever statements these may be. The second premise asserts that *P* is true. But if the conditional statement is true and if the antecedent is true, then the consequent must be true too—which is just what the conclusion says. In short, it is impossible that *Q* be false if both premises are true. This satisfies the definition of a deductively valid argument.

Denying the Consequent

To illustrate this form, let us use an argument that we have already considered implicitly in connection with an earlier example. The argument was, roughly, that PCB must not be highly toxic in small doses because there have not been any deaths due to PCB poisoning among people working in plants that manufacture such chemicals. This argument may be set out explicitly in the form of a conditional argument as follows:

If PCBs are highly toxic in small doses, then there would have been some PCB-related deaths among workers in certain chemical plants.
There have been no PCB-related deaths among such workers.
Thus, PCBs are not highly toxic in small doses.

The name of this form comes from the fact that the second premise is the negation—that is, the denial—of the consequent of the first premise.

You should be able to convince yourself that this is a valid argument. The first premise says that there being some PCB-related deaths among certain workers is a necessary condition for PCB being a highly toxic poison. This premise may similarly be read as saying that PCB being a highly toxic poison is sufficient for there to be some deaths among chemical workers. But the second premise denies that the necessary condition took place. So if these two premises are true, it must be the case that the sufficient condition is not satisfied: that is, that PCBs must not be highly toxic in small doses. Which is just what the conclusion says.

To put it in a nutshell, if PCBs were highly toxic, there should have been some deaths. But there were no deaths. So they must not be highly toxic.

Again the validity of arguments with this structure is apparent from their abstract form alone. As before, let *P* stand for the antecedent of the conditional, and *Q* for the consequent. Then the form of this argument is:

If P, then Q.
Not Q.
Thus, Not P.

The first premise says that if the statement P is true, then Q will be true too. But the second premise says that Q is not true. So P cannot possibly be true either. If it were, Q would be; but Q is not.

Make sure that you get a really good feeling for this form of argument and why it is valid. It will turn out that good inductive arguments are quite similar in structure. That is, good scientific arguments tend to be those in which one reasons "backward" from the falsity of some statement to the falsity (or probable falsity) of a hypothesis that implies that statement.

Affirming the Consequent

As the name indicates, these arguments proceed with the second premise affirming the truth of the consequent. And as we shall see in just a moment, arguments of this type are invalid. But arguments of this form often sound quite convincing. For example:

Jones must have studied hard in college, because if one studies hard one will get a degree, and Jones got a degree.

Here we have a whole argument compressed into a single sentence. The first part is the conclusion. You can tell this by the fact that it precedes the word "because." What follows the "because" are premises. Also, the word "must" is not really part of the conclusion but merely serves to indicate that the statement in which it occurs is supposed to follow necessarily—that is, deductively—from the other statements. If they are true, it "must" be true. Immediately following the "because" is a conditional statement in standard form. Following the "and" is another statement which is in fact just the consequent of the preceding conditional statement. So the argument, when written out in standard form, looks like this:

If Jones studied hard in college, then Jones got a degree.
Jones got a degree.
Thus, Jones studied hard in college.

Before reading any further, stop and see if you can tell why this argument is invalid.

The question to ask yourself is: Is it possible that the conclusion is false if the premises are true? Is there any way this could happen? You should be able to think of lots of ways. Even if it is true that studying hard is a sufficient condition for getting a degree, and Jones got a degree, Jones might not have gotten a degree by studying hard. For example, Jones might be very smart and thus not have needed to study much to graduate. Or Jones might have gone to an easy school where no one has to study hard to graduate. Or Jones might have been a football player at a school that values football above scholarship. Just because studying hard would have gotten Jones a degree and Jones did get a degree, it does not follow necessarily that Jones studied hard. The argument is not deductively valid.

It is possible to see the invalidity of affirming the consequent more abstractly by looking at the general structure. As before, let *P* stand for the antecedent of the conditional statement and *Q* for the consequent. Then the form of the argument is:

If *P*, then *Q*.
Q.
Thus, *P*.

Now the question to ask has nothing to do with Jones or any other subject. It is, simply, is it possible that the two premises represent true statements while the conclusion represents a false statement?

An obviously correct philosophical principle is that if something is actual, it must be possible. An impossible actuality is a contradiction. So one need only ask, are there *any* statements that *P* and *Q* might represent that have the result that both premises of the resulting argument are true but the conclusion false. If you can find *any* such statements, then you know that the same *possibility* exists for any argument of this form. So you know that arguments of this type do not *necessarily* yield true conclusions from true premises—which is to say that they are not deductively valid.

How about the following statements? Let *P* represent the statement, "Indianapolis is in Illinois" and *Q* the statement, "Indianapolis is in the United States." Filling in the abbreviations with these statements yields the argument:

If Indianapolis is in Illinois, then it is in the United States.
Indianapolis is in the United States.
Thus, Indianapolis is in Illinois.

Now, as every auto racing fan knows, Indianapolis is not in Illinois but in Indiana. But it is surely true that *if* it were in Illinois, it would be in the United States since all of Illinois is in the United States. The first premise, therefore, is true. So is the second premise, of course, since Indiana is in the United States. But the conclusion is false. This example proves beyond any doubt that it is *possible* for an argument with this structure to have a false conclusion even if its premises are true. This form cannot be trusted necessarily to yield true conclusions from true premises. Such arguments are not valid.

Denying the Antecedent

This form of conditional argument proceeds with the second premise being the negation—that is, the denial—of the antecedent of the first premise. Like arguments that affirm the consequent, arguments of this type are invalid, but they often sound very persuasive.

Imagine being in a class in which the teacher reassuringly tells everyone that they can all pass if they study hard. Knowing that your friend Jones never studies hard, you may be tempted to conclude that Jones is going to fail this course. The argument that represents this little bit of reasoning may be set out explicitly as follows:

If Jones studies hard, Jones will pass this course.
Jones never studies hard.
Thus, Jones will not pass this course.

Again, you should stop and try to figure out for yourself why this argument is not valid.

The trouble again is that although studying hard may be *sufficient* for passing the course, it may not be *necessary*. There may be other ways to do it. Jones might just be very smart and not have to study hard to pass any courses. Or the teacher may be a personal friend of Jones' parents and not want to fail their child. Lots of things are possible that are not ruled out by the premises. So the argument cannot be deductively valid. The conclusion does not have to be true even if both premises are true.

By looking at the general form of the argument it is fairly easy to see how to construct an argument with the same form in which the premises are obviously true and the conclusion obviously false. The form of the argument is:

If P, then Q.
Not P.
Thus, Not Q.

You can see that arguments of this form must be invalid just by reflecting on the meaning of conditional statements. The conditional statement says what will be true if the antecedent is *true*. It does not explicitly say what will be true if the antecedent is *false*.

For an example that proves that this form of argument cannot be trusted to yield true conclusions from true premises we can use a version of the earlier example.

If Indianapolis is in Illinois, then it is in the United States.
Indianapolis is not in Illinois.
Thus, Indianapolis is not in the United States.

Again, the two premises are clearly true, but the conclusion is false. Arguments with this form do not guarantee a true conclusion from true premises, which is just to say that they are invalid.

Conditional Fallacies

Reasoning that follows an invalid form of argument is a type of "fallacious" reasoning. Invalid arguments themselves are often called fallacious arguments, particularly if they give the superficial appearance of being valid. Thus both affirming the consequent and denying the antecedent are fallacious arguments. You should be sure you can recognize these two forms when you meet them because any attempt to justify a statement using either of these forms of argument is doomed to failure. Justified premises alone are not sufficient for justification. You also need a good argument, and fallacious arguments clearly are not good arguments.

4.4   ARGUMENTS USING "AND," "OR," AND "NOT"

Conditional statements are a natural choice for premises of arguments because they assert a connection between the truth of two statements. This is often just the kind of connection needed for constructing a valid argument. But there are many other, and even easier, ways of connecting two statements and constructing valid arguments. Only a few of these, however, play a prominent role in scientific reasoning. While we are looking at deductive arguments, we should look at these other arguments so that we will be able to use them when the need arises.

Denying a Conjunction

Probably the easiest way of compounding two statements is by *conjunction*. For example, the statement "Smoking causes heart attacks in both men and women," is really a conjunction of two statements. Expanded, the statement is:

Smoking causes heart attacks in men *and* smoking causes heart attacks in women.

This statement asserts the truth of *both* component statements. The standard way of forming conjunctions in English is using the word "and," but there are other ways as well.

Another common way of compounding statements in English is by using the word "or." The result is called a *disjunction*. For example, the statement that PCBs are either highly toxic or harmless is really a disjunction of two statements:

PCBs are highly toxic or PCBs are harmless.

This statement asserts the truth of *at least one* of its component statements.

Now, suppose that you have a conjunction of two or more statements and are told that the whole conjunction is false. That is, you are told that the negation of the whole conjunction is true. What can you infer (validly) about the individual statements in the conjunction? The answer: At least one of them must be false. The conjunction asserts that all the components are true. To deny this you have only to assert that at least one component statement is false.

For example, to deny that smoking causes heart attacks in men and women you must assert either that it does not cause heart attacks in men, or not in women, or not in either. Thus asserting that at least one of the component statements is false is really to assert the *disjunction* of the *negations* of the component statements. You are saying: This one is false *or* that one is false (or maybe both are false).

This can all be said more simply using the notation introduced in the previous section. Indeed, seeing this in symbolic form may give you a greater appreciation for the usefulness of the symbolism. Any conjunction with two component statements can be abbreviated as:

P and Q.

The negation of this statement, whatever it is, would then be:

Not (P and Q).

The claim that from the negation of the conjunction one can infer the disjunction of the negations of the components just means that arguments with the following form are valid:

Not (P and Q).
Thus, Not P or Not Q.

The same type of argument can be used with more than two components. If the original conjunction had three component statements, for example, the conclusion would be the disjunction formed from the negations of each of the three component statements. We shall have several occasions to make use of this valid form of argument in Parts II and III.

Disjunctive Syllogism

The first argument we discussed, the original PCB example, was an argument of this form. So was the first case of a valid argument we examined, the Topeka example. So now we are just giving a name to a form of argument we already know to be valid. In brief, the PCB example was:

Small doses of PCB are deadly or they are harmless.
They are not deadly.
Thus, small doses of PCB are harmless.

The abstract form of this argument is:

P or Q.
Not P.
Thus, Q.

We have already seen that arguments of this form are valid. Here we can see it more abstractly. The first premise says one or the other of the two statements is true. The second premise picks out one and says that it is false. But then the other one must be true. So if both premises are true, the conclusion must be true as well. The argument is deductively valid. We shall meet this form of argument again, under a still different title, in Part II.

4.5   CONDITIONAL STATEMENTS AND VALIDITY

One final point will conclude our brief excursion into deductive reasoning. If you see this point clearly, that is a good indication that you have an exceptionally good grasp of the three fundamental notions: tautology, conditional statement, and validity. If you do not see the point at first, give it a little more time. If you still fail to get it, do not worry. Seeing this relationship will make it easy to grasp one important point in our development of inductive arguments. But not seeing it will not be a great loss. And perhaps by the time it becomes relevant you will find it obvious.

It may already have occurred to you that there is a very close connection between true conditional statements and valid arguments. In fact you may even have found yourself, like some texts in logic, reading the conditional statement "If P, then Q" as "P implies Q." If by "implies" one means "validly implies," then this reading is mistaken, but it is a "deep" mistake. The conditional statement merely says that Q is true if P is true. But to say that P validly implies Q is to say that Q *must* be true if P is true. Validity is a much stronger connection than the conditional connection.

There is, however, one way in which the conditional connection can be just as strong as the validity connection, indeed, equivalent to it. This is if the corresponding conditional statement is a *tautology*. A tautology, you will remember, is a statement so constructed that it must be true. So to say that "If P, then Q" is a tautology is to say that P validly implies Q, and vice versa.

As an example, let us take disjunctive syllogism. The form of this argument is:

P or Q. Not P. Thus, Q.

I have written it out in a line rather than in the standard form for arguments so as to make it easier to see the parallel to the corresponding conditional statement, which is:

If [(P or Q) and (Not P)], then Q.

Now the antecedent of this conditional statement is not just a simple statement, but a complex compound statement: the conjunction of a disjunction with a negation. But this antecedent is just the conjunction of the premises of the disjunctive syllogism. And the consequent is the conclusion of the disjunctive syllogism. To say that the premises of the argument are true is just to say that the antecedent of the conditional statement is true. And to say that the conclusion of the argument is true is just to say that the consequent of the conditional statement is true.

Finally, to say that the *argument is valid* is to say that if the premises are true, the conclusion must be true. To say that the *conditional statement is a tautology* means that it must be true, which means if its antecedent is true, its consequent must be true. But this is the same as saying the corresponding argument is valid.

To sum up, an argument is valid if, and only if, the conditional statement with premises as antecedent and conclusion as consequent is a tautology. That, I admit, is quite a mind full.

CHAPTER EXERCISES

4.1 The following statements all express a conditional relationship between two possible circumstances. Restate each relationship in terms of the notions of necessary or sufficient conditions by rewriting the statement in the form ". . . is sufficient for . . ." or in the form ". . . is necessary for . . . ." You should know, of course, that each statement could be expressed in either of these two ways. So either form may be correct so long as you have correctly identified which condition is sufficient and which necessary. But often there is greater emphasis on one condition or the other, and this may make one form more natural. Pick whichever seems most natural to you.

A. If taxes are cut, the rate of inflation will increase.

B. You may obtain a new U.S. passport by mail only if you already hold a passport issued within the preceding 8 years.

C. You will not be issued a driver's license if you are not at least 16 years old.

D. Only if the pressure is less than standard will water boil at less than 100°C.

E. A proposed amendment to the U.S. Constitution will not become law unless it is ratified by two-thirds of the states.

4.2 Rewrite each of B, C, D, and E of Exercise 4.1 in standard conditional form: "If . . . then . . . ."

4.3 Each of the following short paragraphs contains a conditional argument expressed in everyday language. For each argument do the following:

(a) Identify the conditional premise and write it out in standard conditional form: "If . . . , then . . . ."

(b) Identify the second premise and the conclusion and write them in their appropriate places below the first premise.

(c) Letting *P* represent the antecedent and *Q* the consequent of the first premise, write down the general form of the argument.

(d) Identify the argument by name (for example, affirming the antecedent) and state whether it is valid or not.

Note: In order to make these everyday arguments fit one of the four standard forms you may have to leave out certain things as being irrelevant to the "real" structure of the argument. This is a typical problem in dealing with everyday reasoning. In scientific arguments it will be much clearer what the structure of the argument should be and thus easier to spot what is supplemental or irrelevant.

A.   If there were any truth to the rumors about flying saucers, the Air Force would certainly deny it. So there must be something to the rumors because they certainly do deny it.

B.   I guess Smith is not going to graduate this year because unless he satisfies the language requirement he can't graduate and he is not planning to take any courses needed to meet the requirement.

C.   It looks as though we cannot expect much improvement in the economy this year because although all the economists have said that conditions would improve if there were a tax cut, plans for a cut appear to be dead for the rest of the year.

D.   The supply of natural gas is unlikely to increase this year because the companies will increase the supply only if the price is allowed to rise and the government will not act on their request for a price increase before next year.

4.4 The following questions are more difficult and you should not feel that you ought to be able to answer them. Try. If you can answer them, you are ahead of the game. If not, do not worry.

A.   Consider an argument of the form:

Not P or Not Q.
Thus, Not $(P$ and $Q)$.

Can you convince yourself that this argument is deductively valid? So now you know that any two statements with these forms each validly imply the other. In logic, statements that validly imply each other are called "logically equivalent."

B.   Consider a statement of the form:

Not $(P$ or $Q)$

This is logically equivalent to a form of statement that is a *conjunction* rather than a disjunction. Can you figure out what form of statement this is?

Summary to **PART I**

What follows is a list, together with appropriate definitions, of all the new concepts introduced so far. You should be able to go through this list with no difficulties. If you find something you do not understand, or are unsure of, turn back to the section indicated in parentheses. That is where the concept in question was first introduced and explained. Review that section right away. In the remainder of this text these concepts will be used with the assumption that you already know what they are. You will save yourself backtracking and possible confusion if you take the time now to get them all straight.

Scientific Reasoning (1.2)

Distinguish all the "reasoning" that goes on in the actual course of an investigation from the reasoning contained in the final published report of the research. For our purposes "scientific reasoning" is the reasoning of research reports, not the reasoning of the laboratory.

Statement (2.1)

A sentence that asserts how things might or might not be.

True Statement (2.1)

A statement that asserts how things in fact are.

False Statement (2.1)

A statement that is not true.

Definition (2.2)

A statement that gives necessary and sufficient conditions for the use of some expression. If regarded as a statement, a definition must be counted as true. But definitions need not be regarded as statements at all.

Tautology (2.2)

A statement that is so constructed that it must be true no matter how things are.

Negation of a Statement (2.2)

A statement that is false if the given statement is true and true if the given statement is false.

Contradiction (2.2)

A statement that is so constructed that it must be false under all conditions.

Contingent Statement (2.3)

A statement that both might be true and might be false.

Belief (2.3)

A relation between a person and a statement that holds whenever the person regards, or tends to regard, the statement as being true.

Knowledge (2.3)

A relation between a person and a statement that holds only if (a) the person believes the statement, (b) the person is justified in believing the statement, and (c) the statement is in fact true.

Certainty (2.3)

A relation between a person and a statement characterized by an extremely high degree of belief and/or by an exceptionally strong justification for believing the statement to be true.

Argument (3.1)

A set of statements, some of which are designated as premises and others as conclusions.

Premises (3.1)

Those statements in an argument that are intended to provide the justification for the conclusions.

Conclusion (3.1)

The statement (or statements) in an argument that the argument as a whole is intended to justify.

Justification (3.2)

A relation between two sets of statements that holds only if (a) the statements in the first set are already justified and (b) the statements in the second set are conclusions of an appropriate argument that has the first set of statements as premises.

Valid (or Deductively Valid) Argument (3.3)

An argument so constructed that it is necessary that the conclusion be true if the premises are true.

Good Inductive Argument (3.4)

An argument so constructed that it is highly probable that the conclusion be true if the premises are true.

Conditional Statement (4.1)

A compound statement of the form "If . . . , then . . . ." This statement asserts that the second component is true if the first is true.

Antecedent (4.1)

The component statement introduced by the word "if" in a standard conditional statement. The sufficient condition in a conditional relation between two statements.

Consequent (4.1)

The component statement introduced by the word "then" in a standard conditional statement. The necessary condition in a conditional relation between two statements.

Sufficient Condition (4.2)

The condition described by the antecedent of a conditional statement. The truth of the antecedent is said to be *sufficient* for the truth of the consequent.

Necessary Condition (4.2)

The condition described by the consequent of a conditional statement. The truth of the consequent is said to be *necessary* for the truth of the antecedent.

Necessary and Sufficient Conditions (4.2)

Two conditions that are both necessary and sufficient for each other. The components of a biconditional statement.

Conditional Argument (4.3)

An argument with one of the following four general forms, each with a conditional statement as its first premise.

Affirming the antecedent (valid):
  If *P*, then *Q*.
  P.
  Thus, *Q*.
Denying the consequent (valid):
  If *P*, then *Q*.
  Not *Q*.
  Thus, Not *P*.
Affirming the consequent (invalid):
  If P, then *Q*.
  Q.
  Thus, *P*.
Denying the antecedent (invalid):
  If *P*, then *Q*.
  Not *P*.
  Thus, Not *Q*.

Fallacious Reasoning (4.3)

Reasoning that would be represented by a "bad" argument. For our purposes this is any argument that is neither a valid deductive argument nor a good inductive argument.

Conjunction (4.4)

A compound statement of the form "*P* and *Q*." This statement asserts that both of its components are true.

Disjunction (4.4)

A compound statement of the form "*P* or *Q*." This statement asserts that one or the other (or both) of its components are true.

Denying a Conjunction (4.4)
A valid argument of the form:
   Not (*P* and *Q*).
   Thus, Not *P* or Not *Q*.

Disjunctive Syllogism (4.4)

A valid argument of the form:
   P or *Q*.
   Not *P*.
   Thus, *Q*.

The type of argument used to justify scientific hypotheses is basically the same in all the sciences. However, not all scientific hypotheses themselves are of the same type, and one must approach the justification of different types of hypotheses in different ways. We can get along fairly well on the assumption that there are only two importantly different types of hypotheses to consider. One type involves recognized, and recognizable, *theories*. The claim that the universe will go on expanding forever is such a claim. The other type involves *statistical hypotheses* about some population—for example, "One-half of all American men over 18 smoke cigarettes"—and more complicated *causal hypotheses*—for example, "Smoking causes lung cancer." This division of all scientific hypotheses into just two types is, of course, not perfect. It sometimes takes a bit of squeezing to fit a particular hypothesis into one or the other of these two molds. But the occasional difficulty of fitting a particular hypothesis into one of these categories is more than offset by the simplicity of having only two types of hypotheses to consider.

Part II of this text deals solely with those hypotheses that involve scientific theories. The justification of statistical and simple causal hypotheses is taken up in Part III. For most people, knowing how to deal with statistical and causal hypotheses is of more practical value than knowing how to deal with theoretical hypotheses. But the inductive arguments used to justify theoretical hypotheses are easier to understand because they need not involve probability in any detailed way. So from the standpoint of *learning* how to evaluate scientific hypotheses, it is easiest to start with theoretical hypotheses. Once you have learned to deal with inductive arguments involving theoretical hypotheses, the additional complexity of having to think more carefully about probabilities is not all that troublesome.

Before we can consider the kinds of arguments that might justify hypotheses involving theories, we must first understand what a theory is and thus what kind of statements are at issue. This is the purpose of Chapter 5. With the help of simple examples from physics and biology, we shall distinguish theories from theoretical hypotheses and examine several different uses of the term "model."

PART TWO

Reasoning about Theories

Chapter 6 is devoted to explaining how theoretical hypotheses may be tested and thus justified or refuted. In particular, it teaches you how to analyze a report of a test of a theoretical hypothesis in such a way that you can judge for yourself whether that hypothesis is well justified or not.

Chapter 7 applies what you have learned about theories and the justification of theoretical hypotheses to a particular case, the "World II model," presented in the popular paperback book, *The Limits to Growth*.

Finally, Chapter 8 applies what we have learned to various "popular" theories—for example, those concerning possible visits by extraterrestrial beings—and shows why the reasoning offered to justify such hypotheses is generally fallacious.

5

Theories

Everyone knows that theories play an important role in science, and many people know at least a little about some particular theory. For our purposes, however, it will be necessary to have a good grasp of only those few features of theories that are common to all theories, regardless of subject matter. This is all that is needed to be able to reason intelligently about the justification of hypotheses involving theories in any scientific field—and even in fields that are not ordinarily thought of as strictly "scientific."

As you might expect, the question "What is a theory?" is much discussed by professional philosophers of science. Whole books are devoted to this question. Those who want to delve into this issue should consult the references at the end of this text. I shall again follow the course of presenting the view I think best. I must admit at the outset, however, that my view of the general nature of theories is not the most widely held, even among philosophers. But I do think it is the best view, and, more important, that it provides the easiest way for a nonspecialist to understand scientific reasoning about theories.

The best examples of scientific theories are to be found in the most highly developed sciences: physics, chemistry, and biology. This is to be expected because if anything can be said to be *the* goal of scientific activity, it is the development of theories that are simultaneously as detailed and comprehensive as possible. The most highly developed sciences are the most highly developed simply because they have been most successful in achieving this goal. So it is to these sciences that we should look for examples of what theories in general should be like. "But physics?" you say. "That has to be the most difficult subject one could imagine. If I have to learn physics to learn how to reason about theories, forget it!"

Don't panic. You are not going to be asked to learn physics. You are only going to have to read a little bit about simple Newtonian physics which is presented only as an example of the kind of thing a theory is and of how theories are used in science. Learning physics—that is, to produce solutions to problems in physics—is indeed very difficult. But if it is presented correctly, it is possible for anyone to gain some understanding of what physics, especially classical Newtonian physics, is all about. Moreover, discovering that this is so can be a very liberating experience. If you can understand Newtonian physics you can probably understand most any scientific theory presented in a reasonable manner. So learning a little about physics may give you confidence that you can understand scientific theories and even evaluate arguments for or against theoretical hypotheses. An important component in developing the skill to reason intelligently about scientific issues is simply gaining the confidence that you can do it, even if you are not an expert.

There is also a cultural and intellectual reason for beginning with the example of Newton's physics. Throughout the eighteenth and nineteenth centuries and into the twentieth, Newtonian physics was regarded as the paradigm example of a scientific theory. For many people, Newtonian science *was* science. And many others who sought to establish other sciences on sound scientific principles took Newtonian physics as their model. This was true in chemistry and biology, and also in sociology and political science. Even romantic thinkers who opposed the intrusion of science into new areas took Newtonian science as their chief enemy. So Newtonian physics is not merely *an* example of a scientific theory; it is in many respects *the* example.

Through the example of Newtonian physics we shall learn what a theory is and how theories are used in framing hypotheses. After a brief examination of the phrase "law of nature," we shall apply what we have learned to another theory, Mendelian genetics. We shall then see why Newtonian physics is called "deterministic" and genetics "stochastic." Finally, we shall have to look at some of the various things people mean when they call something a "model." We shall then be ready to consider how theoretical hypotheses can be tested and in some cases justified by appropriate inductive arguments.

5.1 NEWTONIAN PHYSICS

Background

Throughout recorded history people have been curious about the motions of the sun, moon, and stars, and about the motions of objects on earth, such as rocks, arrows, and birds. The first major attempt to give a coherent account of the motions of all physical objects was due to Aristotle and his Greek contemporaries around 350 B.C. Aristotle's writings on the subject were collected in a book whose title translated into English is "Physics," which is what we call the subject today. We shall not delve into Aristotle's conception of physics. Indeed, it might be claimed that our modern conception of science is so different from Aristotle's that what we should say about Aristotle's science would have little application to present science. In any case, everyone who has studied the subject agrees that, while Aristotle's view of things was generally accepted for nearly 2000 years, its general acceptance ended around the second half of the seventeenth century, the time of what we now call "The Scientific Revolution." Our present-day conception of science begins there, and the person who first, so to speak, "put it all together," was Newton.

In Newton's time the main branch of physics was called "mechanics," meaning the study of the mechanical motions of all bodies, including the earth, sun, moon, and stars. Today most people would think it at best quaint to speak of the motion of the moon as a type of "mechanical" motion. We generally use the term to refer to machines, such as automobiles. But in physics departments the general subject is still called "mechanics" as you can verify by looking in any college catalog.

So what we are after is Newton's theory of mechanical motions. Now the theory incorporates a number of concepts, some of which were as clear to Newton as they are to us today. The clear ones are position, velocity, and acceleration.

Position

Every physical object may be located at some position. For example, at any moment, every car in Los Angeles could in principle be pinpointed on a large map. The pins would indicate the positions of each car relative to other objects in the area, including the ocean and the nearby mountains.

Velocity

Velocity is speed in some direction. A car traveling north at 50 miles per hour has a velocity of 50 mph in the northerly direction. A car sitting at a stoplight has a velocity of zero. In general terms velocity may be described as change in position over time. A car traveling 50 mph will change its position by 50 miles in an hour.

Acceleration

Acceleration is change in velocity with time. A car accelerating from zero to 60 miles an hour in 1 minute would be accelerating at an average rate of 1 mile per hour per second. (One would not win any drag races at that rate of acceleration.) A car moving at a constant 50 mph has zero acceleration, as does one standing still.

Newton's mechanics also employs several other concepts that are not so clear. These are:

Mass

We all experience mass as weight. But since the advent of space flight, nearly everyone knows that mass is not the same as weight. An astronaut has the same mass both on earth and on the moon, but weighs less on the moon because the moon is smaller and has less gravity—but that is getting ahead of the story.

Momentum

Momentum is defined as the product of mass times velocity. Whatever amount of mass an object has, its momentum is that number multiplied by its velocity. An object standing still has no velocity and thus no momentum, though it may have a large mass.

Force

This is the killer. Whole books have been written just on Newton's conception of force and others on the concept of force in general. Restricting our attention to just gravitational force helps a little. That is what pulls you down when you jump off a diving board. It also holds you down all the time. If it did not, you would just float out into space.

At this point you may think, "Aha! I was right. Even simple Newtonian mechanics is too difficult to be explained easily. Surely Newton must have given a precise definition of mass and force. If he had not, his physics could never have been the finest example of science for more than two centuries."

This is an intelligent response, but it is based on the mistaken assumption that all the concepts in a really good theory must be clearly and precisely defined. As a matter of fact, physicists and philosophers were still arguing over the definition of force and mass into the twentieth century, as you can learn from the references at the end of the text. But the more important point is that good theories *do not have to* give precise definitions of their key concepts. Indeed, it is almost a necessary (but not sufficient) condition for a theory to be really new and important that it introduce *new* concepts, which, just because they are genuinely new, cannot be defined solely in terms of familiar, well-understood concepts.

Where do we go from here? We go to what Newton said about forces, masses, and the motions of bodies in general. What he said is summarized in three famous statements known as "Newton's laws of motion" and one additional statement known as "Newton's law of universal gravitation." We shall come back to the question of what makes a statement a "law." For the moment, these are just statements.

First Law of Motion: If there is no force on a body, its momentum will remain constant.

This statement tells us that if any object is moving at a constant velocity—that is, at a fixed speed in a straight line—then it will just keep moving at that velocity so long as no forces interfere. A special case of this law occurs if the particle is at rest, that is, not moving at all. Then, according to the first law, it must just stay at rest until some force acts on it.

Second Law of Motion: If there is a force on a body, then it will accelerate by an amount directly proportional to the strength of the force and inversely proportional to its mass.

This statement tells us that the more force you apply to a body, the faster it will accelerate. It also tells us that a given force will accelerate a smaller mass more than a larger one. Applied to automobiles, this is to say that the same engine will give greater acceleration to a smaller car than to a larger one. Obvious enough.

Third Law of Motion: If one body exerts a force on a second, then the second exerts on the first a force that is equal in strength but in the opposite direction.

This is the law about action and reaction being equal. So, for example, the earth and the moon exert equal but opposite forces on each other. The reason the moon accelerates a lot more is that it has a much smaller mass.

Now notice that these laws are all in the form of statements that tell us what a particle should do *if* it is subject to certain forces. But none of these statements tells us what the forces themselves would be. This is the job of the gravitation law.

Law of Universal Gravitation: Any two bodies exert forces on each other which are proportional to the product of their masses divided by the square of the distance between them.

This statement tells us that the amount of force between two objects increases with increasing mass and decreases rapidly with increasing distance. Squaring the distance in the denominator means that if you double the distance between two masses, their mutual force will be only one-fourth of what it was.

Figure 5.1 illustrates these relationships (insofar as they can be pictured) for the simple case of a system of just two particles. You may think of them as being the earth and the moon. We will just call them m_1 (meaning mass number one) and m_2. Each has a location (indicated by their position in the picture) and a momentum (p_1 and p_2, respectively). Each exerts a gravitational attraction on the other, equal and opposite as required by the third law.

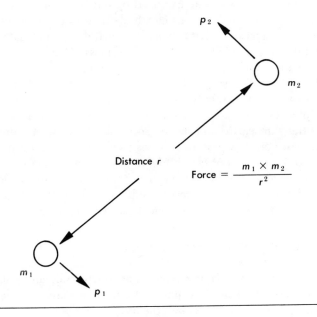

$$\text{Force} = \frac{m_1 \times m_2}{r^2}$$

FIGURE 5.1

Before proceeding, you should make special note of the fact that the above renditions of Newton's laws are given totally in words. No mathematical symbols were used. Yet everyone knows that physics is highly mathematical. What is going on here?

The answer is that in fact there is nothing that can be said using mathematics that cannot be said in words. The problem is that the use of words is so cumbersome that one could never do the calculations required to solve particular problems just in words. For example, all the steps in a simple problem in long division could be written out in words, but you could hardly imagine actually figuring out the answer that way. It takes a special symbolism, really a special language, to do that easily. So if, for example, one wants to calculate how fast a rocket will be traveling after experiencing a certain force for so many minutes, one will need to use mathematics. But to achieve a fairly good intuitive understanding of the fundamental relationships, a special symbolism is rarely necessary. And a reasonable intuitive understanding of what is going on is generally good enough to be able to judge whether a theoretical hypothesis has been justified or not.

While we are talking about mathematics, one additional point should be made. Mathematical reasoning is a type of *deductive* reasoning. That is, all the steps in a mathematical calculation are deductively valid. So when we come to talk about statements that may be deduced from some hypotheses, what is meant is deduction in the sense you have studied in Part I. And this remains true even though you know that in real scientific practice the hypotheses are written in mathematical rather than in ordinary language.

Now let us return to the questions raised by the fact that in Newtonian physics, neither force nor mass seem as clearly defined as position, velocity, and acceleration. One suggestion that may have occurred to you is that we take the four laws as themselves *definitions* of force and mass. These laws do, after all, state certain relations among all the concepts, including force and mass. The immediate difficulty with this suggestion, however, is that definitions do not assert anything about the world, only about how some phrases are to be used. But Newtonian physics is supposed to be a theory about the motions of real objects, such as the sun, earth, and moon. So we are faced with a dilemma. Either we must figure out how theories can be useful in science even though their basic new concepts are not clearly defined or we must figure out how definitions can be used to say interesting new things about the world. Any general account of scientific theories must resolve this dilemma. The account we shall use does so in a neat, though somewhat radical, way.

The Theory of Classical (Newtonian) Particles

At this point let us stop worrying about what Newton said and just worry about what *we* should say about Newtonian physics if we are to understand what is going on with this, or any other, theory.

We shall follow the suggestion that the laws themselves define the basic concepts, but let us carry this suggestion even further. Rather than thinking of the laws as defining the concepts of mass and force individually, we shall take all the laws together as providing a definition of a kind of system *as a whole*. In the case of Newtonian physics, the system in question is a system of physical objects, "bodies," or "particles." So let us take the four laws all together as defining a kind of natural system, which we shall call a classical (or Newtonian) particle system.

Now to call something a "natural system" is not to say very much since, as we shall understand this term, everything in the world is a natural system or part of a natural system. This includes the earth and the moon as well as an automobile. And it also includes plants and animals, and humans too. The main virtue of talking about natural systems rather than just "things" or "events" is that it encourages you to think of the world not merely as a set of individual happenings, but as a complete whole made up of many interacting systems, each of which has its own type of structure. Things do not just happen. They are related to other things in regular, systematic ways. It is the job of scientific theories to define these systematic relations.

The idea that science is concerned with natural systems is not particularly new. Physicists have for centuries studied the solar system. More recently they have turned their attention to "atomic systems" and other "microsystems." Biologists now write about ecosystems, and some psychologists study perceptual systems. Sociologists talk about the social system. Economists and political scientists use the terms "economic system" and "political system" to refer to what most of us simply call "The System." Our account of scientific theories just makes explicit these ways of talking and thinking.

We still have not said just what a scientific theory is. Now we can; that is,

A scientific theory is a definition of a kind of natural system.

It follows immediately that a scientific theory is a single statement, though undoubtedly a very complex compound statement incorporating many statements as part of the definition. Indeed, it will incorporate most of what are standardly called "laws."

Applying our general definition of what a theory is to classical particle mechanics, the theory itself may be stated as follows:

A natural system is a classical (Newtonian) particle system if and only if it is a system of objects satisfying Newton's three laws of motion and the law of universal gravitation.

Instead of simply referring to Newton's laws following the "if and only if," we could have said "if and only if it is a system of objects satisfying the following laws:" and then actually written in the laws themselves. But this would just make the definition very long and would serve no useful purpose here. The important thing to realize is that the definition given above is a statement of the theory of classical particles.

Now we have to face the other part of the dilemma. If a theory is literally a definition, then it is automatically true because it does not tell us anything about how things are. It only tells us how the phrase "classical particle system" is to be used. But are not theories supposed to tell us about the world? Our answer must be a somewhat radical sounding, "No. Not theories themselves." However, I hasten to add, there are other statements, very closely related to theories, that do make assertions about the world. These we will call *theoretical hypotheses*.

Theories and Theoretical Hypotheses

An example of a theoretical hypothesis based on the Newtonian particle theory is:

The solar system is a classical particle system.

This is a genuine contingent statement about a system in the real world. This statement is true if our actual solar system fits the definition of a Newtonian particle system and false if it does not.

In general, for any theory, a theoretical hypothesis has the form:

Such-and-such real system is a system of the type defined by the theory.

So, in our account of theories, all theories have the same general form, that of a definition, and all theoretical hypotheses have the same general form, that of a contingent statement asserting that some designated real system fits the theory. These relationships are pictured in Figure 5.2.

You should explicitly note that our use of the word "theoretical" is special. In everyday speech, and in the writings of many philosophers of science, "theoretical" tends to imply being speculative or not really real. Thus religious fundamentalists reject the theory of evolution saying "That's *just* a theory." And philosophers puzzle over the status of "theoretical entities" like electrons and magnetic fields. It is taken to be a serious question whether, or in what sense, theoretical entities are real.

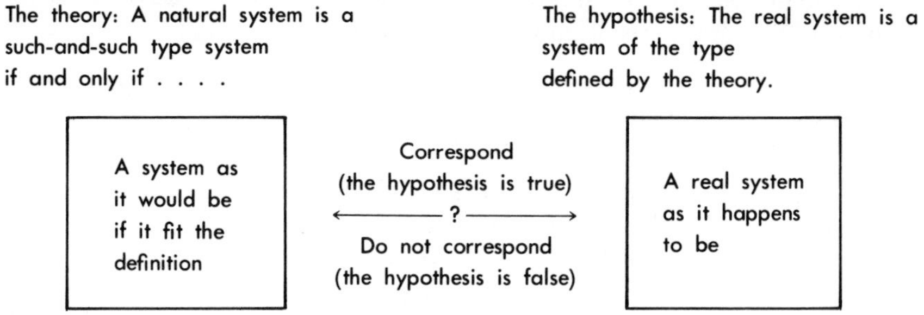

The theory: A natural system is a such-and-such type system if and only if . . . .

The hypothesis: The real system is a system of the type defined by the theory.

| A system as it would be if it fit the definition | Correspond (the hypothesis is true) ←———— ? ————→ Do not correspond (the hypothesis is false) | A real system as it happens to be |

FIGURE 5.2

For our purposes, these associations do not apply. A theoretical hypothesis is simply a hypothesis that asserts that some real system fits some particular theory. It is "theoretical" simply because it applies some theory. And theoretical hypotheses are true or false just like any other contingent statements.

Our only concession to the standard associations of the word "theoretical" concerns the *justification* of such hypotheses. Theoretical hypotheses usually cannot be justified simply by observation. Their justification requires a special form of inductive argument, as we shall see in Chapter 6. For example, anyone can see that the moon appears to move across the sky. That is a claim that anyone can easily justify. But that the motion of the moon should qualify it as part of a Newtonian particle system is not a claim that anyone could justify just by looking at the moon on a clear night. More is required.

For any theory, there are innumerable hypotheses that one could formulate using that theory. These may be very specific, applying to just a single system, such as the solar system, or they may be very general, applying to a whole class of systems. The most general hypothesis ever considered using classical particle theory was propounded by the French physicist and mathematician, Laplace, around 1800. Laplace's claim was:

The whole universe is one big Newtonian particle system.

This includes you and your dog as well as the sun, moon, and the distant stars. You, for example, are made up of little particles. Laplace claimed that these particles form a Newtonian particle system. Needless to say, he created quite a stir. Most people do not like thinking of themselves as just complex systems of particles, let alone ones that fit Newton's laws.

In science, one is often less interested in a theoretical hypothesis itself than in what can be deduced from it. Thus, to take a simple example, consider the hypothesis that a pendulum is a classical particle system. From this hypothesis one can deduce the well-known "law of the pendulum," which gives the relationship between the length l of the pendulum and its period T—that is, the time it takes to make a full cycle across and back. Deducing this relationship is a standard problem in elementary physics texts. The relationships among the theory, the theoretical hypothesis, and the "law" deduced from the hypothesis are pictured in Figure 5.3. In the law of the pendulum the letter g stands for a number that expresses the strength of the gravitational force exerted by the earth on the pendulum. Most problems in physics take the form of having to deduce some characteristic of a system from the hypothesis that it is a system of a specified type, such as a classical particle system. This partly explains why physics is so mathematical and why it is so difficult. These deductions can be very complicated.

You may still have the uneasy feeling that there is something wrong with thinking that theories, by themselves, are just definitions. Definitions are pretty trivial, and theories are not. But you misunderstand. Framing a definition is something anyone can do. But it takes the genius of a Newton to

Theory: The definition of a
Newtonian particle system

Theoretical Hypothesis: A
pendulum is (that is, satisfies
the definition of) a Newtonian
particle system

Deduction from the Theoretical Hypothesis:
"Law of the pendulum," or $T = 2\pi\sqrt{l/g}$

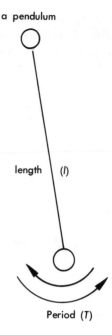

FIGURE 5.3

formulate a definition that will capture the essentials of all physical motion from
the motion of the moon to the motion of an apple falling from a tree. And it
takes genius to realize that one and the same definition applies to both—that
they are both systems of the same type. This is what Newton and his
predecessors—for example, Galileo—had accomplished before the end of the
seventeenth century.

Finally, distinguishing theories from theoretical hypotheses emphasizes a
second way in which science is knowledge expanding. Not only do inductive
arguments have more information in their conclusions than in their premises,
but when the conclusion is a theoretical hypothesis, the conclusion may contain
new concepts as well. These are introduced by the theory, which may define a
totally new kind of system. In everyday life we do not have to distinguish
between our definitions and our hypotheses because we simply take the definitions that are given. But science is in the business of expanding our knowledge,
and one way it does this is by introducing totally new concepts. That is one of
the main functions of theories. Our account of what theories are makes this
function clear and understandable.

5.2 LAWS OF NATURE

The expression "law of nature" has a long history going back to a time
when people generally thought of nature as obeying the laws of its Creator in

something like the way subjects obey the laws set down by their king. Only nature never breaks the law. Today this association of the expression is generally forgotten, but people still tend to think that a law of nature is something special and that it is important to know what makes some hypotheses laws and others not.

From our point of view, there is nothing to be gained by trying to make sense of the notion of a law of nature. In any given situation we can understand everything that needs understanding simply by being clear what the theory is and what theoretical hypothesis is at issue. For us, what are called Newton's laws are just component statements in the definition of a Newtonian particle system. Strictly speaking, they should no longer probably even be called laws of *nature* because nature as we now know it is not a Newtonian particle system. It may be an Einsteinian particle system, but even that is up for grabs.

5.3   MENDELIAN GENETICS

If our account of theories is to be useful, it must apply to sciences other than physics. So we shall now look at the science of biology—in particular, genetics. We shall see that the "classical" theory of genetics, Mendel's theory, can be isolated in the form of a definition of a kind of system and that the various hypotheses employing this theory can be easily distinguished. It is not difficult to get a handle on the essentials if you know what to look for.

Background

It is an obvious but interesting fact, again first discussed systematically by Aristotle, that the offspring of goats look like goats and the offspring of humans look like humans, not goats. Moreover, as Aristotle would have put it, the offspring of Greeks look like Greeks, and the offspring of barbarians look like barbarians. Even individual children tend to resemble their parents and grandparents more than others. Why is this so? How do characteristics of parents get transmitted to their children?

We shall once again have to pass up what Aristotle said. The scientific revolution of the seventeenth century changed biology as well as physics. But the real revolution in biology did not occur until the middle of the nineteenth century with Darwin's discovery of natural selection and Mendel's investigations into the mechanisms of inheritance.

Mendel did most of his experimentation on ordinary sweet peas. Since peas reproduce sexually, his results are relevant to other sexually reproducing organisms, such as humans. For the moment let us follow Mendel and stick to peas. Mendel was following a good scientific strategy by beginning with something simple and easy to study.

Mendel noted that pea plants exhibit a number of characteristics that come in pairs, called "traits." The peas themselves (the seeds of the plant) are either smooth or wrinkled. They are also either yellow or green. And the plants themselves come in two varieties, tall and short. To keep things simple we shall concentrate on just one characteristic, height.

Mendel discovered that when short plants are "cross-pollinated" with other short plants, the resulting seeds always yield short plants. However, tall plants seem to come in two types. Some pairs yield seeds that lead only to tall plants, but others yield a mixture of tall and short "offspring." The short plants, as well as the tall plants that produce only other tall plants, are called "true-breeding" plants, whereas the other tall plants are called "hybrids."

Others before Mendel had discovered that if you cross-fertilize the true-breeding tall plants with the short plants (which are true-breeding), the result is *all tall* plants. But these tall plants are all hybrids. If these hybrid plants are then fertilized together, they yield *both* tall and short offspring. So whatever it is that is responsible for determining whether a plant is short or tall seems to be transmitted through a generation of plants that are themselves all tall. The short plants show up again only in the *next* generation. Mendel's unique contribution was actually to count the different types. He discovered that the ratio of tall to short in the second generation was roughly 3 to 1. This ratio occurred again and again, and with other characteristics besides height. Why?

Figure 5.4 shows Mendel's results. His explanation was the following. Suppose that there are *two* things that determine any characteristic, such as height. Mendel called these things "factors," although he did not define them. (Factors correspond roughly to what we now know as genes.) Suppose further that one factor is associated with being tall and the other with being short. Also, suppose that each individual has some combination of two of these two types of factors. Let H stand for the factor associated with tall plants and h for the factor associated with short plants. Then there are really three different types of individuals: those with two "tall" factors, those with two "short" factors, and those with one of each. These three possibilities are shown schematically in Figure 5.5.

Next, assume that upon "mating" (i.e., cross-pollination) the seeds for the next generation get one factor from each "parent." Which two factors any seed gets are assumed to be a *random selection* from the four pairs available. This is the crucial part of what is usually called "Mendel's law of segregation." In the formation of the seed that becomes a plant in the next generation, the four original factors (two from each parent) "segregate" themselves at random into groups of two, one from each parent.

Finally, suppose that those individuals that have one factor of each type (i.e., the hybrids) exhibit the trait associated with only one of the two, which by definition is called the "dominant" trait. The factor associated with this trait is then called the "dominant factor." That one of the factors should be dominant in this sense is known as "Mendel's law of dominance." The other trait of the pair is called "recessive."

Now let us take a new look at Mendel's experiment represented in Figure 5.4. This time, however, let us represent each individual by its factors alone.

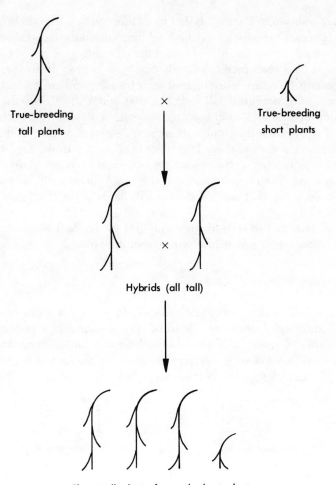

Three tall plants for each short plant

FIGURE 5.4

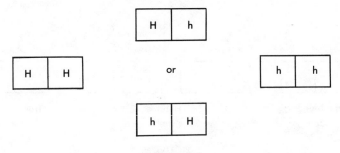

FIGURE 5.5

The result is shown in Figure 5.6. Let us go through this diagram carefully. The original "parents" consist of a true-breeding tall and a true-breeding short. Their union must produce hybrids since the offspring must get one factor from each parent, and each parent has only one type to give. However, with both hybrid parents there are four different ways the offspring can get its two factors: A tall from each parent, a tall from the first and a short from the second, a short from the first and a tall from the second, or a short from each. Since by the law of segregation the combinations are selected randomly, there should be equal numbers of each of these four types of offspring. However, by the law of dominance, only one of these four types will actually be short. The other three—the one true-breeding tall and the two hybrids—will all be tall. Thus the observed result that tall and short plants occur in the third generation in a ratio of 3 to 1.

The above is the kind of story one gets in standard elementary texts in genetics. Now let us put it into our general framework.

Mendelian Inheritance Systems

You should realize, now, that Mendel's theory is not about sweet peas. It is an account (a definition) of a kind of system, namely, a system in which characteristics of individuals are inherited through sexual reproduction. The basic characteristics of such a system are given by the laws of segregation and dominance. Let us first try to state these laws more precisely.

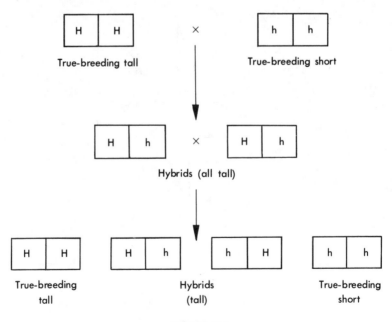

FIGURE 5.6

Law of Segregation: For any characteristic of an offspring of reproduction, the two factors obtained by the offspring are a random selection of one from each parent.

Law of Dominance: A hybrid individual exhibits the trait associated with just one of the two truebreeding types. (This trait and its associated factor are called "dominant.")

We can now state Mendel's theory as a definition of a kind of system.

A system of inheritance of a single characteristic through sexual reproduction is a Mendelian system if and only if it satisfies the laws of segregation and dominance.

As in the case of Newton's theory, we could, of course, actually write out the laws in the definition, but it is simpler just to refer to the laws that have already been stated.

One of the *theoretical hypotheses* that Mendel proposed (though of course not in these words) was:

The system of inheritance of height in sweet peas is a Mendelian system.

This, of course, is not a definition, but a contingent hypothesis, one for which Mendel thought he had ample justification. At the moment we are just concerned with what the hypothesis is and what it means, not with whether it is justified or not. We shall get to that shortly.

As in the case of Newton's theory, there are numerous other hypotheses one can formulate in terms of Mendel's theory. Mendel himself advanced similar hypotheses for at least six other characteristics of sweet peas. It is also now generally claimed that several human characteristics fit the Mendelian definition—for example, eye color and possession of sickle-cell anemia.

5.4 DETERMINISTIC AND STOCHASTIC SYSTEMS

Systems seem to come in two basic types, called "deterministic" and "stochastic." The theories that define such systems go by the same names. Since we have now seen an example of each type of theory, this is a good time to learn what these types are. The distinction will be important when we get to simple causal hypotheses in Part III.

Newtonian particle systems are *deterministic* systems. To explain what this means we need one other concept, that of the "state" of a system. The state of a system, as defined by a particular theory, is described by giving a complete rundown of all the characteristics of the system referred to by the theory. Thus, to give the state of a Newtonian system at a particular time, you must specify the positions and momenta of all the particles at that moment, and give all the forces acting within the system.

To say that a Newtonian system is *deterministic* means that its state at any one moment of time completely determines its state *at all other times*. Thus if you could specify the positions and momenta of every particle in the solar system right now, then, assuming the action of universal gravitation and no interference from outside this system, the position and momentum of every particle would be uniquely determined for all time.

Now you see better why Laplace's contemporaries were unhappy with the idea that the whole universe is just one big Newtonian system. If Laplace is right, all motions of the minutest particle have been uniquely fixed "from the beginning of time." In particular the behavior of everyone who has ever lived, and ever will live, is precisely predetermined. This conflicts strongly with most peoples' feelings of being "free" to make choices and do "unpredictable" things. Thus arises the philosophical problem of "Determinism and Free Will." We don't dare get started on that.

A *stochastic* system, by contrast, is one in which the state at any one time determines only the *probability* of various possible states at later times. Mendelian genetics is a stochastic theory because the factors possessed by both parents do not in general uniquely determine the factors possessed by any offspring. They only determine the probability of the offspring having one of the available combinations. For example, suppose eye color is a Mendelian characteristic in humans with the two traits being "blue" and "brown." Then if brown is dominant, and if both you and your mate (or a prospective mate) are hybrids, your chances of having a blue-eyed child would be only 1 in 4. The eye color of your (potential) child is not a deterministic result of the state of you and your mate.

Whether stochastic systems are any more "free" than deterministic ones is also a question we don't dare get into here.

5.5 THEORIES AND MODELS

The word "model" occurs with increasing frequency in scientific literature, both professional and popular. Unfortunately the word has several, quite different uses. The only way to deal with this situation is to become familiar with the different uses so that you can quickly sort out what is being said on any particular occasion.

Scale Models

Model cars and model airplanes are scale models of real things. Such models are also used as teaching aids in the sciences, and sometimes in research. On the demonstration tables in a chemistry lecture room, for example, you may see scale models of molecules and crystals. Anatomy departments use models of various human organs (e.g., the heart) that you can take apart to see the inner workings.

Most of the time when scientists talk about models, however, they are *not* referring to scale models. But knowing that this is one meaning of "model" will enable you to pick out those cases in which the word is used in this way.

Models and Analogies

The use of analogies in science is neatly illustrated by early attempts around 1900 to develop a theory of the structure of atoms. From the then-recent discovery of X rays, radioactivity, and other phenomena, it was clear that atoms contained both positive and negative electric charges. The problem was to figure out how these charges were arranged in an atom in such a way that would explain how atoms could exist as stable entities and not simply fly apart.

One of the first attempts to describe the actual structure of atoms was based on what has been called "a raisin pudding model" of the atom. What this means is that there are *analogies* between the atom and raisin pudding, analogies that can be used in investigating what the atom is like. In general, to say that there is an *analogy* between two things is to say that they are alike *in some respects*, though of course not in all. In the case of the atom, it was thought that the positive charge is spread out like pudding, and that the electrons move around in the pudding like raisins when the pudding is stirred.

Later experiments by the English physicist Ernest Rutherford seemed to show that the positive charge was not spread out, but highly concentrated. Rutherford, and later Niels Bohr, proposed a different model. They suggested that we think of the atom as being analogous to the solar system. An atom, they said, is like a miniature solar system. All of the positive charge is in the center, just as the sun is in the center of the solar system, and the negative charge is carried by electrons, which move around the center like planets in orbits. So Rutherford and Bohr were taking the solar system as a model for the atom and saying that its structure is analogous to the structure of the solar system. This model is still used today in elementary discussions of atomic physics.

This use of models in science can be described in general terms as follows. There is a type of system, such as atoms, about which not much is known. However, there are other systems, such as solar systems, about which a lot is known. In 1900 there were already good theories of solar systems (e.g., Newton's). Someone then suggests that maybe the unknown type of system is like the known one in certain important respects. This in turn suggests questions that one should ask about the unknown system: How fast are the electrons moving around in their orbits? Are the orbits circular or elliptical? and so on. The model (i.e., the known system) also suggests ways of answering the questions. In this way one is led to discover how the previously unknown system does in fact work. In particular, one discovers ways in which it is *not* like the model system.

So it is clear that models as the basis for analogies do play an important role in scientific research—that is, in the creation of new theories. The question for us, however, is whether analogies play any role in the *justification* of the new theoretical hypotheses. For example, does the analogy with the solar system provide any justification for the hypothesis that an atom is a Bohr-type system—where a Bohr-type system is the type of system defined by the *theory* Bohr proposed?

This is again a matter of debate among philosophers of science. The safe answer is no. There is no really good account of inductive arguments that are based on analogies, although several have been proposed. There is, however, a good account of how to justify theoretical hypotheses themselves. So we shall say that each theoretical hypothesis must stand on its own. However useful some model may have been in creating the theory behind a theoretical hypothesis and however well justified our hypotheses about the model itself, any new theoretical hypothesis about another type of system must be justified by information about that type of system. It cannot be justified by information about the model.

Models and Theories

Theoretical hypotheses, as we have seen, all have the same general form: Such-and-such real system is in fact a system of the type defined by so-and-so's theory. Now if such claims are taken literally as meaning that the system in question fits the definition *exactly* and *in all details,* then few, if any, interesting theoretical hypotheses are true. Most are known to be false right from the start. This may come as a shock, but it is really not all that surprising if you think about it.

In biology, for example, not even the characteristics of sweet peas fit the Mendelian pattern perfectly. And eye color in humans, which is often cited as an example of a Mendelian trait, obviously does not fit exactly. There are, for example, more than the two colors, blue and brown. Some people have eyes that are gray, blue-green, or even green. There are even cases of people with one brown eye and one blue. So the real systems (e.g., humans) are more complicated than Mendel's definition allows.

The same is true even in physics. No real system behaves exactly like a Newtonian system ought to, although many do behave very similarly. There are several reasons for this. One is simply that mass and velocity, for example, are not related in just the way Newton prescribed. When very high velocities (i.e., velocities near the speed of light) are involved, the discrepancy between Newton's theory and real systems becomes obvious. Presumably this is taken care of in Einstein's "relativistic" mechanics. Even Einstein's theory, however, treats only mechanical forces. It leaves out nuclear and other even more esoteric forces. But in any real system these other forces are present, at least to some degree, and they do influence the behavior of any system. So no real system behaves exactly like an Einsteinian system either.

So it is only in a few areas of physics that scientists have theories that yield hypotheses whose deviation from the truth is not currently detectable in actual experiments. In other areas of physics, and in all the other sciences, the theories used are *known* not to yield hypotheses that are exactly true.

But is not science interested in true hypotheses? Of course. But theories that yield hypotheses whose deviations from the truth are undetectable are generally beyond our powers to construct. A theory that yields hypotheses with known, and not too large, deviations from the truth is far better than no theory at all. Such a theory provides at least a basis for further investigations and may eventually lead to the discovery of a better theory—that is, one yielding hypotheses that more effectively describe the systems of interest.

In many areas of science, but especially in psychology and the social sciences, there is a tendency to use the word "model" rather than "theory" in order to emphasize the fact that the corresponding hypotheses are obviously not true, and in many cases quite far from true. In these contexts a "model" is, like a theory, a definition of a kind of system. It is just that it is well known that the real systems of interest are only approximately like the defined systems, and then maybe only in a few respects. These "models" (read "theories") are used partly because there are none better and partly because they do provide *some* insight into the workings of the real systems, even if they obviously do not tell the whole story. In Chapter 7 we shall examine one such model in detail, the "World II model" presented in *The Limits to Growth*.

Sometimes when people talk about a particular model, they seem to be referring not to the definition, and not to any real system, but to "the system defined by the definition." Indeed, people will refer to "properties of the model" when it is quite clear that they are not referring to properties of a definition. They seem to be talking about a system, not a definition. You might well wonder, "What system is this?"

The answer is that such talk is just a convenient way of referring to what a real system *would be like* if it *were* to fit the definition exactly. It *would have* all of the characteristics specified in the definition. But in English, as in most languages, the subjunctive form is cumbersome. It is much easier to speak *as if* one is referring to some definite thing. So one in effect creates a hypothetical entity, "the system defined by the definition," and discusses its properties, which, of course, are just the properties specified in the definition. This is a convenient linguistic device, and we shall use it occasionally in this text. Just do not be misled into thinking that in addition to definitions and real systems, the world somehow also contains things called "models," which are systems that actually do have all the properties specified in the definition. Such things are purely fictitious.

Now that we have straightened out the differences between "theories," "models," and "theoretical hypotheses," we can move on to questions of justification. But keep in mind that theoretical hypotheses are not exactly true. The justification offered for a theoretical hypothesis will never justify believing that it is exactly true, but at best only that it is *approximately* true. This is another way in which the conclusions of scientific reasoning are "not certain."

CHAPTER EXERCISES

5.1 By writing the letter T, H, or N in the indicated space, identify each of the following statements as being:

T: A theory or a statement that follows deductively from a theory itself.

H: A theoretical hypothesis or a statement that follows deductively from a theoretical hypothesis itself.

N: Neither T nor H (i.e., some other kind of statement).

A.   A shooting star is part of a classical particle system.   _____

B.   In Mendelian inheritance systems, individuals with one dominant and one recessive factor exhibit the dominant trait.   _____

C.   In a Newtonian particle system each particle attracts each other particle.   _____

D.   In humans, eye color is transmitted by a Mendelian inheritance system.   _____

E.   An apple falling from a tree will accelerate at the rate of 16 feet per second per second.   _____

F.   A characteristic in a Mendelian inheritance system satisfies the law of segregation.   _____

G.   The force between the earth and the moon is proportional to the product of mass of the earth and the mass of the moon.   _____

H.   Isaac Newton had blue eyes.   _____

I.   In humans, blue eye color is a recessive trait.   _____

5.2 Describe, in general terms, three different things that could be meant by the phrase "Bohr model of the atom."

5.3 Suppose you come across a reference to "the Keynesian model of the economy" in a newspaper article about the effect of cutting taxes. What is the most plausible interpretation of the meaning of the phrase "Keynesian model of the economy"?

5.4 In 1950 physicists studying the structure of the nucleus of atoms were in a similar position to people studying the structure of atoms in 1900. In 1950 nuclear physicists were talking about "the liquid-drop model of the nucleus." Describe in general terms what they must have meant by this phrase.

5.5 Imagine a system for which there are some states that uniquely determine all later states, but other states that determine only probabilities for later states. Would you call such a system deterministic or stochastic? Explain your choice. Can you support your answer with an example from either Newtonian physics or Mendelian genetics?

The following questions concern either Newtonian mechanics or Mendelian genetics. You should not feel that you have to be able to answer these questions, but being able to do so shows that you have a good grasp of the basic ideas.

5.6 It is known that some serious birth defects are genetically determined and that some even follow a simple Mendelian inheritance pattern. Suppose you were to learn that you are carrying a recessive factor for some serious defect that, because of the law of dominance, you yourself do not have. If you were to have a child with someone genetically like yourself in this respect, what are the chances that your child would exhibit the unwanted trait? If you were to have a child with someone who is not carrying the recessive factor for this trait, what are the chances that your child would exhibit the unwanted trait? (In Part IV we shall consider how this difference in probabilities might be relevant to your choice of mates.)

5.7 In modern treatments of classical mechanics, Newton's first law does not appear because it is a direct deductive consequence of the second law. Try to explain in words why this should be so.

5.8 Galileo is famous for being among the first to claim that any two bodies should fall at the same rate, regardless of their weight. The legend is that he demonstrated this claim by dropping large and small cannonballs simultaneously from the top of the Tower of Pisa. Galileo's claim is a logical consequence of Newton's theory applied to bodies near the earth. Try to explain why Galileo's claim is indeed true for Newtonian systems. (*Hint:* You need only consider the second law together with the law of universal gravitation.)

5.9 This question is just to give you something to think about. By 1920 most scientists were convinced that everything, including humans, was made up completely of atoms. Then in the 1920s physicists discovered that atoms are stochastic, not deterministic, systems. That was one of the exciting things about the new "quantum mechanics." Well-known physicists then gave popular lectures in which they claimed that physics had demonstrated the existence of "free will." Do you think that the fact that atoms are stochastic systems shows that free will exists? (*Hint:* Think about Mendelian genetics.)

6

Testing Theoretical Hypotheses

Scientists talk about testing a theory, or model, by seeing whether or not it "fits the facts." In this chapter we shall see what this means and how it is done. Our object will be to learn how to judge for ourselves whether it has been done well or not.

In general, testing a theory involves applying it to some real system, which for us means formulating the theoretical hypothesis that says that the theory fits. Judging whether the theory fits is really judging whether or not the hypothesis is true. So for us the objective in testing a theory is easily understood as trying to determine whether a corresponding hypothesis is or is not true. And determining that it is requires producing an argument with justified premises that justifies the hypothesis.

So far we have only seen theoretical hypotheses in the role of *premises* of deductive arguments. The deduction of the "law of the pendulum," for example, involved a valid argument in which one premise was the hypothesis that a pendulum is a Newtonian particle system. But justifying a theoretical hypothesis itself requires that *it* be the *conclusion* of an argument in which some *other* statements are premises. We already know that the appropriate argument in this case will be an *inductive* argument. Now we must learn about the detailed structure of these arguments. This means discovering what kinds of statements may serve as premises.

As in the previous chapter we shall develop our account of the testing and justification of theoretical hypotheses in the context of a simple example from physics: a test of a Newtonian hypothesis made possible by the appearance and reappearance of Halley's comet in the seventeenth and eighteenth centuries. We shall then apply what we have learned to analyze an experiment that Mendel used to test his hypotheses regarding sweet peas. The exercises that follow provide you with the opportunity to apply these same techniques to a variety of other examples. Detailed answers to some of the exercises are given in the back of the book.

You should regard the physics and genetics examples as models for doing other problems. Here I mean "model" in the sense of something that provides useful *analogies* you can use in analyzing tests of other hypotheses. Thus it is important that you get a good grasp of what is going on in the Newtonian and Mendelian examples, not so much because these cases are so important in themselves but because they provide you with helpful guidelines for dealing with new cases.

6.1 TESTING A NEWTONIAN HYPOTHESIS: HALLEY'S COMET

Newton's theory was first published in 1687. In 1695 a young astronomer, Edmond Halley, began applying Newton's ideas to the motions of comets. Comets are a very interesting phenomenon, historically, because they had always been viewed as mysterious and even ominous. If even comets behave as Newtonian particles, then surely all of the heavens are a Newtonian system, and not just the earth, moon, sun and other planets. Halley began investigating a comet that he himself had observed in 1682. Halley proceeded to deduce what should be the orbit of the comet if it and the sun together constitute a Newtonian system. He realized that there must be some influence from the gravitation of the largest planet, Jupiter, but ignored this as being relatively small and too difficult to calculate.

Using data recorded in 1682 on the path and motion of the comet, Halley deduced that the comet should travel in a large elongated ellipse and that it should take about 75 years to make one complete circuit. He concluded that the comet must have been around many times before. Sure enough, there had been comets reported at roughly 75-year intervals going back to 1305. Halley speculated that they were all the same comet traveling in an elliptical orbit as given by Newton's hypothesis. He then proceeded to calculate the time of the next return, which turned out to be, given the uncertainties in his data, sometime in December 1758. Figure 6.1 will help you to keep in mind the relevant details of this example. It is obviously not drawn to scale.

Halley published his work on comets in 1705, but it attracted only minor attention. He died a respected scientist in 1743, fifteen years before the predicted return of the comet. The comet reappeared on Christmas Day, 1758, within the predicted time interval. The whole European intellectual community then took notice. Even the French, some of whom still clung to the earlier theories of Descartes, were persuaded of the value of Newton's theory. It was then that the comet was named "Halley's comet."

Intuitively it seems quite clear that the success of Halley's prediction did provide substantial, if not conclusive, justification for the hypothesis that the comet and the sun constitute a Newtonian system. Why is this so? What made this a good test of Halley's hypothesis? How does the experimental verification of Halley's prediction provide an argument that justifies this hypothesis?

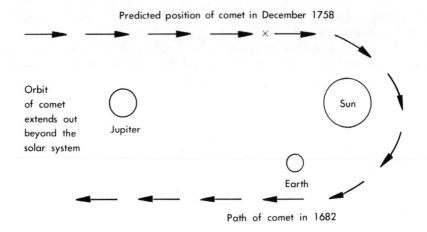

Predicted position of comet in December 1758

Orbit of comet extends out beyond the solar system

Jupiter

Sun

Earth

Path of comet in 1682

FIGURE 6.1

6.2   THE BASIC ELEMENTS OF A GOOD TEST

The preceding section presents the facts of Halley's discovery in the way that is typical of reports involving the justification of theoretical hypotheses. In such reports not much distinction is made between the experiment and the result of the experiment nor between the experiment and the justification of the hypothesis. These things must be distinguished, however, if we are to develop a general account that can be applied to a variety of cases. In this section we shall concentrate on the distinction between the experiment or observations that provide a *test* of the hypothesis and the *outcome* of the test.

Experiments, Results, and Justification

Before we examine the details, it will be helpful to see the basic distinctions set out in a preliminary way. This will give you some idea of how the various pieces are supposed to fit together so that you can better appreciate the details as they are presented.

The basic idea behind any experiment or observation designed to *test* a theoretical hypothesis is that things should be arranged so that in the end you will be sure to get information that you can then use to argue either that the hypothesis is true (or approximately true) or false (or not approximately true). You ought to be able to tell at the beginning, before the experiment is actually performed, whether or not things have been so arranged. In Halley's case, for example, it was clear in 1705 that Halley had set up a good test of the hypothesis that the comet and the sun formed a Newtonian system.

When the experiment terminates, something will have happened, or failed to happen. This is the *result* of the experiment. You cannot know what this is, of course, until the experiment is over. In Halley's case the results were not in until 1758, fifteen years after Halley himself had died. In this example the *experiment* was simple. Just wait until 1758 and look to see whether the comet reappears. The *result* of this experiment, of course, was that it did. It might not have. That it would was a contingent statement.

Now to *justify* either the hypothesis or its negation, one constructs an argument. The premises of this argument will be statements that relate both to the experiment *and* to its results. Indeed, on our account, such arguments will have just two premises, one relating to the experiment and one describing its results. The conclusion of the argument will be either the hypothesis or its negation.

The Basic Elements

The Hypothesis. The most important element, of course, is the theoretical hypothesis whose truth or falsity is at issue. The whole point of designing an experiment is to be able to come up with a justification either for the hypothesis or for its negation. In the example, Halley's hypothesis was that the comet and the sun form a Newtonian particle system—with only minor influence from the planets, particularly Jupiter.

The Prediction. One might use the word "prediction" to refer either to the *possible event* whose occurrence is predicted or to the *statement* that describes this possible event. We shall generally use it to refer to the statement and not the event described.

In an experimental test of some hypothesis, the prediction always describes the occurrence of a *possible state*, or part of a state, of the system under investigation. To assert the prediction, then, is to say that the state occurs at the indicated time. This is clearly a contingent statement—that is, a statement that might be true or might be false. In the example, the prediction was that the comet would again be visible from the earth in December 1758. It turned out to be true.

Typically, as in Halley's case, the prediction describes some possible *future* state of the system under study. That is why we use the word "prediction." However, it is not necessary for the prediction to describe a state of the system that lies in the future relative to the experiment. In archaeology or evolutionary biology, for example, one studies systems that operated long ago. All the events in the history of these systems are in the past relative to present investigations. In such investigations, then, the "prediction" would still describe some part of the state of those systems, but a state that occurred long ago.

Initial Conditions. The term "initial conditions" may also be used to refer either to the occurrence of a state of the system or to a statement describing this occurrence. Again we shall generally mean the latter. Typically, the initial conditions describe the occurrence of some state at the *beginning* of the experiment—thus the term "initial" conditions. In the Halley example the initial conditions describe the positions and velocities of the comet that were recorded when the comet appeared in 1682. Halley used this information in his deduction of the shape of the orbit and the time of return. In general, it is not necessary that the initial conditions describe some state that occurred before the event described by the prediction. But they must describe some *other* state.

Auxiliary Assumptions. In designing any experiment one will need to make additional assumptions about the system being studied and about its immediate environment. Halley, for example, assumed that the planet Jupiter would not have much influence on the comet's path. He must have made other additional assumptions as well, as we shall soon see.

This brief rundown of the basic elements of a test leaves many important questions unanswered. Most of these questions, however, will be covered in the following discussion of the way these elements fit together to characterize a good test and to help provide justification for theoretical hypotheses.

One thing you should note immediately, however, is that isolating these elements requires little knowledge of what actually went on at the time. You do not have to know anything about the details of Halley's calculations. You need to know only what he assumed and what he concluded. Nor do you need to know anything in particular about the kinds of instruments used to observe comets. You need to know only that there were some and that they were good enough to yield a prediction accurate to within a month in 76 years. These are things that you can discover even from relatively popular reports of the experiment without having any detailed knowledge of the subject. Thus, if the question of whether the hypothesis is justified or not can be answered in terms of just these elements, questions of justification are indeed within the range of any careful person.

6.3   TWO NECESSARY CONDITIONS FOR A GOOD TEST

We shall now see how the basic elements of an experiment can be put together to formulate two statements, each of which must be true if a test of a theoretical hypothesis is to be a *good test*. Both conditions being true will be *sufficient* for the test to be a good test.

Keep in mind that the point of a test is to test the *hypothesis*, not the prediction. It is the hypothesis that will be justified (or refuted) when the experiment is done. Determining the truth or falsity of the prediction is merely a means to this end.

First Condition for a Good Test

It is clear from the example that the prediction is in some way based on the hypothesis. If this were not so, it would be hard to see why the success of the prediction should provide any justification for thinking that the hypothesis is correct.

The way the prediction depends on the hypothesis is through a *deductive* argument. The prediction is *deduced* from premises that include the hypothesis and it could not be deduced from the remaining premises if the hypothesis were removed.

It is important to realize also that the prediction could not be deduced just from the hypothesis all by itself. Just from the hypothesis that the comet and the sun form a Newtonian system, nothing at all follows about where the comet will be at any particular time or, indeed, what particular shape its orbit will have. The orbit could be much closer to a circle, like the orbits of the planets, and the system would still be a Newtonian system.

In order to determine the shape of the orbit and thus how long it would take the comet to return, Halley had to know some details about the state of the system at some time. This was provided by the observations made in 1682. The comet's position and velocity could be determined for the brief time it was close enough to the earth to be observed. But remember that Newtonian systems are *deterministic*. One need know only the complete state of the system at one particular time in order to determine its position at all other times. That is why Halley was able to calculate back to all the earlier recorded appearances of the comet and forward to the next appearance. So among the premises of the deductive argument leading to the prediction is the set of statements we have called the *initial conditions*.

Finally, it is possible that both the hypothesis and the initial conditions are true, but the prediction nonetheless fails to come true. This means that the prediction does not follow deductively from the hypothesis and initial conditions alone. (If you do not see why the previous sentence is true, you should review Section 3.3.) Halley might have been wrong about the negligible influence of Jupiter. If the gravitational attraction of Jupiter were great enough to deflect the comet on its way back out into space, it would not have returned when Halley predicted because he assumed there was no influence. But worse yet, there may have been other objects out in space that were unknown to anyone at the time but were big enough to deflect the comet if it came sufficiently close to them. The planets Neptune and Pluto, for example, had not yet been discovered when Halley made his calculations. If, unknown to Halley, one of these planets were in the path of the comet, it could have been deflected sufficiently that it might never come back. The possibilities for failure of the prediction are endless.

Here there seems to be a problem. It is hard to imagine all the things that might go wrong in such a way that the prediction fails even though the specific hypothesis at issue, and the initial conditions, are both true. In order to be sure that nothing did go wrong, one would have to know a lot more than even we today could know. What then do we put in for the auxiliary assumptions that will ensure that the prediction will follow deductively from all the premises? What we put in, in effect, is simply the assumption that nothing outside our knowledge will interfere. In short, we make the auxiliary assumptions just as comprehensive as they need to be to ensure that the prediction does follow validly from the conjunction of the hypothesis, the initial conditions, and the auxiliary assumptions.

This may strike you as fudging. Surely science is not as vague as all that. But the fact of the matter is that science is as vague as all that—at least in some respects. One of the values of learning to understand scientific reasoning is learning just where science is vague and what effect this has on the justification of scientific hypotheses. We shall uncover some of the dangers inherent in this particular vagueness as we proceed.

At this point it is again useful to introduce a bit of symbolism. Let us abbreviate the theoretical hypothesis simply by the letter H. Whatever theoretical hypothesis we happen to be talking about will be called simply H. Similarly, whatever initial conditions are at issue will be called IC. The auxiliary assumptions will be called AA. Finally, we shall let the letter P stand for the prediction.

Using this symbolism, our progress to date can be summed up by saying that the following must be a deductively valid argument.

H.
IC.
AA.
Thus, P.

Now if this argument is indeed valid, that means that it is *impossible* for the premises to all be true and the conclusion false. But then surely it is true that if the premises are true, the conclusion is true too. But this means that the following conditional statement is true:

Condition 1. If [H and IC and AA], then P.

As indicated, our first condition for any test of a hypothesis to be a *good test* is that this conditional statement be true. And this statement will be true so long as the above argument is valid. (If you are having difficulty following this paragraph, reviewing Section 4.5 may help.)

One way of understanding Condition 1 is as requiring that the hypothesis be able to *explain* the prediction, if the prediction indeed comes true. But this assumes that to explain why the prediction came true, if it does, requires being able to deduce it from the hypothesis (plus initial conditions and auxiliary assumptions). Many philosophers of science would now question whether deduction is necessary for explanation. So I will not insist on this reading of Condition 1, but it may help to make it intuitively more understandable.

Perhaps one reminder about conditional statements is in order. In saying that Condition 1 is true, we are not saying that the hypothesis is true or that the prediction is true. We are only saying that there is a conditional relationship between the hypothesis (plus initial conditions and auxiliary assumptions) and the prediction. We are trying to identify characteristics of a good test of a hypothesis—a test that might in the end provide us with a justification for the hypothesis. Obviously we cannot start out assuming that the hypothesis is true, much less that the prediction will be true. But in requiring that Condition 1 be true, we are not making any such assumptions. We are only assuming that a conditional relationship holds between the statement of the hypothesis and the statement of the prediction. (If you are unclear about this last point, you might want to review Sections 4.1 and 4.2.)

Finally, it is helpful to realize that Condition 1 will be satisfied by almost any experiment you hear about. This is because the way scientists typically hit upon predictions to use in testing hypotheses is by seeing what follows from the hypothesis itself. They set out investigating its deductive consequences. When they find an interesting consequence, then they think about designing a specific experiment that yields specific initial conditions, auxiliary assumptions, and a specific prediction. So the way they put together an experiment leads automatically to Condition 1 being satisfied. The only time you have to worry much about Condition 1 is when the hypothesis itself is sufficiently vague that there may be some dispute as to just what it implies. Then Condition 1 serves the important function of reminding you that this question must be settled at the outset. Otherwise you simply do not have a good test of the hypothesis.

Second Condition for a Good Test

There is one feature of the example we have not yet used. This is that the prediction of the comet's return was a remarkable feat. Halley's prediction was accurate to within one month in roughly 76 years. He said it should reappear in December 1758, and it did. That is an accuracy of 1 month in 912 months, or roughly one in a thousand. No wonder people were impressed. Most people at the time regarded the behavior of comets as being quite unpredictable, let alone predictable with such precision.

To see that this is not just an accidental feature of this example, imagine that Halley had used Newton's theory to calculate how long it would take a small cannonball to fall from the top of the Tower of London. With the hypothesis that the cannonball and the earth are a Newtonian system, together with the known height of the Tower, plus obvious auxiliary assumptions, one can easily deduce the duration of the fall. Suppose Halley then loudly proclaimed his prediction, marched to the top of the tower, and dropped the cannonball. Of course his prediction would have been correct within the accuracy of devices then available for measuring short periods of time. But would anyone have taken notice? Of course not. Galileo had discovered the law of free fall early in the seventeenth century. Every scientist knew it. That Newton's theory permitted the same prediction seems to count for little.

One way of describing this feature of a good test of a theoretical hypothesis is to say that the prediction must be such that it would be unlikely that anyone should succeed in getting it right unless they used the hypothesis in question. Or, to put it more intuitively, anyone not using Newton's theory would have had to have been fantastically lucky to have picked the right month just using whatever else was known about comets in 1705.

Using the previous abbreviations, we can express this intuitive idea more precisely as follows:

Condition 2.   If [Not *H* and *IC* and *AA*], then very probably Not *P*.

In words, if Halley's hypothesis is false, but the stated initial conditions and auxiliary assumptions are true, then it is very unlikely that the prediction will turn out to be true.

At this point we must remember the difference between a hypothesis being *true* and its being only *approximately true*. Condition 2 is not satisfied if we read *H* as meaning that the hypothesis is exactly true. But it is satisfied if we read it as meaning that the hypothesis is approximately true. The Halley example provides a good illustration of why this is so.

Although this was not discovered until the twentieth century, we now know that the solar system and related objects do not, strictly speaking, form a Newtonian system. They form an Einsteinian system. However, for objects like planets and comets, which move much slower than the speed of light, the difference between a Newtonian system and an Einsteinian system is minute. Halley's data on the motion of the comet in 1682 were only precise enough to allow him to make his prediction accurate to 1 month. The differences between a Newtonian comet and an Einsteinian comet are far less than that. So whichever theory had been applied, the prediction would have been the same. But this means that Condition 2 is false if *H* is read as meaning that the system is exactly a Newtonian system. The antecedent is true, but the consequent is false; that is, Not *H*, IC, and *AA* are all true, but the prediction is still *P*. It is not very probable that *P* will be false.

On the other hand, if we read Not H as saying that *H* is "not approximately true," then Condition 2 can be satisfied. By saying that *H* is not approximately true we mean that it is far enough from correct that it could yield a detectably different prediction with the same initial conditions and auxiliary assumptions.

Of course we should now go back and replace *H* by "approximately *H*" in Condition 1. But that is fine. The same prediction, accurate only to within 1 month, follows if we make this change.

Writing "approximately *H*" is cumbersome, so we will continue simply to use the letter H to represent theoretical hypothesis. However, whenever *H* appears in the statement of Conditions 1 and 2, or as the conclusion of an argument with one of these conditions as a premise, it must be read as "approximately *H*." The degree of approximation will be determined by the precision with which the experiment can be performed. It is not usually important to know just what the precision is, only that it is not perfect.

Returning to Condition 2 itself, it is important that you get a good intuitive feel for what it means. Don't just memorize the symbolic formulation. The symbolic form is meant to make it easier for you to see the essential logical structure of Condition 2. It is not intended as a substitute for understanding the meaning of specific statements of that form.

Try thinking of Condition 2 this way. Suppose that the real system is actually a good bit different from the type of system defined in the theory. Then ask yourself, "Is the prediction in this case very likely to come out true anyhow?" or "Would it have to be a big coincidence if the real system, whatever it is like, just happened to produce the precise result predicted by a mistaken hypothesis?" If your answers are "No" and "Yes, it would be a coincidence," then Condition 2 is satisfied.

Or try this. If *H* is not even approximately true, then it should not in general be much help in making specific predictions. So if you take the given initial conditions and auxiliary assumptions and use *H* to predict something quite specific, then you should not expect to be right.

Condition 2 is also a conditional statement, but not quite like any we have seen before because of the qualification "very probably" in the consequent. As you might suspect, this makes a big difference. Most of what you know about regular conditional statements does *not* hold when there is a "probably" in the consequent. We shall have to go more deeply into the concept of probability in Part III, but here we can get by using our everyday intuitions. In the example, we know that if Halley's hypothesis had been very far wrong, then it would have had to have been a gigantic cosmic coincidence for the comet to reappear during the predicted month 76 years later.

It is important to realize that Condition 2 is completely independent from Condition 1. In particular, just because Condition 1 is satisfied, that does not mean that Condition 2 must also be true. Not everything that follows deductively from a theoretical hypothesis is automatically something that would be unlikely if that hypothesis were false. It follows from Newton's hypotheses, for example, that an apple breaking loose from its branch on a tree will fall to the ground and not just float away. So the prediction that the apple will fall satisfies Condition 1. But it does not satisfy Condition 2. It does not take Newton's theory to be able to predict that result. This is not a result that anyone would regard as a coincidence if it occurred. Even Aristotle would have predicted the same thing.

If Condition 2 does not follow from Condition 1, how do we determine whether or not it is satisfied in any particular case? Unfortunately, there is no simple general answer to this question. In principle, the truth of Condition 2 depends on two things: the prediction and what other hypotheses might be true if the hypothesis being tested is not right (or not at least reasonably close to being right). However, even for scientists it is difficult to say much about what other hypotheses might be true. So most of the weight of Condition 2 falls on the nature of the prediction.

Usually it is sufficient if the prediction is very precise. In this case you can be quite confident that any other theory would yield a different prediction with the same initial conditions and auxiliary assumptions. However, being precise is not necessary. If what is predicted is quite unusual or completely new, for example, something no one had ever thought of before, that would be sufficient to say that the success of the prediction would be very improbable if the hypothesis in question is seriously mistaken. In the exercises for this chapter you will find examples of both of these ways of satisfying Condition 2. In most cases you will meet, the report of a test will contain some indication of how "surprising" the success of the prediction was to other scientists, or to those conducting the experiment. This will give you some basis for judging just how improbable the prediction was when it was made, and thus for judging whether Condition 2 is satisfied or not.

Summary: What Is a Good Test of a Theoretical Hypothesis?

A good test of a theoretical hypothesis, H, requires an experiment or set of observations that involve the hypothesis, H, initial conditions, IC, auxiliary assumptions, AA, and a prediction, P, such that the following two statements are *true*.

Condition 1.   If [*H* and *IC* and *AA*], then *P*.
Condition 2.   If [Not *H* and *IC* and *AA*], then very probably Not *P*.

Moreover, whether these two statements are true or not is something that can, and should, be determined at the outset of the experiment. Neither requires that any of the component statements actually be true. Both conditions express only conditional relationships. Figure 6.2 pictures some of these relationships.

Finally, note once again that you can usually tell just from a report of some experiments or observations whether these two statements are true. And to do so usually does not require any detailed special knowledge of the subject in question.

6.4   ARGUMENTS FOR AND AGAINST THEORETICAL HYPOTHESES

The determination that a proposed test would be a good test does not require that the experiment actually be performed. It might be designed and then for some reason never performed. For example, the scientist might not be able to get a grant to carry out the required experiments. However, if the

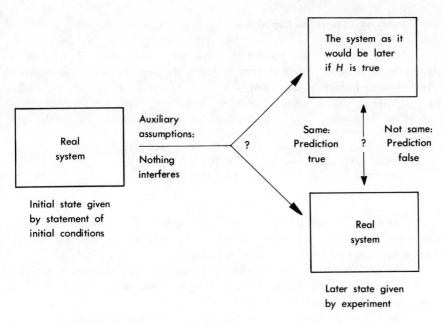

FIGURE 6.2

experiment is performed, then one needs only one additional bit of information given by the result of the experiment to be able to justify either the hypothesis or its negation. This bit of information is simply whether the prediction turned out to be true or not.

Thus there is one constraint on the prediction that we have not yet explicitly mentioned. It must be a statement that can reliably be determined to be true or false using methods that do not themselves presuppose that the hypothesis in question is true. A prediction that cannot be reliably checked using independent experimental means is useless for the purpose of justifying the hypothesis. It must therefore be presumed that the prediction was originally chosen so that at the end of the experiment one of the following two statements is available as a justified premise for use in an argument: "The prediction is true" or "The prediction is false."

Be careful not to misunderstand. The reason for the experiment is not just to justify the *prediction*. The point of the experiment is to reach some conclusion about the *hypothesis*. The determination of the truth or falsity of the prediction is merely a means to this end.

Now since a prediction is, after all, a statement, it must be either true or false. Once we have determined which, we can proceed to consider what we should conclude about the hypothesis.

Justifying a Theoretical Hypothesis

In the example the prediction came true: The comet showed up on schedule. So we know that P is true. To justify the hypothesis we now construct the following *inductive* argument:

First Premise:	If [Not H and IC and AA], then very probably Not P.
Second Premise:	P (is true).
Preliminary Conclusion:	Not [Not H and IC and AA]
Preliminary Conclusion:	H or Not IC or Not AA
Additional Premise:	IC and AA (are true).
Final Conclusion:	H.

Let us go through this argument once carefully. Thereafter we can give a shortened form or simply refer to it without actually writing it down. But it is important that you know just what is going on.

The first premise is, as you should recognize, just Condition 2. So if you have already determined that the test is a good test, which you should have done, then this premise is already itself justified.

The second premise just gives the result of the test. We presume that the prediction was indeed something that could be reliably checked. The second premise records the fact that P turned out to be true.

The step from the first two premises to the first preliminary conclusion is the *inductive* step. It is the only inductive step in the argument. We shall reserve comment on this step until we have gone through the others.

The step from the first preliminary conclusion to the second is a purely deductive inference. It is simply a matter of denying a conjunction. (If you have forgotten why this is a valid inference, go back and review Section 4.4.) The only step that has been left out is the step from Not (Not H), to H, that is, the inference that the negation of the negation of a statement is the statement itself.

At this point in the argument we must make explicit our commitments to the truth of both the initial conditions and the auxiliary assumptions. The initial conditions are usually not much of a problem. Like the prediction, they describe some state of the real system that can be fairly reliably determined with existing methods of measurement and experimentation.

The auxiliary assumptions are more problematic. Some can be justified. Halley had good reason to believe that Jupiter's gravitational force on the comet would be tiny compared with that of the sun. But there are all those other auxiliary assumptions needed to make the inference to the prediction a valid deduction. Let us put off this problem a bit longer.

The second preliminary conclusion together with the additional premises form the premises of a disjunctive syllogism with three initial disjuncts. (Here again you may want to review Section 4.4.) You need only remember that IC is the denial of Not IC every bit as much as Not IC is the denial of IC. Similarly for AA.

And we are done. We have justified the hypothesis.

Before discussing the inductive step in more detail, let us collapse the argument into just two premises and a single conclusion. Nothing fundamental is changed; it is just easier to write down and look at.

If [Not *H* and *IC* and *AA*], then very probably Not *P*.
P and *IC* and *AA*.
Thus (inductively), *H*.

In words, very roughly and intuitively: Unless the hypothesis is fairly close to being right, it is not very likely that the prediction would succeed. But it did. So the hypothesis must be fairly close. From now on, all references to the argument justifying the hypothesis will be to this "short form" of the argument, unless otherwise noted.

A *good inductive argument,* as indicated in Chapter 3 (Section 3.3), is an argument in which the premises make it very probable that the conclusion is true. This is the same as saying that the premises make it very *improbable* that the conclusion is *false.* Now just ask yourself how it could happen that the conclusion is false if the prediction were true and both the initial conditions and the auxiliary assumptions were likewise true. This could happen only if by some great coincidence the real system just happened to evolve in such a way as to fulfill the prediction of a false hypothesis. This is possible, which is why the argument is inductive and not deductive. But the prediction was one that would make Condition 2 true, and so this coincidence must be very unlikely. Thus it is possible, but unlikely, that our conclusion is incorrect.

Remember our qualification that *H* in this argument means "*H* is approximately true" and not simply "*H* is true." So now you see why it is that, in science, one can never be justified in claiming that any theoretical hypothesis is exactly true. In order for Condition 2 to be justified, it can only contain the statement "Not approximately *H*" in its antecedent, which is then denied if the prediction is true. So the conclusion can only be "approximately *H*," not simply *H* itself.

Here we have, in one argument, two distinct forms of uncertainty in scientific reasoning. One is the uncertainty due to the fact that the conclusion says only that the hypothesis is approximately true. The other is due to the argument being inductive. Although both premises are true, it is possible that even our conclusion that *H* is approximately true is false. *H* might be nowhere near true and still the prediction could have come true by coincidence.

Take another look at the short form of the argument. You will note that, except for the words "very probably," the form of the argument is just like the valid deductive form, *denying the consequent.* That the antecedent is a conjunction makes no difference. Our conclusion is the denial of the whole conjunction, which is the antecedent of the conditional statement. In scientific reasoning, the good inductive arguments almost always have this form.

How Not To Justify a Theoretical Hypothesis

Knowing how theoretical hypotheses are justified is important. Knowing how they are *not* justified is almost as important.

The most common fallacy in scientific reasoning is to suppose that just because a hypothesis "explains" some event, that event provides grounds for believing that the hypothesis is correct. But if you examine the argument on which this reasoning is based, you see that it is fallacious. The premise that the hypothesis explains the prediction is just our Condition 1. So to conclude that the hypothesis is (probably?) true, one must be reasoning as follows:

If [*H* and *IC* and *AA*], then *P*.
P.
Thus, [*H* and *IC* and *AA*.]
Thus, *H*.

The last step of this argument makes use of the definition of a conjunction (see Section 4.4), which says that a conjunction asserts the truth of each one of its components.

The first step of the argument you should recognize as an example of affirming the consequent. The fact that the antecedent of the conditional statement is itself a compound statement (i.e., a conjunction) makes no difference to the basic structure of the argument. The argument concludes, invalidly, that the antecedent is true because the consequent is true. You should know enough about this fallacy that we need say no more here. (If not, review Section 4.3.)

At this point it is often said that although affirming the consequent is not a valid deductive argument, it is a good inductive argument. That is, the truth of the premises do not guarantee that the conclusion is true, but they do make it "probable" that the conclusion is true. But this is just plain wrong, as the following simple example proves.

It is a medical fact that a very small percentage of men who contract syphilis develop an advanced form called paresis. Moreover, having syphilis is the only known way of getting paresis. Suppose, then, that you read an account of a man, Jones, that records the fact that he had once had syphilis. If you think that affirming the consequent is a good inductive argument, you might reason as follows:

If Jones has paresis, then Jones had syphilis.
Jones had syphilis.
Thus, Jones has paresis.

But given these premises, it is not "probable" that the conclusion is correct because only a very small percentage of syphilitics develop paresis. You do not have to know anything special about "probability theory" to see that this is so.

In short, affirming the consequent is not a valid deductive argument, and it is not a good inductive argument either. It is simply a bad argument. It justifies nothing. If you want to justify a theoretical hypothesis, you must use Condition 2 as your first premise, and reason "backwards" by the inductive version of denying the consequent.

One final point. If you have an eye for philosophical puzzles, it may have just occurred to you that perhaps the inductive version of denying the consequent is not a good inductive argument either. You may even have thought of an example to show this. Something like the following perhaps.

Suppose you are playing the slot machines at Las Vegas or Atlantic City, and you win a lot of money. That, you know, is very improbable. So, you argue:

> If I play the slot machines, then very probably I will not win a lot of money.
> I did win a lot of money.
> Thus (inductively), I didn't play the slot machines.

But of course the conclusion is false. And arguing in this way is likely to get you false conclusions. So arguments of this form are not in general good inductive arguments. But is this not the general form of the argument we used to justify Halley's hypothesis about his comet? What is wrong? Try to figure it out before reading further.

You have missed a subtle point. I did not say that arguments of this form are *in general* good inductive arguments. I said that *if* you have a good test, and *if* the prediction turns out to be true, *then* you may use this form of argument to justify the hypothesis. But you must first have a good test. And that requires Condition 1 as well as Condition 2. The above example does not involve a good test of the conclusion because the corresponding Condition 1 is not true. That is, it is not true that you will win a lot of money if you do not play the slot machines. Certainly winning a lot of money does not follow deductively from the hypothesis that you don't play the slot machines. So remember, first you must have a good test. Then you can go on to consider whether the hypothesis is justified. Without a good test, you can't even consider this question.

Refuting a Theoretical Hypothesis: The Phlogiston Theory

To call a test of some hypothesis a good test does not mean that it ended up providing justification for the hypothesis being tested. It is not necessarily good from the standpoint of someone who wants to justify the hypothesis. A test is good in the more neutral sense that it is guaranteed to provide *an answer*. The answer may be that the hypothesis is justified. But it may also be the reverse. The hypothesis may be refuted. It all depends on whether the prediction was true or not. That a test is a good test also does not mean that the prediction came out true. It only means that there was a clear prediction whose truth or falsity became known at the end of the experiment.

In Halley's case, of course, the prediction came true and the hypothesis was justified. In science, as in everyday life, one tends to remember only the winners. And history is the same. The losers tend to be forgotten. Some losers, however, are remembered, if only because they had a period of considerable success before their downfall. A famous loser is the phlogiston theory.

Fire, like motion and inheritance, has always fascinated people. In the Western world, recorded speculation about the nature of fire also goes back to the Greeks. To them we owe both the myth of Prometheus and the view that the world is made up of a few separate "elements," namely, earth, air, fire, and water. All four are present in the process of combustion. The common sense view of combustion is that something leaves the burning object, leaving only ashes behind.

By the eighteenth century, this "something" had a well-established name, "phlogiston." It explained most of the obvious facts about combustion. Heating drives off the phlogiston into the air; cooling makes it less volatile; smothering holds it in.

There are no clear statements generally referred to as "the laws of phlogiston." But we may take these statements of the general properties of phlogiston to define what we would call "a phlogiston system." A phlogiston system is anything containing a substance with those properties.

In 1775 a French chemist, Lavoisier, published a report of his recent careful experiments with mercury. Using a magnifying glass (a burning glass), he heated a measured amount of mercury carefully enclosed in a known amount of air, all in a glass container. A red powder, or ash, formed on the surface of the mercury. But at the end of the experiment, the volume of the air had *decreased*, and the mercury with the red powder on its surface weighed more, not less, than the original mercury alone.

The *hypothesis* at issue in Lavoisier's experiment is that mercury is a phlogiston system. The *prediction* is that the mercury and ash should weigh less than the original mercury, and that there should be a greater volume of air since it now contains the phlogiston that was driven off. The *initial conditions* consist of the information about the original weight of the mercury, the original volume of the air, and the heating of the mercury. It is difficult to discern any specific *auxiliary assumptions* from the above account. But we know there must be some, if only in the form: Nothing unknown interferes.

Now that we have identified the basic elements, we can go on to ask the central question: Did Lavoisier's experiment constitute a good test of the phlogiston hypothesis? To answer this question we need only check to see whether both conditions are justified.

Condition 1 is satisfied. If mercury is a phlogiston system, then under these conditions it ought to give up some of its phlogiston into the air. We know that this conditional statement is true because we can deduce P from the conjunction of H with IC and AA.

Condition 2 is also satisfied. Lavoisier took great pains to ensure that everything was accounted for. The amount of phlogiston in the air should have exactly corresponded to the amount that left the mercury. Everything was precisely controlled and carefully measured.

But the prediction failed. In fact, the results were just the opposite of what was predicted using the phlogiston hypothesis. In this case we construct our argument using Condition 1 as the first premise and the falsity of P as the second. As before, we will go through the details of the argument once carefully.

The argument proceeds as follows:

First Premise:	If [*H* and *IC* and *AA*], then P.
Second Premise:	Not *P* (i.e., *P* is false).
Preliminary Conclusion:	Not [*H* and *IC* and *AA*].
Preliminary Conclusion:	[Not *H* or Not *IC* or Not *AA*].
Additional Premise:	*IC* and *AA* (are true).
Final Conclusion:	Not *H*.

Every step in this argument is deductively valid. The first premise is just Condition 1, which we have determined is true. The second premise is the result of the experiment; that is, the prediction failed. By the valid step of denying the consequent we reach the first preliminary conclusion. This conclusion is the denial of a conjunction, which we know leads validly to the next preliminary conclusion, a disjunction of the negations of the original component statements (again, see Section 4.4 if you are not sure about this step). Finally, we must commit ourselves to asserting both *IC* and *AA* if we are to reach the conclusion that *H* is false. The final step is a valid disjunctive syllogism once again.

Do not forget our earlier qualification. "Not *H*" in the conclusion means more than simply that *H* is false. It means that *H* is not even approximately true. This means, roughly speaking, that *H* is far enough from being true that its deviation from the truth could be clearly detected by this experiment. Another experiment might not have been so sensitive to this deviation from the truth.

As in the case of the previous argument justifying a hypothesis, we can collapse this argument into just two premises and a conclusion.

If [*H* and *IC* and *AA*], then P.
Not *P* and *IC* and *AA*.
Thus (deductively), Not *H*.

In words, very roughly and intuitively: If the hypothesis is close to being right, the prediction should be true. But since it was not, the hypothesis cannot be all that close. We shall call this the "short form" of the argument refuting a theoretical hypothesis.

You may be wondering why we bothered determining whether Condition 2 is true if in the end we were not going to use it in an argument. Indeed, since the argument refuting *H* was completely deductive, why do we even need to ask whether there was a good test. All that it takes to refute a hypothesis is some deductive consequence of the hypothesis that we know is false.

This makes good logical sense, but not good scientific sense. It is indeed true that any false consequence of a hypothesis refutes it. But one does not, so to speak, usually find false consequences of significant scientific hypotheses lying on the sidewalk. You have to look hard to find them. Satisfying Condition 2 means looking hard, because any prediction that satisfies Condition 2 must be one that is very likely not to come true if the hypothesis is not close to being true. Which is to say that it is a prediction that is very likely to refute H if H is indeed not more or less true.

Indeed, there is a neat interplay between the two conditions in scientific research. Most experiments are done because scientists want to justify their hypotheses. To set up experiments so as to make this possible, one has to satisfy Condition 2. But that automatically makes it likely that any hypothesis will be refuted if it is mistaken. So trying to justify any hypothesis automatically ensures that it will be likely to be refuted if it is wrong. Of course the hypothesis *should* be refuted if it is wrong. So everything is as it should be.

It is true, then, that if *all* you want to know is that there is justification for rejecting some hypothesis, then all you need know is that Condition 1 is true and the prediction was false. But if you want to understand what the whole experiment was about, you will want to think about Condition 2 as well.

One sidelight. It is sometimes thought that any objectivity there is in science is due to scientists being objective people. And this means being dispassionate regarding the truth or falsity of one's hypotheses. But this is a mistake. The objectivity of science, imperfect as it is, is not a function of the objectivity of the scientists. It is a function of the "logical" rules of the game. And these are embodied in Conditions 1 and 2. So there is no reason why scientists should not hope to justify their hypotheses and be very disappointed if they are refuted. The rules of the game ensure that the harder one tries to get a good justification, the greater the risk of refutation—unless the hypothesis is indeed on the right track.

How Not To Have a Theoretical Hypothesis Refuted

There is a way to avoid the conclusion that the hypothesis is false even if the prediction fails. This is by denying the auxiliary assumptions instead. To see the logic of this strategy, let us look again at the last part of the argument refuting a hypothesis and see what happens if one refuses to assert the truth of the auxiliary assumptions. Then the last step is as follows:

Not H or Not IC or Not AA.
IC.
Thus, Not H or Not AA.

This is a valid argument. The conclusion, however, is that either the hypothesis is false or the auxiliary assumptions are false. Given this choice, one might rather say that something unknown happened to make the prediction false than grant that the hypothesis is false. So one avoids having the hypothesis refuted by withdrawing one's assent to the original auxiliary assumptions.

In the process of scientific investigation, this is a legitimate move—up to a point. It can only be used as a temporary "holding action." In the first place, if a theory is going to be useful, it has to lead to some interesting hypotheses that get justified. Avoiding refutation is not enough.

Secondly, one cannot just say that *something* went wrong. That eventually must be justified. It is not justified by the failure of the prediction. The falsity of the prediction (plus the truth of the initial conditions) only justifies the disjunction, "Not H or Not AA." To get from here to "Not AA" one would have to say that H is true. But that is just what we were trying to justify in the first place. It is not justified by the experiment, so it can't be used to justify the conclusion that something unknown went wrong. One has to discover, independently, just what went wrong. Unless this can be done, the refutation must be allowed to go through.

In the case of phlogiston, supporters of the phlogiston theory tried many different ways of explaining away Lavoisier's results, including even the suggestion that phlogiston has negative mass! In the end, none of these explanations could themselves be justified, and the phlogiston theory had to be abandoned.

6.5 TESTING A MENDELIAN HYPOTHESIS

As a further illustration of the use of our schema for evaluating tests of theoretical hypotheses, let us examine an experiment that Mendel himself performed to test his hypotheses on sweet peas. If you find that you have forgotten some of the details of Mendel's theory, a quick review of Section 5.3 may be in order.

Why the Original Experiments Do Not Constitute a Good Test

Before going on to examine the "backcross" test, we should say why a further test is necessary. Do not the original experiments provide adequate justification for Mendel's hypotheses? As we shall discuss shortly, really solid justification of any far-reaching hypothesis requires more than one successful test. But that is not the issue here. The question is whether the experiments outlined in Section 5.3 provide *any* justification for Mendel's claims.

There is no question that the original experiments are "explained" by an application of Mendel's theory. Condition 1 for a good test is satisfied. The theory was *designed* to account for these results; that is, the 3-to-1 ratio of dominants to recessives in the third generation. But that is just the problem. If the theory was designed to use in explaining these results, what sense does it make to say that these results would be very improbable if the theory did not apply? Any theory Mendel would have proposed would have fit this case, whether or not the corresponding hypotheses were true in general. Fitting this case was, for Mendel, a necessary requirement for any theory to be seriously entertained. So there seems no way to argue that Condition 2 is true. And this means, of course, that the results of the experiment cannot be used to justify the hypothesis because that requires using Condition 2 as a *justified* premise. Moreover, to argue that the hypothesis is true because it "explains" this case would be simply to commit the fallacy of affirming the consequent. In short, another experiment is necessary if the claim that sweet peas are Mendelian systems is to be justified.

The Backcross Test

Mendel's theory itself suggests another experiment that does provide a good test of his hypothesis. Why not try crossing the second generation hybrids "back" with the true-breeding short plants of the "parental" generation. A quick review of Figure 5.6 will make it obvious what kind of a mating this backcross represents. It seems that no one had ever done this experiment before, and if they had, no one ever counted to see what ratio of dominants to recessives would occur. Let's figure out what ratio Mendel's hypothesis predicts. The experiment is represented symbolically in Figure 6.3. The numbers 1, 2, 3, 4 are attached to help keep straight the various possible combinations of factors in the offspring.

The *hypothesis* is that height in pea plants is transmitted by a Mendelian system. The *initial conditions* are that all the offspring are the result of "mating" hybrids with true-breeding short plants. The *auxiliary assumptions* are unspecified.

To deduce the *prediction* we note that there are four different possible ways an offspring could get its complement of two factors. If the hypothesis is true, the law of segregation applies. It says each of these possibilities is equally

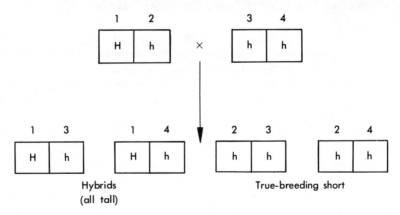

FIGURE 6.3

likely. So on the average there should be equal numbers of each type. Now as you can see, two of the possibilities yield hybrids, which by the law of dominance will be tall. The other two are true-breeding short plants. So the ratio of tall to short plants in this sort of experiment should be one to one. That is, on the average, half should be tall and half short. This is the *prediction*.

We have deduced the prediction from the hypothesis (plus initial conditions and auxiliary hypotheses). So Condition 1 is satisfied. What about Condition 2? It also seems well justified. Since no one had ever counted to see what the ratio from such an experiment ought to be, Mendel had no way of knowing what it might be. Any ratio between zero and one was possible. But applying the theory leads to a precise prediction. That ratio should be 1 to 1. If Mendel's hypothesis were not more or less true, there would be no particular reason for thinking that this should be the ratio rather than any other. If Mendel's hypothesis were false, it does not seem likely that the true hypothesis, whatever it might be, should give just the same prediction. But that is just what Condition 2 says. So both conditions are satisfied, and the backcross test is a good test.

Mendel reports producing 208 plants by this type of cross-pollination. Of these, 106 came out tall and 102 short. That is surely close enough to call the prediction true. We need not bother actually going through the argument. The conclusion is that the hypothesis is justified as being at least nearly true. Height in sweet peas is at least approximately a Mendelian trait.

Testing Stochastic Hypotheses

Mendel's theory is a stochastic theory. This means, you may recall (see Section 5.4), that the type of "parents" do not uniquely determine what factors any offspring will have. The parents' factors only determine the *probabilities* for the various possible combinations of factors in the offspring. Since this is true of any individual offspring, it is true of any number as well. Thus when we deduced our prediction, what we said is that "on the average" half the offspring should be tall. This is not the same as saying that "exactly" half will be tall. And indeed, in the experiment the result was not that exactly half were tall. It was 106 out of 208, or a little greater than half. But it was only a *little* greater, so we called it a successful prediction. And it was.

What this all means is that for stochastic theories, Condition 1 should also have a "very probably" in it. It should read: If [*H* and *IC* and *AA*], then very probably the ratio of tall plants will be near one half. Moreover, to determine whether such a prediction is fulfilled may require some more sophisticated reasoning. What if Mendel's result had been 110 tall and 98 short? Or 120 tall and 88 short? When does the ratio stop being "pretty close" to half?

We shall have to deal directly with such questions in Part III. For now, we shall just treat the testing of stochastic hypotheses informally. This means we can for the moment only deal with clear cases, like Mendel's, where there is no question but that the prediction was fulfilled. But the majority of cases you will ever come across will be roughly this clear-cut. So there is no need to worry here about using more sophisticated reasoning to evaluate tests of stochastic hypotheses. When we have finished Part III you will know what the more sophisticated reasoning is like.

6.6   WHY ONE SUCCESSFUL TEST IS NOT ENOUGH

Up to now we have spoken as if one successful test—that is, a good test in which the prediction is true—were enough to call a theoretical hypothesis "justified." But in science, one successful prediction, even if very surprising, would not generally be taken as settling the issue. Several, at least, are required. Why is this so?

There are a number of reasons. One is simply the fear that maybe the success of the prediction was, after all, only a coincidence. Maybe, for example, comets do not generally behave like parts of a Newtonian system, but something out in space deflected this one so that it came back just when a Newtonian comet would have. Coincidences do sometimes happen. However, it is hard to take seriously the idea that there should be several such coincidences in a row. So several successful tests would pretty well settle the issue.

Another reason has to do with the *scope* of the hypotheses one wants to assert. Newtonians did not want simply to say that the earth, moon, and sun form a Newtonian system. They wanted to say that every object in the solar system is part of a Newtonian system, including comets, and cannonballs, and pendulums, and apples falling from trees. This partly explains the interest in Halley's test. Predictions of the positions of the moon and major planets had fairly well established that they were proper Newtonian objects. Predictions of eclipses of the sun and moon with great precision provided especially good tests of these hypotheses. But one might still wonder about comets, which seemed to come and go with no special regularity. Maybe comets are not Newtonian objects. Halley's test put an end to that possibility. And there did not seem to be any other major mechanical phenomena to which Newton's theory could not be successfully applied.

In cases where the only worry is the possibility of a coincidence, we might say that one successful test provides "substantial" justification and that several provide "full" justification—it being understood that even full justification does not *absolutely* rule out the possibility that the hypothesis is seriously mistaken. This precaution should not cause you any difficulties. In most reports of new scientific findings it is usually made clear whether the experiment in question is the first major test of some hypothesis or a later test that provides additional justification for taking the hypothesis to be correct.

Experimental result / Status of hypothesis	Prediction true. Use Condition 2 to conclude that H is approximately true.	Prediction false. Use Condition 1 to conclude that H is not approximately true.
The hypothesis is approximately true.	Very likely	Very unlikely
The hypothesis is not approximately true.	Very unlikely	Very likely

FIGURE 6.4

6.7 SUMMARY: GOOD TESTS AND GOOD ARGUMENTS

The relationships between good tests and arguments for or against theoretical hypotheses can be summarized in a table something like that used to characterize valid arguments in Part I (Section 3.3). Figure 6.4 shows these relationships for tests and inductive arguments.

Figure 6.4 is to be read from left to right. For any theoretical hypothesis there are two possibilities. Either it is correct within detectable limits or it is not. If one sets up a good test hoping to justify believing that the hypothesis is approximately correct, then one of two things must happen: Either the prediction succeeds or it fails. In either case the conditions that define a good test provide you with the first premise of an argument to use in drawing the appropriate conclusion. If the test is a good test, then, no matter whether the hypothesis is approximately true or not, the test is very likely to lead you to reach the correct conclusion and very unlikely to lead you to draw the wrong conclusion. You cannot ask for anything better than that.

CHAPTER EXERCISES

Each of the following exercises consists of an account of some scientific investigation followed by a series of questions. Your task in general is to analyze the scientific episode with an eye to seeing what conclusions, if any, are justified. The questions are intended to help you organize your analysis. The first several exercises contain many questions, which give you a great deal of help in organizing your analysis. Later exercises have fewer. In these you are more on your own. Answers to some of the exercises are given at the end of the book.

Your general strategy for approaching each of the exercises should be roughly as follows:

First, make sure what hypothesis is at issue. This means determining what kind of system is defined by the corresponding theory and deciding what real system is claimed to be a system of this type.

Second, identify the other elements that would go into a good test. The prediction is most important. Then the initial conditions and auxiliary assumptions if given.

Third, check to see if both conditions for a good test are satisfied, particularly Condition 2. Was the prediction something unlikely to be true if the hypothesis is not close to being right?

Fourth, determine whether the prediction was true or not.

Fifth, determine whether the justified conclusion is "Approximately *H*" or "Not Approximately *H*." The earlier exercises will ask you to write out the short form of the appropriate argument. In general this is not necessary so long as you know what the argument is and which result of the experiment justifies which conclusion. Once you have determined that both conditions for a good test are met, you can move directly from the success or failure of the prediction to the appropriate conclusion.

6.1 THE DISCOVERY OF NEPTUNE

In the early 1800s astronomers were still working out tables and charts giving the positions of the various planets. In this they were aided by Newton's theory. But then the outermost planet, Uranus, caused some difficulties. Its observed position differed from what it should have been applying Newton's theory. And the difference was much too great to be attributed solely to inaccuracies in measurement. Implicitly following the logic of the argument refuting a hypothesis, they came to the conclusion that either Newton's hypothesis was wrong—Uranus is not part of a Newtonian system—or "something" was interfering with the orbit. By that time there had been so many successful predictions from Newton's hypotheses that they were justifiably reluctant to conclude that the hypothesis was wrong. This meant discovering just what was interfering. "Something" is not an answer that can be accepted for long. Around 1843 the English astronomer J. C. Adams and the French astronomer La Verrier independently calculated that the observed orbit of Uranus could be explained if there were an additional planet beyond Uranus whose gravitational force on Uranus produced the deviations from the earlier Newtonian predictions—which of course assumed no such planet. Assuming that the new planet was Newtonian in all respects, Adams and Le Verrier were able to calculate just where it should be at a particular time—given the observed positions of Uranus. Astronomers at several observatories began looking. The planet, named "Neptune" by Le Verrier, was observed, as predicted, in 1846.

> A. The theory in question is of course Newton's mechanics. The hypothesis is that the solar system is a Newtonian system. Identify the initial conditions and the prediction. What are the auxiliary assumptions?

B. Using the standard symbols, *H*, *IC*, *AA*, and *P*, write out Condition 1. Is Condition 1 true? Why?

C. Again using the standard symbols, write out Condition 2. Is it justified? Why?

D. What was the result of the experiment?

E. What conclusion is justified? Write out the short form of the appropriate argument.

6.2 THE MISSING PLANET: THE STORY OF VULCAN

Later in the nineteenth century it was discovered that the orbit of the innermost planet, Mercury, also fails to fit the Newtonian theory by amounts that can be reliably measured. Fresh from their discovery of Neptune, astronomers immediately assumed that there was yet another planet, still closer to the sun than Mercury. They named the new planet Vulcan and calculated just where it should be, but it wasn't there. Although several explanations of the failure to find it were offered, none worked out. The supposed planet was never found.

Answer questions A through E as given in Exercise 6.1. Don't just give the answers mechanically. Think about it a bit.

6.3 MEMORY TRANSFER

In the 1950s biologists and psychologists began speculating that memory works by changing and storing certain chemicals in the brain. In short, memory is a kind of chemical system. In the 1960s it occurred to some psychologists at the University of Michigan that if memory is a matter of stored chemicals, one might be able to transfer memory from one organism to another just by transferring chemicals from the brain of one to the brain of another. Experiments were done on both worms and rats. In the rat experiments a group of rats were taught to get a small drink of milk by pressing a lever on the opposite side of their cage from where the milk was dispensed. This is difficult for rats to learn. The average time it took a typical group of rats to learn it was about 25 hours. Then chemicals from the brains of the trained rats were extracted and injected into the brains of other similar rats. Still other rats were injected with material from the brains of other untrained rats. The injected rats, they reasoned, should learn faster if they received injections from the previously trained rats. The rats that received injections from trained rats learned the trick in an average of only 3 hours. The others averaged around the usual 25 hours.

A. The theory defines learning and memory as a kind of chemical system; the details are not too well specified. What is the hypothesis in the experiment?

B. Identify the prediction. Then locate at least some initial conditions. Can you think of any specific auxiliary assumptions?

C. Does the experiment satisfy the first condition for a good test? Why? The second condition? Why?

D. Was the prediction successful? What conclusion is justified?

6.4   CONTINENTAL DRIFT

During the past several decades there has been a revolution in the science of geology. Twenty-five years ago very few respectable scientists believed that the continents moved. According to the current hypothesis, however, the crust of the earth consists of a system of land masses that have broken off from a single large land mass and drifted to their present positions. Application of this hypothesis to the continents of Africa and South America has always seemed quite plausible because some of the existing shorelines obviously "fit" together quite well. Some of the best tests of claims about continental drift, however, have become possible only since the development of accurate methods of determining the age of rocks, methods based on the examination of the products of radioactive decay.

The following two paragraphs are taken from an article ("The Confirmation of Continental Drift") in *The Scientific American,* April 1968. The particular hypothesis at issue is that Africa and South America are part of the system of land masses that have broken up and drifted apart. The investigators write:

> Of special interest to us at the start was the sharp boundary between the 2,000-million-year-old geological province in Ghana, the Ivory Coast and westward from these countries, and the 600-million-year-old province in Dahomey, Nigeria and east. This boundary heads in a southwesterly direction into the ocean near Accra in Ghana. If Brazil had been joined to Africa 500 million years ago, the boundary between the two provinces should enter South America close to the town of Sao Luis on the northeast coast of Brazil. Our first order of business was therefore to date the rocks from the vicinity of Sao Luis.
>
> To our surprise and delight the ages fell into two groups: 2,000 million years on the west and 600 million years on the east of a boundary line that lay exactly where it had been predicted. Apparently a 2,000-million-year-old piece of West Africa had been left on the continent of South America.

Figure 6.5 indicates the relevant geography.

> The theory and hypothesis have been given. You should be able to identify the prediction, initial conditions, and auxiliary assumptions. Then determine whether the two conditions for a good test are satisfied, see if the prediction was successful or not, and give the justified conclusion.

6.5   CHILDBED FEVER

An episode of a recent television series dealt with the investigation of Ignaz Semmelweis into the causes of "childbed fever" during the mid-nineteenth century. In the years 1844 to 1846 the death rate from childbed fever in the First Maternity Division of the Vienna General Hospital averaged roughly *10 percent.* The rate in the Second Division, in which women were attended only by midwives rather than doctors, was only around *2 percent.* Semmelweis tried in vain for 2 years to discover why the rate should be higher in the "better" division. Then one of his colleagues received a small cut on the finger from a student's scalpel during an autopsy. The colleague died exhibiting symptoms exactly like those of childbed fever. Semmelweis wondered whether the disease might be caused by something in "cadaveric matter" which was being transmit-

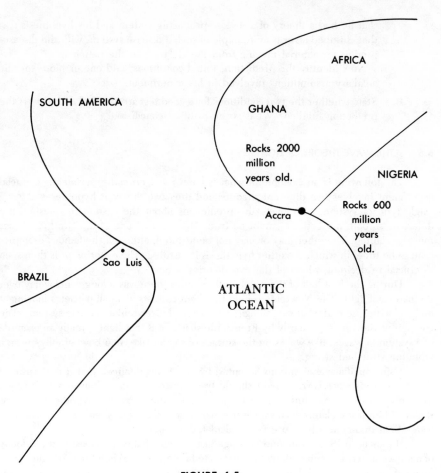

SOUTH AMERICA

BRAZIL

Sao Luis

AFRICA

GHANA

Rocks 2000 million years old.

NIGERIA

Rocks 600 million years old.

Accra

ATLANTIC OCEAN

FIGURE 6.5

ted to the women during childbirth on the hands of the doctors and medical students who spent every morning in the autopsy room before making their rounds. Semmelweis reasoned that if this idea were right, the death rate could be cut dramatically simply by requiring the doctors and students to wash their hands in a strong cleansing agent before examining their patients. He then insisted that no doctor enter the ward without first washing his (there were no women doctors at that time) hands in a solution of chlorinated lime, which Semmelweis assumed was strong enough to remove whatever it was that caused the disease. It worked. The death rate in the First Division for 1848 was *less than 2 percent*.

A.   There is a theory behind Semmelweis' work, though it is not very explicit. It is an early version of what later became "the germ theory of disease." Semmelweis did think that disease, and childbed fever in particular, could be produced by "something" present in corpses that was carried on the hands of doctors and entered the bloodstream of their patients. Let us say that he had a theory of a disease-producing system and his hypothesis was that childbed fever is an example of such a disease system, with the disease-producing material moving from the cadavers to the patients.

Now identify the prediction, initial conditions, and one or more specific auxiliary assumptions involved in his experiment.

B. State whether the two conditions for a good test are satisfied and why. Is the prediction fulfilled? What is the justified conclusion?

6.6   THE WAVE THEORY OF LIGHT

The following is an example of what is known as a *crucial experiment*. A crucial experiment involves *two* different theories and thus two different hypotheses about a single type of system. Moreover, the predictions about the system in question are different, so both predictions cannot be true. Thus at least one of the two hypotheses must be refuted. The other may or may not be justified, although the famous examples tend to be those in which the other hypothesis is justified. The experiment is then seen as crucial in deciding which of the two theories is best.

During the first half of the nineteenth century there was a long controversy over the nature of light. The Newtonians held that light consists of small particles moving at high velocities, so that what we call light rays are really Newtonian particle systems. The competing view, held primarily by French physicists, was that light is really an example of a system of waves, like waves on the surface of a calm lake, in a bowl of jelly, or on a vibrating stretched string.

Using auxiliary assumptions accepted by all, it was deduced that if the claims of the Newtonians were correct, light should travel *faster in water* than in air, and by a precisely calculated amount. Similarly, using the same auxiliary hypotheses, it was deduced that if the claims of the wave theorists were true, light should travel *slower in water* than in air, also by a precisely calculated amount.

It took until 1850 for anyone to design instruments that could measure the velocity of light accurately enough to detect the predicted differences. When the experiment was finally done, it was found that the velocity of light is *lower in water*, and by the amount claimed by the wave theorists.

A. Indicate in general terms the nature of the two theories at issue. State the two hypotheses that are to be tested simultaneously in this experiment. Call one *PH* and one *WH*. Specify the initial conditions and auxiliary assumptions simply as *IC* and *AA* since you are given no detailed information about them.

B. State the prediction *PP* associated with the particle theory.

C. State the prediction *WP* associated with the wave theory.

D. Does the experiment constitute a good test of the particle hypothesis? Explain why or why not.

E. Does the experiment constitute a good test of the wave hypothesis? Explain why or why not.

F. What was the fate of the two predictions?

G. What is the justified conclusion concerning the particle theory? Give the short form of the appropriate argument.

H.   What is the justified conclusion concerning the wave theory? Give the short form of the appropriate argument.

6.7   THE DIGESTIVE SYSTEM

The following experiment was also a "crucial experiment." At the beginning of the twentieth century it was known that the pancreas (a small organ near the stomach) secreted digestive juices into the duodenum (the tube connecting the stomach with the small intestine). This happened whenever partially digested food entered the duodenum from the stomach. The question was whether the signal that started the pancreas working was transmitted by the nervous system or by a chemical substance carried by the blood. At the time there were known examples of both nerve-stimulated organs and chemically stimulated organs. No other stimulating mechanism was known.

To decide the issue, two physiologists, W. M. Bayliss and E. H. Starling, cut all the nerves going to and from the duodenum of an experimental animal. All blood vessels to and from the duodenum were left in place. They also inserted tubes that allowed the flow of digestive juices from the pancreas to be detected.

The result of this experiment was that when food entered the duodenum of the experimental animals, the pancreas secreted digestive juices in the normal way.

A.   The two theories define nerve-stimulated organs and chemically stimulated organs respectively. Call the hypothesis that the pancreas is a nerve-stimulated system NH. State the corresponding CH.

B.   State the predictions, NP and CP, associated with the nerve hypothesis and chemical hypothesis respectively.

C.   Does the experiment constitute a good test of the nerve hypothesis? Explain why or why not.

D.   Does the test constitute a good test of the chemical hypothesis? Explain why or why not?

E.   What was the fate of the two predictions?

F.   What is the justified conclusion concerning the nerve hypothesis?

G.   What is the justified conclusion concerning the chemical hypothesis?

6.8   THE EXPANDING UNIVERSE

Now you are on your own. In one page or less, analyze the description of the issue in order to determine what conclusion (or conclusions) are justified. Obviously you should follow the general pattern established in the previous questions.

One of the most interesting discoveries of the twentieth century is that the universe is expanding; that is, the galaxies are all moving away from each other. This discovery stimulated the creation of numerous theories defining kinds of systems in which such expansion would take place. Two of these models were widely regarded as possibly giving the structure of the real universe. One was an "explosion model" (the "big bang" theory) in which all matter is originally concentrated in one place and explodes outward. The other is a "steady-state model," in which subatomic bits of matter are created out of nothing and eventually move outward, leaving each region of space with the same total amount of matter for all time.

If the universe is an exploding system, it follows, using accepted auxiliary hypotheses and initial conditions, that the density of matter (e.g., the number of galaxies per cubic light year) gets less and less the farther away from the original explosion one gets. If the universe is a steady-state system, on the other hand, the density of matter should be exactly the same everywhere.

To decide which of these two models fits the real universe, all we need to do is measure the density of matter in regions of space other than those near our own galaxy. Recently, radio telescopes have made it possible to do this. The measurements show a clear decrease in density at great distances from our galaxy.

6.9   MUTATIONS

In the 1940s, during the beginning of what we now call "molecular biology," geneticists Salvador Luria and Max Delbruck solved a major problem concerning the production of mutations in living organisms. It was known that certain viruses, bacteriophages ("phages" for short), can attack and kill bacteria. However, in some bacteria cultures some of the bacteria survive—apparently resistant to the phages. Moreover, the descendants of the resistant bacteria continue to be resistant, so the resistance is a genetic trait. The question was whether the mutations in the genetic material that lead some of the bacteria to be resistant are produced by the attacking phages or just arise spontaneously, at random, in the bacteria population. To decide the issue, Luria and Delbruck produced a number of bacteria cultures, each from a small number of bacteria. They then introduced phages into each of the cultures. If the phages produced the mutations leading to resistance, then all of the cultures should have had roughly the same number of resistant bacteria since each was roughly the same size when the phages attacked. But if the mutations arise by chance, the number of resistant bacteria should differ from culture to culture depending on when in the short history of the growth of the culture the mutation appeared. If early, there would be lots of resistant bacteria produced before the phages were introduced. If late, there would only be a few. Early or late should be a matter of pure chance. When they examined the cultures after introducing the phages, the number of resistant bacteria varied greatly from culture to culture.

6.10   THE WAVE THEORY OF LIGHT

In the course of studying the wave theory of light, a French physicist deduced the following startling conclusion: If a tiny spot of light were to be projected on a small circular object, like a coin, and then against a screen, the shadow of the object on the screen should exhibit a small bright area right in the middle of the shadow (see Figure 6.6). On the hypothesis that a beam of light is a Newtonian particle system, the edges of the shadow would be somewhat blurry, but there would not be a bright spot in the middle. Advocates of the Newtonian theory rushed to do the experiment in hopes of refuting the claims of the wave theorists. But to almost everyone's surprise, when the conditions of the experiment were satisfied, there was a bright spot in the middle of the shadow, just as the French physicist had predicted.

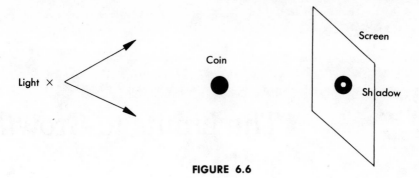

FIGURE 6.6

6.11

Find a discussion of the results of some experiment that is relevant to the justification of some theoretical hypothesis. You may find an example in a newspaper or newsmagazine. The Sunday supplement to your local newspaper is a good bet. Or you might try some popular sources that specialize in scientific findings, e.g., *Scientific American, Psychology Today, Science,* or *Science News.* When you have found something that you find interesting and substantial enough to work on, take one page to analyze the experiment along the lines suggested in the above exercises.

One of the side effects of doing this exercise is that you get an idea of the different levels of science reporting in various popular sources. Some sources tell you everything you need to know in order to judge whether the stated conclusion really is well justified or not. Others give so little information that you cannot tell at all. You are forced to take their word for it. Most sources fall somewhere in between.

This exercise may be turned into a longer project by looking for other sources of information on the same theory. You might be able to uncover a whole history of experiments relating to the theory, each of which can be analyzed. That way you may discover other experiments justifying hypotheses using the theory. Or you may discover cases in which such hypotheses were refuted—or were refuted and then "saved" by changes in auxiliary assumptions or in the theory itself. You can then look to see if these changes were themselves later justified or not, and so on.

The Limits to Growth

Up to now we have discussed only examples taken from the established sciences. Except for the example of genetics, which may be of great practical interest to people contemplating having children, these examples have few direct practical implications. Their interest is primarily intellectual. They feed one's curiosity and imagination. Now we are going to examine at some length an example with implications that may be almost too directly related to practical concerns. The implications are so far-reaching that many people would prefer not even to think about them. But they must be thought about, even if it hurts. And the techniques we have developed for dealing with scientific theories and hypotheses can help us to reason more effectively about these issues.

The problem is complex. In the 1960s it was called "The Population Explosion." In the 1970s it became "The Energy Crisis." But these are just two aspects of the same problem. The problem is *growth*. How long can the world's resources sustain current rates of growth in both population and industrial output? If growth must stop, how can this be achieved? Will it come about naturally? Will the "natural" course be a human disaster? What can be done to avoid a possible disaster? What can an individual do?

These issues were recently brought into the area of public discussion in a "scientific" framework by a little book entitled *The Limits to Growth*. In this chapter we will use what we have learned so far to evaluate the arguments of this book and those of some of its critics.

7.1 THE WORLD II MODEL

The introduction to *The Limits to Growth* informs us that the analysis of the book is based on a "model of the world." The model is named "World II," being a modification of an earlier model called "World I." One does not have to read far to discover that what the authors mean by a model of the world is a definition of a kind of system that they are going to use to represent the whole world physical-social-economic system.

116

So they are using the word "model" in the sense of "theory." The strategy in such a situation is first to get clear about what kind of system is being defined. Then one can go on to ask whether hypotheses using this model are justified. In particular, "Is it true that our real world is approximately like a World II model?" For the next several sections of this chapter we shall be concerned solely with the model itself. Then we shall ask how well it fits the real world.

The model, we are informed, is very complex, being written in the form of many mathematical equations. Moreover, the equations have been programmed into a computer, which does all the calculations. But that should not scare us. Neither should we be overly impressed. Deductions from a definition are simply deductions, no matter whether they are done by a person or a computer.

The reason the authors use the word "model" rather than "theory" is that the definition of World II, though complex, is obviously much too simple to apply very precisely to the real world. The model has only five basic quantities. Each is intended to represent just one aspect of the real world. The five aspects are:

1. Population
2. Industrial output
3. Pollution
4. Food production
5. Nonrenewable resources

In the model these are just overall quantities. There is nothing in the definition of World II systems that represents the *distribution* of any of these quantities. In the real world, most of the population is in Asia, Africa, and South America, but most of the industrialization is in Europe and North America. In World II systems, there is just population, period. Similarly for the other quantities. There is in fact nothing in the model that represents different geographical areas or different political boundaries.

In the model, as in the real world, each of the basic quantities is represented by *two* numbers. The *amount* of that quantity at a given time, and the *rate of change* of the amount at that time. For example, in 1970 the *amount* of population (i.e., the number of people) in the world was about 3.5 billion. The *rate of change* in the amount was a growth of about 2 percent a year. The other quantities are harder to measure. Note that the distinction between amount and rate of change is like the distinction between position and velocity (rate of change in position) or between velocity and acceleration (rate of change in velocity). This is a fundamental type of relationship that exists in almost all systems.

The reason the model is complex is that it includes relationships among all five quantities—really ten if you distinguish amounts and rates of change. The number of interactions is quite large. Figure 7.1 gives you a rough idea of how complex things are. In this respect the kind of thinking that went into develop-

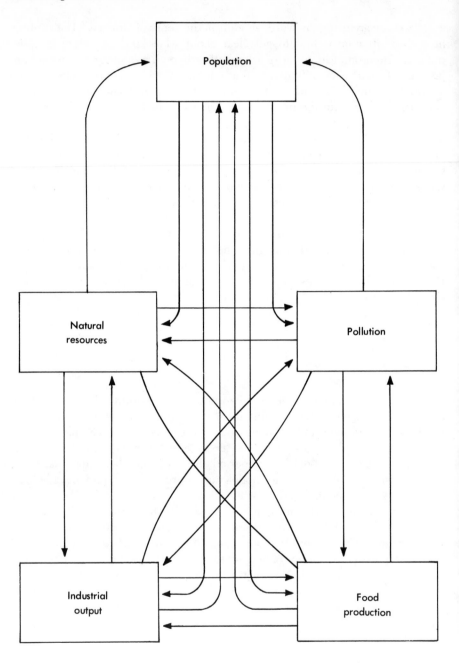

FIGURE 7.1

ing the World II model is much more advanced than thinking that concentrates just on population or just on energy. All these things interact in the real world. If population grows, so must food production if the standard of nutrition is to remain constant. But more food means more machines, more fuel and electricity, more fertilizer, and so on. And more people mean more manufactured goods, at least in industrialized areas. But more people and more industrial production means more pollution, which may lead to there being less food and maybe even fewer people. And so on, and on.

Although the authors do not say so explicitly, it is easy to discern that the World II model is *deterministic*. If you add to the definition of a World II system specific values of the amounts and rates of change for the five basic quantities, the computer will deduce for you these ten values for all later times. In principle this procedure is just like taking the observed positions of the 1682 comet and using Newton's theory to deduce when the comet should return, if it is indeed part of a Newtonian system.

What the authors did, then, is to take the best estimates of these ten values for the real world in the year 1970 and give these values to the computer as initial conditions. The computer then calculates the values for all later times and draws them out on a graph. The result for the amounts of the five basic quantities is shown in Figure 7.2. The authors refer to this result as the "standard run."

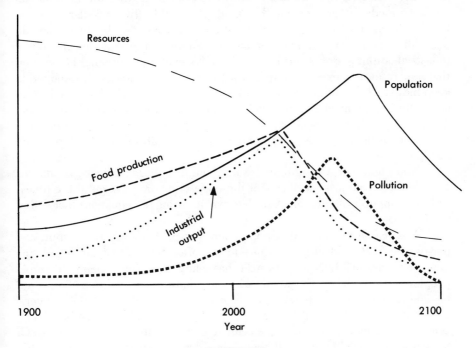

FIGURE 7.2

What happens in a World II system is that population, industrial output, and food production continue increasing until sometime after the year 2000. Then industrial output drops because of depleted resources. Food production, which depends on industrial production for machines, fertilizers, and so on, then drops. Because of natural lags in the system, both population and pollution continue to grow for a while. Finally the large population simply cannot be sustained and the system "collapses," with a resulting sharp decline in the population.

One of the advantages of having the definition programmed into a computer is that it is then relatively easy to play around with a variety of different initial conditions and see what the system does. For example, if one puts in the information that the store of natural resources is two or even four times the 1970 estimates, the system still collapses. So assuming that new technology can open up new energy sources, for example, does not make all that much difference. At best it only delays the collapse by a few years.

If our world is like a standard World II system, the future is a disaster. That collapse occurring sometime before 2100 would involve the death of roughly 5 billion people from starvation and disease within a relatively short span of 10 to 20 years. If this is indeed what is going to happen, those of us alive today should be thankful that we will have died a nice natural death by then. But our children may not be so lucky and our grandchildren certainly not.

The standard run exhibits a World II system in what the authors call "collapse mode." However, another mode is possible. In particular, suppose that by the year 1985 people have children only at the "replacement rate" of one child per person, or two per couple. Suppose also that the amount of industrial output per person is kept at the 1985 level. In addition, it is specified that new resources are created (presumably by new technology) fast enough to keep industrial output per person at the 1985 level. Similarly, food production and pollution are fixed at 1985 levels per person. With these conditions specified, a World II system does not collapse, but reaches a stable equilibrium sometime during the twenty-first century. This "equilibrium mode" is pictured in Figure 7.3.

Note, however, that nothing is said about *how* these particular conditions are achieved. One just specifies that these are to be the values at these times, and the computer blindly calculates later values using the equations that define the model.

The reason the amount of population, per-person industrial output, and per-person food production continue to rise after 1985 is that the population now, and presumably in 1985 as well, has many more young people than old people. But it is the young people who have most of the children. So even if beginning in 1985 each person produces only one child, it will take several generations until the number of people being born equals the number dying. When that happens, we shall have achieved zero population growth, or ZPG. Similarly, industrial output will have reached zero industrial growth, or ZIG.

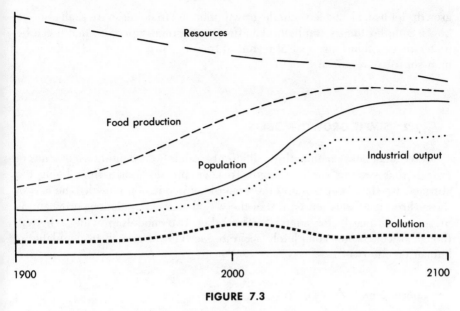

FIGURE 7.3

The fact that the population continues to grow even if births are occurring at the replacement rate is an example of a "lag" in the system. Other lags are also built into the definition of World II systems. The effects of pollution, for example, typically exhibit considerable lags. It may take 20 years of air pollution to raise the number of deaths due to lung cancer. Mercury released into the environment may take several years to work its way up the food chain from tiny organisms, to small fish, to larger fish, to humans.

Partly because of the lags built into the definition of World II systems, it makes a difference when conditions leading to equilibrium are imposed. If instead of 1985, these conditions are not imposed until the year 2000, the system does not exhibit a stable equilibrium. It does not exhibit so drastic a collapse either, but before 2100 the population begins to drop off due to food and resource shortages. The lesson, of course, is that if the real world is like a World II system, we cannot wait until the middle of the twenty-first century to do something. It must be done by the end of this century.

These are sobering conclusions if they are correct—that is, if the model gives even an approximate account of the nature of the system in which we live. But how could this be? The world has gone along for centuries with no indications of global catastrophes. Why suddenly should we have only 100 or so years left to go on as we have been all along? The answer to these questions lies in the nature of the growth pattern exhibited both by populations and by industrial output. These patterns are built into the World II model in a set of complex interactions. But the basic pattern is quite simple and can be understood apart from its inclusion in such a complex model. As a means to a better understanding of the World II model and the phenomenon of population growth, let us look at a few simple growth models. In doing so we shall also see how a complex theory can be broken down into simpler parts so that it can be understood without any special technical training. Just a little "scientific common sense" is required.

7.2   SIMPLE GROWTH MODELS

We shall now examine three different models (i.e., definitions) of kinds of growth that systems may undergo. To keep the discussion from being too abstract, we shall keep in mind the example of population growth. The first of these three models is not typical of the way that real populations grow and is presented primarily for contrast. The other two models provide a better, though of course not completely accurate, account of growth in populations, human or otherwise.

Linear Growth Models

Imagine a population consisting of exactly 100 people and organized in such a way that the number is allowed to increase by exactly 10 people each year. These people might be members of an exclusive club, a communal farm, or workers in a small factory or other business. Beginning with 100 and increasing by 10 per year would produce the pattern given in Table 7.1. These figures are easily represented in graphical form in Figure 7.4. The fact that the graph for this kind of growth is a straight line is responsible for the name "linear growth model."

Linear growth may be defined *qualitatively*—that is, with no explicit use of mathematical expressions—as follows:

> A system exhibits *linear growth* if and only if it increases by a *constant amount* each time period.

The amount added and the time period are related, of course, in that the amount is specified as some amount *per unit of time*. In the example, the constant amount added was 10 per year.

For those with sufficient background and interest, the *quantitative* definition of a linear growth model is:

TABLE 7.1

End of Year Number	0	1	2	3	4	5
Size of Group	100	110	120	130	140	150

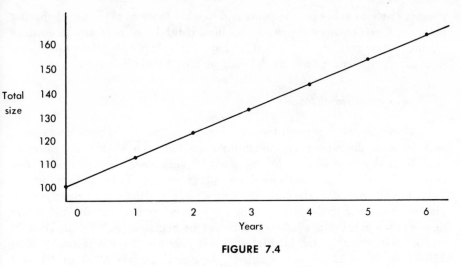

FIGURE 7.4

A system exhibits *linear growth* if and only if the total amount $N(T)$ at any time T is given by the equation

$$N(T) = N(0) + (A \times T)$$

where $N(0)$ is the amount at the beginning and A is the amount added per unit of time.

In our illustration, $N(0) = 100$ and $A = 10$, so, for example, $N(6 \text{ years}) = 100 + (10 \times 6) = 100 + 60 = 160$. For our purposes, a good understanding of the *qualitative* relationships will be sufficient.

The definition of linear growth may be applied to anything that can be added in equal units, not just population. Thus if you save $10 a week, your savings will grow linearly at that rate.

Now let us ask ourselves: Do populations under "normal" conditions fit the definition of a linear growth model? Think about how populations normally grow. Fundamentally there are only two things to consider. Some people die and other people are born. Now if the population of 100 in our example were a naturally growing population, say a group of people on an isolated island, then to add 10 people the first year, there would have to be at least 10 births. If there had been 1 death, then there would have to have been 11 births, and so on. Suppose, to be definite, that there were 2 deaths and 12 births, and that this continued year after year.

Now imagine this population 10 years later. As you can easily calculate, there are now 200 people. But if there are still only 2 deaths and 12 births per year, something dramatic and unusual is happening. At the beginning, when there were only 100 people, the *death rate* was 2 per 100 people; 10 years later it is only 1 per 100—that is, *half* the earlier rate. Similarly for the *birthrate;* it too has been cut in half. This is possible, but hardly what one expects for normal populations. Everything else being equal, a population twice as big will have roughly twice as many people dying and roughly twice as many people having children. One's chances of dying or having a child do not normally go down as the population gets larger. But that is what must be happening if the population is to satisfy a linear growth model. Let us now try another type of model.

Exponential Growth Models

These models represent the idea that the *rate* at which births and deaths occur remains the same even though the population as a whole is increasing in size. To understand this model one must be very clear about the difference between a fixed *amount* and a fixed *percentage* (or rate). Let us concentrate on the notion of *birthrate*. The birthrate for a given population, for a given year, is just the number of births that occur in that population that year, divided by the number of people in the whole population at the beginning of the year. Thus, in our previous example, the birthrate the first year was (an unrealistically high) 12/100, or .12, or 12 percent. Similarly, the *death rate* was 2/100, or .02, or 2 percent. This means that the overall *growth rate* was 12 percent − 2 percent = 10 percent, or 10/100.

These relationships among birthrate, death rate, and growth rate are sufficiently important to be set off symbolically. Let B stand for the birthrate, D the death rate, and R the growth rate. Then

$$\text{growth rate} = \text{birthrate} - \text{death rate}$$
$$R = B - D$$

This relationship is fundamental to all discussions of a "population explosion" and similar issues.

Now let us consider an example similar to the previous one. We shall again begin with a population of 100, but this time we shall assume that it grows at a constant *rate* of 10 percent per year. The data for the first 5 years are given in Table 7.2. Here it is clear that adding a *constant percentage* each year means adding a *greater number* each year. This is simply because the population is bigger each year and taking the same percentage yields a bigger number than it did the previous year. The graph for this growth model is given in Figure 7.5. Exponential growth may be defined qualitatively as follows:

TABLE 7.2

End of Year Number	0	1	2	3	4	5
Number Added		add 10	add 11	add 12	add 13	add 15
Size of Group	100	110	121	133	146	161

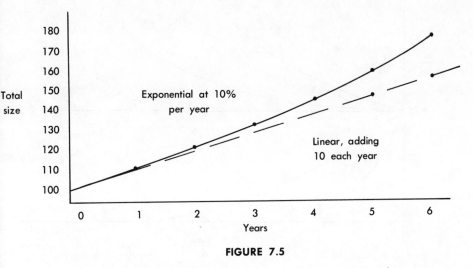

FIGURE 7.5

A system exhibits *exponential growth* if and only if it increases by a *constant percentage* each time period.

In our example the constant percentage was 10 percent per year.

For fixed time intervals, the quantitative definition of exponential growth is:

A system exhibits *exponential growth* if and only if the total amount $N(T)$ at any time T is given by the equation

$$N(T) = N(0) \times (1 + R)^T$$

where $N(0)$ is the amount at the beginning and R is the rate of growth.

In the equation that defines exponential growth, the time T appears as what in mathematics is called an exponent—thus the name "exponential growth."

In older discussions of population growth, exponentially growing populations were often said to be growing geometrically while quantities exhibiting linear growth were said to be growing only arithmetically. This terminology is misleading as well as out of date, but since you may run across it, it is well to know what it means.

Most people have had some explicit contact with exponential growth in the form of interest on savings or on loans. If you keep a sum of money in a savings account at 6 percent interest, your savings will grow exponentially at the rate of 6 percent per year, even though you make no further deposits. If you do not pay back a 12 percent loan, the amount you owe will grow exponentially at the rate of 12 percent a year.

For our purposes, you need not know how to apply the equation for exponential growth. You need only have a good grasp of the basic relationships in rough quantitative terms.

In addition to exponential *growth*, there is also exponential *decay*. This

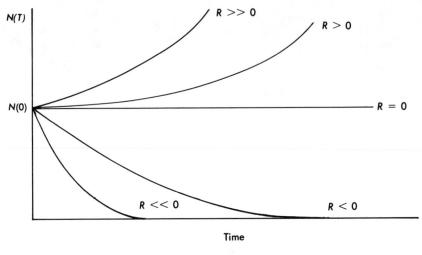

FIGURE 7.6

occurs whenever *R* is less than zero. In the case of populations, decay occurs whenever the death rate exceeds the birth rate. (Remember that $R = B - D$.) Collapse in World II systems is an example of rapid exponential decay. Radioactive wastes from nuclear generators exhibit very slow exponential decay.

Between growth and decay there is stability. If $R = 0$, then the system exhibits zero growth (also zero decay), and the amount remains constant at its initial value. These relationships are pictured in Figure 7.6. ("$R > 0$" means that *R* is greater than zero.) Just by comparing the behavior of a simple exponential system for various values of *R* makes it obvious why ZPG is so appealing.

So far we have just looked at a fairly abstract characterization of exponentially growing systems. It is worth spending a little time developing a good intuitive feeling for exponential growth. The world's population has grown exponentially in the past. So there is at least the possibility that it might continue to do so—with disastrous results. Whether it will or not is just the question of whether our world is approximately like a World II system. We shall face that question shortly.

Doubling Times

It is hard to get an intuitive feel for growth rates if one thinks simply in terms of percentages, especially when the percentages are small. The difference between a growth rate of 1 or 2 percent and one of 4 or 5 percent does not seem very great. This is why demographers (i.e., people who study the growth and structure of populations) often use the concept of a *doubling time*. It is an interesting fact about exponential as opposed to linear growth that the length of

TABLE 7.3

Growth Rate per Year (percent)	Number of Years To Double in Size (T_d)
1	70
2	35
3	23
4	18
5	14
6	12
7	10
10	7

time it takes an exponentially growing population to double in size depends only on the growth rate and not at all on the original size. The relationship between the growth rate R and the time it takes a quantity to double T_d is given by the following simple relationship (which you should understand):

$$T_d = .7/R \quad \text{or} \quad 70/R \text{ percent}$$

Thus any quantity growing at a rate of 7 percent a year will double in 10 years. This holds for populations as well as money in the bank. Some doubling times for a range of possible growth rates are given in Table 7.3.

In 1970 the growth rate in the United States was around 1 percent. This means that someone born that year could expect the population of the country to double during his/her lifetime (70 years). That same year the growth rate in Costa Rica was around 4 percent, which means that its population would double by the time that person reached 18. So the difference between a growth rate of 1 percent and 4 percent is quite striking.

Small Rates and Large Amounts

Another way of gaining an intuitive feel for the nature of exponential growth is to consider the amounts that correspond to typical growth rates, and compare these amounts to something you know, like the population of familiar cities.

As a general rule, scientific reasoning tends to be more quantitative than everyday reasoning. One of the benefits of studying scientific reasoning is that it helps you develop the habit of thinking somewhat more quantitatively. But in developing this habit it is important not to try to be too precise; otherwise you will simply get lost in all the numbers and be no better off than you would be just thinking in more qualitative terms, such as large or small, fast or slow, and the like. The way not to be too precise is simply to "round off" relevant numbers to something you can remember and manipulate in your head.

TABLE 7.4

Unit	Approximate Population
Medium U.S. city	1 million
Large U.S. city	5 million
Largest U.S. city	10 million
Total U.S.A.	200 million
Total U.S.S.R.	250 million
Total India	500 million
Total China	700 million
Total world	4,000 million

In round numbers, a medium-sized U.S. city has a population of about 1 million (i.e., 1,000,000). This includes, for example, San Francisco, Baltimore, Detroit, Cleveland, Dallas, Houston, and Milwaukee. A large city like Chicago or Los Angeles has a population near 5 million. And the largest city, New York, is pushing 10 million. The total population of the United States is about 200 million; the Soviet Union is somewhat larger. India has a population of roughly 500 million, and China is somewhat larger. The total world population is roughly 4 billion, (i.e., 4,000,000,000). These numbers are summarized in Table 7.4.

The total growth rate for the world has been around 2 percent per year. But 2 percent of 4 billion is 80 million. That is equivalent to 8 cities the size of New York or 80 the size of Detroit. It is almost half the total U.S. population. And that is the number of people added in *one year*. Every subsequent year the number added is a little larger. Note also that at this rate, the world population will double in just 35 years.

Even considering just the United States, which until very recently has had a relatively low growth rate of around 1 percent, a year's growth corresponds to two Detroits. Most recently, the U.S. birthrate has dropped to nearly the replacement rate, but most experts expect it to rise again fairly soon.

The general point is that even a small percentage of a very large number is a large number. This helps to explain why exponential growth is so deceptive. When the total amounts are not too large, the amount added each year is not so great. But when the total amount gets large, the amount added is large. And the next year the addition is a percentage of this greatly enlarged total. So as time goes on the growth in terms of actual numbers added becomes tremendous, even though it continues to be a small percentage of the total.

Superexponential Growth Models

It is instructive to look at the pattern of growth exhibited by the world's population in the past. It is especially instructive to begin, say, with the year zero, and then look for the date at which the population was double that at year zero. Then look for the date the population doubled again. And so on. In this way one can get a rough idea of the changes in the growth rate that have occurred over time. Of course the estimates of the total population in the distant past are rough, but they are good enough to get a good idea of what has happened. These data are given in Table 7.5.

Table 7.5 reveals that the growth of the world's population since the year zero has not followed the growth pattern of a simple exponentially growing system. In an exponentially growing system, as we defined it, the growth rate, and thus the doubling time, is a fixed constant. But Table 7.5 shows a *decreasing* doubling time. This means that the growth rate itself has been *increasing*. So what we have had is exponential growth, but with an increasing growth rate. Such growth is sometimes called "superexponential."

In qualitative terms, the definition of a superexponential system is:

A system exhibits *superexponential* growth if and only if it increases by an *increasing rate* each time period.

There is no standard quantitative definition of superexponential growth since there are many ways the rate can increase. Nor is there any point to drawing a graph of superexponential growth. It looks like the graph for exponential growth, except that it curves upward faster than a corresponding simple exponential curve. Needless to say, superexponential growth is even more deceptive than simple exponential growth.

Table 7.5 also provides some insight into why it is that the phenomenon of population growth has only recently become an object of special concern. It is only during the past 50 years that the doubling time has been shorter than the average lifetime. When the doubling time is longer than an average person's lifetime, it is difficult to perceive any difference between linear growth and exponential growth. Thus our historical expectations, our institutions, and perhaps even our genes, have been geared to dealing with linear growth, at most. It is only in the past 50 years that civilization has actually experienced the difference between linear and exponential growth. If we are indeed on an exponential curve with no built-in brakes, the question becomes whether we can change our expectations and institutions fast enough to avoid a crash.

TABLE 7.5

World Population	Year	Doubling Time
250 million	0	
500 million	1650	1650 years
1 billion	1850	200 years
2 billion	1930	80 years
4 billion	1975	45 years

7.3   THE POPULATION BOMB ARGUMENT

A decade or more ago, biologists and others offered a simple argument for a worldwide policy of zero population growth. The argument was perhaps best summarized in a popular book called *The Population Bomb.*

The first premise of the argument is that at any particular time there is a maximum number of people that can be sustained on the earth at some given standard of living. For the purposes of the argument the standard of living could be defined simply in terms of the number of calories available per person per day, and might be quite low. And we may assume full use of all technologies available at that time. This maximum number, whatever it is, is called the *carrying capacity* of the earth at that time.

Next it is assumed that the carrying capacity itself must have some upper limit, for example, when there is one person for every square meter of land on the earth. Finally it is assumed that, left to itself, the world's population will continue to grow exponentially (if not superexponentially).

It follows that at some time the population must approach the maximum carrying capacity. Since by definition the population cannot sustain itself at numbers greater than the maximum carrying capacity, the growth rate must be reduced to the growth rate of the carrying capacity, which eventually is zero. The argument thus far is illustrated in Figure 7.7.

Now since the growth rate is the difference between the birthrate and the death rate, there are only two ways the growth rate can decrease. Either the birthrate is lowered or the death rate goes up. So the meeting of the population and the carrying capacity can have one of two resolutions, a "birthrate resolution" or a "death rate resolution." It is then claimed that if we let nature take its course, it will impose the death rate solution in the form of famines, disease, and so forth. The argument concludes with the assertion that the only way to

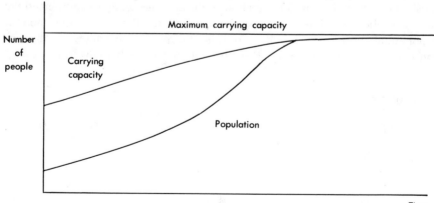

FIGURE 7.7

avoid this obviously undesirable result is for us consciously to reduce the birthrate to the level of the lowest achievable death rate, thus producing ZPG.

Even granting all the premises, however, the conclusion need not be taken seriously unless one believes that we are in fact approaching the carrying capacity at the moment and that we cannot increase it fast enough to stay ahead of the population. If you add this premise, then, granting the other premises, the argument seems conclusive. But what reason is there to believe this premise?

This is the intended contribution of the World II model. As you can now see, the basic behavior of World II systems is like that of the system assumed in the simple population bomb argument. Population grows exponentially until it outstrips the more slowly growing food supply. Industrial production grows exponentially until it uses up the available resources. The World II model, however, uses as initial conditions current data on the five basic quantities and attempts to define various interactions among these quantities. Thus it does give some estimate of where we are now and how long we have to go if we do not impose severe restrictions.

The question remains whether the World II model gives a sufficiently accurate account when applied to the actual system in which we all live. In other words, is there a good justification for believing the hypothesis that the real world is reasonably close to being a World II system? So far, no such justification has been presented. We have by and large just been looking at what the model says about the system *it defines*. The existence of a collapse mode and an equilibrium mode, for example, are *deductions* from the *model itself*. Any system satisfying the definition will have these two modes. But this does not help us to decide whether *our* system satisfies the definition. That is another question.

7.4 TESTING THE WORLD II MODEL

It is time we raised the question of justification. What justification is there for thinking that our world is anything like a World II system?

Are There Any Good Tests of the World II Model?

At this point you would expect to be told about some good tests of the hypothesis that the real world is more or less like a World II system. The fact of the matter, however, is that there have been none simply because there cannot have been any. Before reading on, stop and try to figure out for yourself why this is so.

Let us just apply what we have learned. To have a good test of any hypothesis there must be a prediction that satisfies both Condition 1 and Condition 2.

Finding a prediction that satisfies Condition 1 is not difficult. For example, since the publication of *The Limits to Growth* in 1972, there has been no serious effort to achieve ZPG or ZIG on a worldwide scale. Given these initial conditions, applying the model leads to the prediction that population, industrial output, food production, and pollution should have increased and the stock of nonrenewable resources should have decreased. This is what in fact has happened. But it would be fallacious reasoning to conclude that therefore the model fits. To justify that conclusion we need Condition 2.

The difficulty is that it is impossible to use the model to make short- or even medium-range predictions with enough precision to satisfy Condition 2. The model, as we have seen, is highly simplified relative to the known complexity of the real world. So our hypothesis can only be that the real world is very roughly like a World II system. But almost anything that might reasonably be expected to happen between 1975 and 1995 would qualify as being roughly like what the model would lead us to predict. So what the model leads us to predict over a 20- or 30-year period is not at all improbable, even if the world in general is not at all like a World II system. Condition 2 cannot be justified for these predictions. So even if the world in 1980 or 1985 or 1990 does fit within the wide limits set by the model, we cannot justify the conclusion that the real-world system is in general like a World II system.

This does not mean that there are no predictions that could provide a good test of the hypothesis. For example, if the world is not close to ZPG and ZIG by 2025, then the hypothesis leads to the prediction that there will be a serious collapse before 2100. This prediction satisfies both conditions for a good test. But it would do us no good to learn when the collapse is upon us that we are indeed living in something like a World II system. If we are, the model tells us that we need to take action before the year 2000. So we would like to know before then whether the model fits or not. But, as we have seen, there is no way to justify either conclusion in this short time. The short-term predictions are so rough that they are sure to be correct in any case.

At this point it is well to remember that failure to justify a hypothesis does not mean that it is false or that one is justified in believing it to be false. Neither belief is justified. We just have no way of telling whether the hypothesis is correct or not.

Other Models

The Limits to Growth was the first attempt to present to the public at large a model that was intended to be global in its application and takes into account interactions among several basic quantities. There are other models that might be applied, but most of these were not designed to be applied on a global scale, and they have not been presented to the public in so dramatic a fashion as the World II model. Yet such models have been applied on a large scale and have been used in debates of these issues. So it is worth getting at least some idea of what these models are like.

As you should expect, these alternative models appeal to features of the world that are not even represented in the World II model, namely, social, political, and economic mechanisms. The authors of *Limits to Growth* claimed that introducing such mechanisms could only lead to predictions of consequences that are *worse*, not better, and things are bad enough as it is. This is a cynical view of existing social, political, and economic mechanisms, and may not be justified. But it is difficult to refute too.

Regarding population growth in particular, it has been claimed that something like the following chain of causal mechanisms operates in the real world:

Increasing population, *leads to*
Increasing cost of raising children, *which produces*
A desire for fewer children, *which results in*
Fewer children, *which yields*
A leveling off of the population growth rate.

It has been argued that this model explains the pattern of population growth in France between 1830 and 1930, which grew fairly rapidly and then leveled off. But even if this is true, we know that the existence of some systems that fit a model does not provide a good test of the hypothesis that the model fits the whole world system. We need some justification for believing that such mechanisms will operate on a global scale in the twenty-first century. For this we need some precise, fairly short-term predictions about the behavior of the global system. But these models do not provide such predictions any more than the World II model does.

A similar causal chain has been proposed for the creation of new resources:

Scarcity of resource X, *leads to*
Increase in the price of X, *which stimulates*
Research to develop a substitute for X, *which produces*
A substitute for X.

Again one can cite cases in the past where new technology produced alternative resources. For example, 50 years ago no one had even thought of the idea of using nuclear power to generate steam to drive electric generators. But such examples do not provide good tests of the hypothesis that the world system will operate this way in a global scale in the twenty-first century.

These models are not only unjustified, but have many shortcomings. For example, a desire for fewer children will not lead to a decrease in births unless reliable methods of contraception are known, approved, and implemented. In many places with the highest birthrates at present, these conditions are not being met. Similarly, what guarantees that research will produce a needed substitute *just when it is needed?* Many research projects have failed and were eventually abandoned.

This situation is fairly typical of "global" models in the social sciences.

They are as a general rule simply not precise enough to generate hypotheses that could be tested in the near future using data that one might reasonably be able to obtain. The rational attitude in these cases can only be one of suspended judgment.

7.5 WHAT SHOULD WE DO?

If the nature of the world system were simply an object of scientific curiosity, like the nature of galaxies, we could just wait and see whether someone manages to develop a model that is sufficiently complex to be tested in a reasonable time, and hopefully tested successfully. But that is not our situation. How the world system operates will be a matter of life and death, if not for us, then for our children or grandchildren. And if our system is like a World II system, then the outcome depends on what *we* do before the year 2000. So we are now faced with the decision either to rely on "natural" mechanisms or to try deliberately to impose constraints leading to ZPG or ZIG on a worldwide scale. Moreover, it is a decision that cannot be avoided. Doing nothing is, if only by default, making the decision to rely on the "natural" mechanisms.

Our problem in this case is not one of knowledge or justification. It is a question of taking a course of action in spite of not knowing what the consequences will be. We are faced with a problem that falls in the category of "decision making with uncertainty." In Part IV of this text we will learn what this means and then return to the present "decision problem."

One point, however, can be made here. An important part of "rational decision making" involves assessing the *possibilities* even if one has no knowledge of which possibility will be realized or even which is the most probable. Thus, even if it cannot be justified in time, the World II model serves a valuable function. It makes clear one of the possibilities we must consider. It shows that the possibility that we are living in a World II system is a serious possibility. Thus not to take it into account when making "global" decisions could not be rational, as we shall see.

CHAPTER EXERCISES

7.1 Each of the following questions has one and only one best answer. Pick the answer you think best.

A. According to the World II model described in the text:
   (a) The world system has to collapse by 2100.
   (b) The world system has to reach equilibrium by 2100.
   (c) The world system may do either, depending on what we do during the next 25 years.

B.   If a population is increasing according to an exponential growth model:
  (a) The number of people added each year is fixed.
  (b) The number of people added each year is increasing.
  (c) The percentage of people added each year is increasing.

C.   Imagine two populations, X and Y, each increasing exponentially. Suppose that the growth rate of X is greater than the growth rate of Y.
  (a) The doubling time of X is greater than that of Y.
  (b) The doubling time of X is less than that of Y.
  (c) The doubling time of X may be either greater or less than that of Y.

D.   Imagine two populations, X and Y, each increasing exponentially. Suppose that the growth rate of X is greater than the growth rate of Y.
  (a) X must have a higher birthrate than Y.
  (b) X must have a lower birthrate than Y.
  (c) X may have a higher or a lower birthrate than Y.

7.2 Give a qualitative and informal statement of the law of growth for each of the three growth models discussed in the text: (A) linear, (B) exponential, and (C) superexponential.

7.3 Given below are five sets of birthrates and death rates. Assuming the same initial population, $N(0)$ in each case, make a rough sketch $N(t)$ as a function of time for each case. Put all five on a single graph, labeling each curve with the appropriate letter (A, B, C, D, or E).

A. $b = 2\%; d = 2\%$
B. $b = 4\%; d = 2\%$
C. $b = 3\%; d = 4\%$
D. $b = 4\%; d = 3\%$
E. $b = 2\%; d = 4\%$

7.4 Which of the five sets of conditions in Exercise 7.3 yields the *shortest* doubling time (A, B, C, D, or E)?

7.5 Define linear decay; that is, give the definition of a linear decay model.

7.6 Suppose that you have an outstanding charge on a regular "revolving" charge account. The interest rate on such accounts is usually 18% a year. If you forgot about the account, that is, made no payments and no new charges, how long would it be until your bill doubled in its total amount?

7.7 Imagine seeing a house for sale at $50,000. That is much more than you can afford right now. Suppose you say to yourself, "In 10 years I'll be able to buy a house like that." But suppose that prices for the next 10 years are subject to inflation at the rate of 7% a year—about the average in the late 1970s. How much will your "dream house" cost in 10 years?

7.8 The story is told of a king that saw a golden chess set in the marketplace. Inquiring after the price, he was told by the old merchant that the set was very special and that it could be sold only under special conditions. "If you want it," the old merchant said, "you must wait 64 days, one day for each square. On the first day you must bring me one penny. On the second, two pennies; the third, four pennies; the fourth, eight pennies; and so on, doubling the number of pennies each day until all 64 squares are covered. Then the set is yours." If the king accepts the bargain, what will be the result? Start calculating the total cost of the chess set until you see what will happen. You will need a calculator to get all the way to the end.

7.9 Looking at Table 7.5 you will see that the most dramatic increase in the growth rate of the world's population has taken place during the past 50 to 100 years. What do you think accounts for this dramatic increase in the growth rate? (*Hint:* Remember that $R = B - D$.)

8

Fallacies of Theory Testing

In Part I we identified conditional arguments that affirm the consequent or deny the antecedent as *fallacious arguments*. They seem superficially to be valid, and are often passed off as such, but they are not. Something similar happens with tests and arguments for theoretical hypotheses. There are recognizable patterns that seem superficially to be all right, but do not in fact provide adequate support, whether deductive or inductive, for the stated conclusions. The general mistake in these fallacious patterns is the failure to satisfy the second condition for a good test. But the failure gets disguised in various ways. It is thus helpful to have in mind some clear examples of these fallacies so that you can spot them when they come by in real life. The examples provide you with "models" (i.e., analogies) that you can use in analyzing other cases.

8.1 THE DELPHI FALLACY: VAGUE PREDICTIONS

In classical Greece, a Greek of means when faced with a difficult decision might well have made a trip to Delphi to consult the oracle, who was widely believed to have the power to foresee the future. The Greeks' faith in the ability of the oracle was partly based on the fact that the oracle seemed to be successful in making correct predictions. The question we shall consider here is whether the oracle's successful predictions provided justification for the belief that it could indeed "see" the future.

In a typical case a rich man might simply have asked the oracle whether his family would continue to prosper or not. The reply might have been: "I see grave misfortune in your future." If, a year later, the only son and heir had fallen into a well and drowned, that would have been regarded as the fulfillment of the oracle's prediction. At this point the man, his faith in the oracle increased, might have planned a new trip to Delphi hoping to hear a more favorable prediction this time. But would his faith have been justified?

Although the reasoning here does not involve any sort of recognized scientific theory, we can still treat the question of justification as we would in the case of any theoretical hypothesis. This, however, requires that there be a theory, and it may not be obvious just what the theory could be. In such cases it is best to begin with the theoretical hypothesis and work back to the theory. It turns out that it is not very important what the theory is, so long as it corresponds with the hypothesis.

In the present example, the hypothesis may be stated in terms we have used for theoretical hypotheses as follows: The oracle is a future-seeing system; that is, it has special powers that enable it to see the future. Behind this hypothesis is a vague theory of future-seeing systems. It is vague because no one ever tried to define explicitly what a future-seeing system might be like. But this need not bother us too much. We know that even the most venerated scientific theories, like Newton's, are vague to some extent. They do not explicitly define all the terms used in specifying the type of system at issue. So the difference between Newton's definition of mechanical systems and the Greeks' definition of future-seeing systems may be regarded as merely a difference in degree. The latter is just a lot more vague than Newton's.

Our next question is whether there was a good test of the hypothesis. The *first* condition for a good test seems to have been fairly well satisfied. Let H stand for the hypothesis that the Delphic oracle is a future-seeing system. The prediction, of course, was that a grave misfortune would befall the family. In order to deduce the prediction, P, as required by Condition 1, we need to add some initial conditions and auxiliary assumptions. The main initial condition, IC, is that the oracle said that misfortune would occur. Now to deduce P we need only add the auxiliary assumption, AA, that the oracle was honest, that is, that it reported truthfully what it saw in the future. So if the oracle could see the future, and said it saw misfortune, and is honest, then there should have been a misfortune. That is:

If [H and IC and AA], then P.

At this point one might suspect that the Greeks were simply guilty of affirming the consequent: observing the truth of the prediction and then concluding that the hypothesis is correct. But whether they in fact reasoned this way or not, the real question is whether they need have. If Condition 2 is satisfied, their conclusion was justified.

Having worked out Condition 1, it is easy to state Condition 2:

If [Not H and IC and AA], then very probably Not P.

Was this condition true? Well, suppose that the oracle was no better than the rest of us at knowing the future. Would it then be very improbable that the prediction would have come true? Not at all. The nature of the misfortune was left very indefinite. So was the time until it would occur. Thus, if one were simply to pick any rich Greek family at random, it would not be at all unlikely that something would happen in the next year or so that would count as a "grave misfortune." There could be a drought that would kill all his olive trees. His wife might have died in childbirth. A new city council might have imposed ruinous taxes. A minor war might devastate his lands. Any of these not uncommon occurrences would have fulfilled the prediction. In short, one would not have needed special powers to have a good chance of having this prediction fulfilled. Anyone could have done it. So the conclusion that the oracle had special powers was not justified because the needed premises for a good inductive argument (i.e., Condition 2) was not true.

The general lesson is this. If a prediction is sufficiently vague, then it stands a good chance of being fulfilled no matter what happens and no matter what theoretical hypotheses happen to be true or false. In particular, it is not very probable that the prediction will come out false if the hypothesis in question is false. But this condition must be met if one is to have a good inductive argument for the conclusion that the hypothesis is true. Without it, the claim remains unjustified. The oracle was a con artist.

Interestingly enough, the ancient Greeks, as well as other more primitive people, did have an intuitively sound grasp of scientific method even if they did occasionally allow themselves to be deceived. They did realize that if someone really could predict the future, that must indicate some special power. We don't now talk about special powers except the power of truth. And we don't attribute this power to people but to hypotheses. If a theory provides us with true predictions about the future, we conclude that it does capture part of the real structure of the world. Which is to say, the hypothesis that it does is true. Our *test of truth*, however, is fundamentally that used by the most primitive of peoples.

8.2 THE JEANE DIXON FALLACY: MULTIPLE PREDICTIONS

If you regularly shop at supermarkets or walk past newsstands, you know that there are numerous publications that feature "predictions" by contemporary "seers"—Jeane Dixon being among the best known. As several biographies attest, at least some people believe that Ms. Dixon has special abilities to foresee the future. Moreover, it is explicitly argued that her powers are proved by her many past successes. Let us consider whether this is so.

Ms. Dixon, and others in her profession, regularly issue whole sets of predictions. Thus you may in January see the headline: "Jeane Dixon's Predictions for the Coming Year." The predictions concern all types of subjects, but most deal with the lives of famous people in entertainment, politics, and the arts. Taken individually, these predictions may be more or less vague, but few are so vague as to guarantee success. For example, it is widely claimed that Ms. Dixon "predicted" the assassination of John Kennedy. What she predicted was that he would die in office—not quite so precise, but not hopelessly vague either. In any case, does the fact that this and other predictions have been correct provide justification for believing in her reputed special powers to see the future?

Let us for the moment ignore the fact that some of the predictions are a bit vague; we already know how to deal with that. The main complication, then, is that there are numerous predictions to consider. How should we treat them? Given what we have learned so far, there are really only two possibilities. We can form a *conjunction* of all the predictions; or we can form a *disjunction*. Let us consider both possibilities.

Let us form the conjunction of all of Jeane Dixon's predictions for the coming year. So we have a "superprediction" $P = P_1$ and P_2 and . . . and P_n. As in our discussion of the Delphic oracle, the hypothesis is that Ms. Dixon is a futureseeing system. The initial condition is simply that she has made all these predictions. In order that we be able to deduce the prediction, we must be a little careful in formulating the auxiliary assumptions. Let us assume here that she only makes predictions when she is fully confident that she is right, and that she honestly says what she sees. With these ingredients we can state Condition 1:

If [H and IC and AA], then P.

Now there is an immediate problem. Even Ms. Dixon's most ardent supporters admit that she makes mistakes—at least a few each year. So among the predictions for the coming year there will no doubt be at least one that comes out false. But that means that the whole superprediction, P, is false since we regard a conjunction as false if any of its components are false. But assuming that our initial condition and auxiliary assumptions are correct, we should immediately conclude that she cannot see the future. This, as we know, follows *deductively*. However, those who admit that Ms. Dixon makes mistakes do not draw the conclusion that she has no special powers. We might simply assume that they do not know that denying the consequent is a valid deductive form. But let us try being more generous and reconstruct the superprediction not as a conjunction but as a disjunction. One false prediction will not make the disjunction false.

So now our superprediction is $P = P_1$ or P_2 or . . . or P_n. There is no problem with Condition 1. The disjunctive superprediction still follows from the antecedent of the conditional. So let us look at Condition 2. If it is satisfied, the claim is justified—inductively. Condition 2 is:

If [Not H and IC and AA], then very probably Not P.

Now is it true that probably Not P if Ms. Dixon has no special powers? No, because simply by making a fairly large number of predictions she can be quite confident that at least one will be right. And if one is right, then the whole disjunction is true. Moreover, she is not just guessing. Living in Washington, D.C., she can hear a lot of gossip that most of us do not. And she can read newspapers and make inferences just like the rest of us. In short, it is not very probable that Not *P* even if Ms. Dixon has no special powers at all. Condition 2 with the disjunctive superprediction is not true, and thus cannot be used to justify the claim that Jeane Dixon can see the future.

But perhaps we are still being somewhat unfair. Perhaps treating the set of predictions as a disjunction is also not the best way to reconstruct the logic of this example. The only other possibility is to treat the theoretical hypothesis as a *stochastic* hypothesis. That is, the hypothesis might be not that she can see the future clearly but that she has a *higher probability* of being right than anyone without her special powers. But we do not yet know enough about how such stochastic hypotheses are tested. That comes in Part III.

8.3 THE PATCHWORK QUILT FALLACY: NO PREDICTIONS

Chariots of the Gods?

There now exists a whole rack of paperback books expounding the theme that the prehistoric development of the human race on earth was influenced in various ways by early visitors from outer space. For our example of the patchwork quilt fallacy, let us take the most famous of these, Erich Von Daniken's *Chariots of the Gods?*

In general terms, the hypothesis is that the earth was visited by intelligent aliens from somewhere else. Whether from another planet in our solar system or from beyond our solar system is not specified. In the literature this hypothesis is often referred to as the hypothesis of "extraterrestrial visitation," or ETV hypothesis. To fit this hypothesis into our general framework, we can simply define any inhabited planet that has been visited by intellectual aliens from somewhere else as an "alien intelligence interacting system." The ETV hypothesis, then, is just that the earth is such a system. What further features such systems have remains unspecified.

Von Daniken claims that the ETV hypothesis is true and, moreover, that the "visitations" took place very early in human history. In support of this claim he produces numerous examples of archaeological or anthropological findings that are difficult to explain. For example, many different cultures have myths about "gods" who came from the heavens and performed great feats. How come so many cultures have similar myths? In Egypt and in Peru there are huge temples, pyramids, and other structures constructed out of cut stones weighing many tons. How could men have possibly constructed such things with only the simple tools then available? Similarly, the Plane of Nazca in Peru exhibits what resemble roads cut in patterns. It is difficult to imagine why or how these were constructed. Again, on Easter Island on the Pacific Ocean there are 25-ton statues erected a fair distance from the source of the rock from which they were carved. Who could have carved them? How could they have been moved and set up on an island that could not support more than a few thousand people? For Von Daniken the answer is easy. These things are all explainable as the result of actions of ancient astronauts.

Now let us apply our account of tests and arguments to unravel Von Daniken's argument. In our terms, he seems to be taking the general ETV hypothesis and adding to it specific details about the nature of the alien visitors—whatever is necessary for them to have had the capacity to leave behind the findings cited. Moreover, he must be assuming specific initial conditions and adopting further auxiliary assumptions that, together with his detailed hypothesis, lead to there being just those things to be explained. Schematically, Von Daniken has created a hypothesis H, initial conditions IC, and auxiliary assumptions AA, such that Condition 1 is true for the "prediction" P. The prediction consists of statements describing all those archaeological and anthropological findings. Included in P, for example, will be statements describing the pyramids in Egypt.

Now you see the reason for the name "patchwork quilt fallacy." The hypothesis, initial conditions, and auxiliary assumptions are pieced together in such a way that they logically imply the known facts. Like a real patchwork quilt, the statements that make up the quilt do hang together. The quilt does have some overall pattern. But statements, unlike quilts, must be justified. We already know that the prediction is true. So the only question is whether the conditions for a good test are satisfied.

We have seen that Condition 1 is true. That is,

If [H and IC and AA], then P.

The antecedent was *designed* so that this conditional statement would be true. One may suspect that Von Daniken is trying to get his readers to agree to the antecedent on the basis of the truth of the consequent. We should not be so easily led into fallacy.

What about Condition 2? Here we must make a slight modification. In most standard scientific experiments, the initial conditions and at least some of the auxiliary assumptions are themselves justified by independent observation or experimentation. Halley, for example, used records of the observations made in 1682 as the initial conditions for his prediction of the return of the comet. In cases like Von Daniken's, however, the initial conditions and auxiliary hypotheses are all in the distant past. They are in as much need of justification as the hypothesis itself. So in stating Condition 2 we want to know what happens not just if the hypothesis is false but if the whole "quilt"—that is, the hypothesis plus initial conditions and auxiliary assumptions—is false. Condition 2, then, is:

If Not [H and IC and AA], then very probably Not P.

Here the negation applies to the whole antecedent, not just to H.

Now we must ask whether Condition 2 is itself justified. You may have had the feeling that something is wrong because the "prediction" is not really a prediction at all. It is merely a statement of known facts. Now we see why this is indeed a problem. How could we determine whether these facts are improbable if the hypothesis (really the whole antecedent of our modified Condition 2) is false? Indeed, what would it mean to say that these known facts are improbable? It is not like predicting the return of a comet within one month 76 years later. We all know *that* is something that anyone in 1705 would have been unlikely to get right unless Newton's theory applies to the motion of comets. But what about those facts that anyone can look up in a good encyclopedia?

Rather than puzzling over the meaning of Condition 2 in Von Daniken's case, let us just go back to the general characterization of what a good test should be like. That will tell us whether to interpret Condition 2 as being justified or not. In general, a good test should be arranged so that it is likely that we will conclude the hypothesis is correct *if it is correct*, and likely that we will conclude it is false *if it is false*. (This was summarized in Section 6.7.)

Now let us look at Von Daniken's procedure. He begins with facts that are presumably well known, such as the temples in South America, the pyramids in Egypt, and so on. He then fills in the details of a general hypothesis, the ETV hypothesis, in such a way that it logically implies the existence of the known facts—assuming initial conditions and auxiliary hypotheses that are also (at least implicitly) constructed for the purpose. And that is basically it.

Now ask yourself: Is this procedure a good test? That is, is it a procedure that is likely to lead you to conclude that the hypothesis is true if it is true and false if it is false? No, it is surely not. Whether Von Daniken's hypothesis is true or false, all it takes to follow his procedure is some imagination and ingenuity. With a little imagination and ingenuity, anyone could put together a set of statements about ancient astronauts that would imply some given set of known facts. The amount of imagination or ingenuity required is about the same whether the story one pieces together is true or not. Von Daniken, then, has not provided us with a good test of his hypothesis, and so we cannot regard it as being justified. This means that we cannot understand Condition 2 as being true.

Realizing that Condition 2 is not satisfied helps to explain what is going on in books like Von Daniken's. The long discussions of many "unsolved mysteries of the past"—the statues on Easter Island or the temples in Peru—are intended to convince the reader that the existence of these things *would have been improbable* unless his ancient astronauts really did exist. So in a way he really is trying to convince us that Condition 2 is true. But hypotheses about what would have been improbable from the standpoint of conditions that existed thousands of years ago are at least as difficult to justify as hypotheses about supposed ancient astronauts. To justify such hypotheses would require tests that are very difficult if not now impossible to construct. In any case, Von Daniken has not presented any such tests. At best he has told a good story in a way that can lead the unwary reader to think that such hypotheses are justified and thus that Von Daniken's hypothesis is also justified. But it is not.

At this point it is helpful to recall our discussion of Mendel's test of his hypothesis (Section 6.5) and compare Von Daniken's procedure with Mendel's. We noted that Mendel used the results of his original experiments to construct his theory. The theory concerning "factors" and the way they combine was constructed in order to explain the 3-to-1 ratio of dominants to recessives. Similarly, Von Daniken's theory was devised to explain many puzzling archaeological facts. But we did not appeal to the original experiment to justify Mendel's hypothesis about pea plants. What justified Mendel's hypothesis was the backcross test, and others like it. This was another experiment that involved a precise prediction whose truth or falsity was unknown to anyone at the time and thus was unlikely to come out as predicted if Mendel's theory did not apply. Von Daniken has given us nothing corresponding to Mendel's backcross test. It is as if Mendel had tried to justify his hypothesis solely on the basis of the originally observed 3-to-1 ratio and had not gone on to test his hypothesis with further experiments.

There is one final point about Von Daniken's book that must be made. This is that one cannot always trust an author to be correct even about things claimed to be well-known facts. One cannot even rule out deliberate lying or omission of relevant data. In his discussion of the statues on Easter Island, for example, Von Daniken cites the work of the anthropologist/adventurer Thor Heyerdahl. In the book cited, Heyerdahl reports actually observing some modern-day inhabitants of Easter Island moving and setting up one of the ancient statues. Using only primitive methods, it took fewer than 200 men to move it, and only a dozen to set it upright. Yet Von Daniken claims there is no known way that the early inhabitants could have done this. Clearly something is wrong here.

The moral for our purposes is that it takes more than good reasoning not to be misled into reaching unjustified conclusions. You have to learn when to be suspicious of the "data" and when to check further on your own. And you have to take the time to do it. There are no good rules that you can use here except to remember that people who are sloppy in their reasoning are likely to be sloppy about their facts as well.

Conspiracy Theories

The patchwork quilt fallacy often occurs in attempts to justify a "conspiracy theory." When something happens that is very difficult to comprehend or explain, it is very tempting to try to explain it as the result of a deliberate conspiracy on the part of some persons who had reasons for bringing it about. Currently the best known conspiracy theories focus on the Kennedy assassinations, which are still widely debated. One of the most discussed conspiracies is that in which the conspirators were some members of the Mafia together with some factions within the CIA. There was ample motive since the Mafia would have liked to reestablish the gambling and other criminal operations it had going in Havana before Castro took over. And anticommunists in the CIA would have liked to eliminate the hemisphere's only communist government. Moreover, John Kennedy withheld crucial support for their operations at the Bay of Pigs and Robert Kennedy prosecuted Mafia figures when he was attorney general.

Given the basic Mafia/CIA thesis, it only takes a lot of time and a little ingenuity to patch together more detailed claims (auxiliary assumptions and initial conditions) so that one can explain (i.e., deduce) all the known facts about these assassinations. If we call the conjunction of all the statements describing these well-known facts P, then Condition 1 is satisfied. That is, it is true that

If [H and IC and AA], then P.

But to argue for H, one needs the second condition. However, just as in the Von Daniken example, one cannot say that it is improbable that P if the hypothesis is false, because the hypothesis has been constructed so that it implies P. Even if H is false, there is no chance whatsoever that we should discover that P is false and use Condition 1 to conclude that H is false. The facts that the conspiracy hypothesis was designed to explain do not provide a good test of this hypothesis. And one cannot justify the claims of conspiracy merely by showing that Condition 1 is true.

It does not follow, of course, that conspiracy theories are never true, or that they cannot sometimes be justified. It just takes more than "circumstantial" evidence. If witnesses to the conspiracy are willing to talk, or there are letters or even tape recordings identifying key figures in the conspiracy, then one may prove it did take place. But one really needs such "direct" evidence. A "circumstantial" argument cannot be good enough.

8.4 THE *AD HOC* RESCUE: FAILED PREDICTIONS

The popular literature on unidentified flying objects (UFOs) contains numerous cases in which people have claimed actually to have made contact with the extraterrestrial visitors. Such contacts are now commonly called "close encounters of the third kind." In some of these cases, the people involved also claimed to have arranged a time and place for another meeting. But in real life, unlike the movies, the visitors fail to reappear. The people involved, however, generally refuse to change their story. In those cases where reporters and other officials were present at the appointed time and place, it is claimed that the alien visitors were scared off by the presence of so many strangers. Having "explained" why the visitors failed to reappear, the people continue to insist that they did have the original encounter. Let us look into the logical details of this "*ad hoc* rescue."

The basic hypothesis is that ETV is true. The earth is a system in interaction with intelligent aliens. The initial conditions are that the original meeting took place and that a further meeting was arranged at a specified time and place. One of the auxiliary assumptions must be that the visitors keep their appointments. The prediction, of course, is that they will appear at the designated time and place. Putting this all together, we find, as usual, that Condition 1 is satisfied, that is:

If [H and IC and AA], then P.

Condition 2 is also obviously satisfied since it is very improbable (indeed impossible) that the visitors will reappear, if the ETV hypothesis is false. However, since the initial conditions and the auxiliary assumptions are not justified by any independent evidence, we must use the modified form of Condition 2; that is:

If Not [H and IC and AA], then very probably Not P.

So the reporters and officials are correct in thinking that here they have a good test of the hypothesis. They do.

The prediction, however, fails to be true. In this case one should apply Condition 1, deny the consequent, and conclude by a valid deductive argument that the antecedent is false. That is, one concludes:

Not [H and IC and AA].

In straightforward scientific cases, that is, if the initial conditions and auxiliary assumptions are themselves well justified, one simply asserts IC and AA, and concludes, again deductively, Not H.

But, as we have already seen (at the end of Section 6.4), it is possible to avoid refuting the hypothesis. One refuses to assert IC and AA and simply converts the conclusion to a disjunction; that is:

Not H or Not IC or Not AA.

So far, so good. No fallacy yet.

The next step *is* fallacious. It is asserting that AA is false and that H is true. In the example, it was denied that visitors in general keep their appointments, and it was maintained that the original encounter took place. There is no justification for these claims. All that is justified is the conclusion stated above.

In most cases of this type, the fallacy is compounded still further. A new auxiliary assumption is introduced to replace the one that was denied. In the example it was that the aliens keep their appointments *unless* there are too many strangers present. This claim, too, is not justified. However, if one then combines this new auxiliary assumption, call it AA*, with the original hypothesis, H, and a revised initial condition, IC*, including the statement that there were too many strangers present, then one obtains a revised Condition 1, which is true! That is, it is true that

If [*H* and *IC** and *AA**], then Not *P*.

And Not P is what happened. The actual facts are thus "explained," or rather, the failure of the original prediction is "explained away."

But we know that being able to explain the facts does not by itself justify any hypothesis. To think that the truth of Condition 1 with the revised initial conditions and auxiliary assumptions provides any justification for the hypothesis is simply to affirm the consequent, and we know that is a fallacy.

To clinch the case, just look at Condition 2 with the revisions included. It is:

If Not [*H* and *IC** and *AA**], then very probably *P*.

This statement is not only unjustified, but false. If the aliens are scared off by strangers, then it is not very probable that they will appear. Again there is no justification for any claim other than that one or more of *H*, *IC*, or *AA* are false.

The *ad hoc* rescue and the patchwork quilt fallacy are related in that if one continues using the *ad hoc* rescue, after a while one has a patchwork of statements resulting from the numerous revisions of the auxiliary assumptions, initial conditions, and perhaps even the original hypothesis itself. But no positive claim is ever justified. One has only a trail of refuted conjunctions.

Again it is useful to compare this example with typical scientific cases. As we have seen, scientists may avoid having a hypothesis refuted by questioning the auxiliary assumptions instead. But this can only be a stopgap measure. One has eventually to come up with an independent justification for any proposed explanation of what went wrong. In the case of phlogiston this could not be done, and the whole theory was eventually abandoned. In the example we have been discussing the *only* evidence for a change in the auxiliary assumptions was that the prediction failed. No independent way of checking that the supposed visitors dislike strangers is suggested. This is what makes the rescue *ad hoc*, meaning "for this purpose only." The only purpose of the change in assumptions is to save the hypothesis. Nothing further is suggested that might justify the change.

8.5 JUSTIFICATION BY ELIMINATION

There is one other type of argument that is often used in attempts to justify theoretical hypotheses, but it cannot in principle provide the desired justification. It is easy to see in general why this is so because the argument is *deductively valid*, and we know that the justification of a theoretical hypothesis requires an *inductive* argument. But just knowing in principle that the proposed strategy for justification cannot work is not enough to enable you to spot instances of the argument and to see just where the failure of justification occurs. We shall look at the general structure of the argument first and then work through an example in which it is used in an attempt to justify the ETV hypothesis—unsuccessfully, of course. Finally, we shall look at a fictional example in which the same form of argument is used successfully to justify a conclusion. In this case, however, the conclusion is not a theoretical hypothesis.

The Strategy of Justification by Elimination

Justification by elimination is based on the *disjunctive syllogism* (Section 4.4), which is a deductively valid type of argument. For cases in which the first premise offers three alternatives, the form of the argument is:

P or *Q* or *R*.
Not *P* and Not *Q*.
Thus, *R*.

Just to refresh your memory, the first premise says that at least one of the three statements, *P*, *Q*, or *R*, is true. The second premise says that both *P* and *Q* are false. So *R* must be true, which is the conclusion. Arguments of this type are valid. They are called "arguments by elimination" because the second premise "eliminates" some of the alternatives, leaving the conclusion. In most cases the argument is set up so that the second premise eliminates all but one of the original statements.

I shall first explain in general terms how this type of argument is used in attempts to justify theoretical hypotheses, and then go through an example.

One begins, always, with some "facts" that are taken as well established. Let us just suppose that the statements describing these facts are put together in one big conjunction, which we can call simply *F*. These facts, of course, are not just any old facts, but some for which one would like to find an explanation.

Someone proposes a theory which, when applied to the situation, yields a hypothesis that, with appropriate initial conditions and auxiliary assumptions, could explain the facts, *F*. In short, the following conditional statement is true:

If [H_1 and IC_1 and AA_1], then *F*.

I have used the subscripts because this is not the only hypothesis that could explain the facts. Indeed there are several others. Suppose there are just three in all, although there could be any number of different theories considered. So in addition to the above conditional statement, there are two others, namely:

If [H_2 and IC_2 and AA_2], then *F*.
If [H_3 and IC_3 and AA_3], then *F*.

In short, we have three different possible explanations of the known facts, *F*. The question, of course, is which explanation, if any, is the right one. That is to say, which hypothesis, if any, is correct.

Now suppose that someone wishes to argue that H_3 is the correct hypothesis. But instead of talking about a good test of H_3, this person proceeds to offer arguments refuting both H_1 and H_2. This is done, in principle, by trying to find other statements, say G_1 and G_2, respectively, implied by these hypotheses. G_1 and G_2 are then shown to be false. So each of the alternative hypotheses are refuted by a standard refuting form of argument, denying the consequent:

> If [H and IC and AA], then G.
> Not G (and IC and AA).
> Thus, Not H.

Since this has been done for both H_1 and H_2, it is then concluded that H_3 is correct. And here the argument must be:

> H_1 or H_2 or H_3.
> Not H_1 and Not H_2.
> Thus, H_3.

This argument, of course, is valid. Its second premise is justified by the supplementary arguments that refute both H_1 and H_2. Why does this procedure not justify H_3? Try to answer this yourself before going on.

For the conclusion of a deductively valid argument to be justified, of course, all the premises must be justified. But as presented above, at best only the second premise of the argument by elimination has been justified. Nothing has been said about the first. All we knew when we started is that each of the three hypotheses could explain the facts, F. But we were not given any justification for the claim that one of these explanations had to be correct. Why can't they all be false? In sum, the above procedure justifies H_3 only if the first premise is itself justified.

Now a disjunction of theoretical hypotheses must still be regarded as itself a theoretical hypothesis. Thus if the first premise is to be justified, it would have to be subjected to a *good test*. Such a test is imaginable, but it is never encountered in real life. If someone is going to put together a good test, it would usually focus on one or another specific hypothesis, not a disjunction. And if one has done this successfully for some hypothesis, there is no need for any argument by elimination. One has a justified hypothesis.

So, as a general rule, when you spot what looks like an elaborate argument by elimination, identify all the alternative hypotheses being considered, and then ask whether any justification has been given for believing that they cannot all be false. If not, then the conclusion has not been justified, no matter how many chapters are spent refuting some of the alternatives that are considered.

So far we have been talking in the abstract. Now let us consider a typical example.

Extraterrestrial Visitation Revisited

Some of the best examples of arguments by elimination are in the popular UFO literature. Indeed, whole books can be seen as one elaborate argument by elimination, with various chapters devoted to the elimination of various alternative hypotheses. What follows is a composite taken from many books.

The basic data in any discussion of the ETV hypothesis consists of reports of UFOs—literally, *unidentified flying objects*. Reports range from the modest to the outlandish—that is, from reported "encounters of the first kind" to reported "encounters of the third kind." A modest report would be one of an object, usually with lights, that hovered for 5 or 10 minutes and then moved off silently with great speed. Such observations are frequently made from automobiles, often with more than one passenger. Reports of trouble with the car's ignition and electrical system are common. In less modest reports, the object is described as being like a "spaceship," long, with windows down the side, or saucerlike, with a dome and windows. Still less modest reports contain descriptions of the inhabitants of the vehicle, either through a window or actually outside. Finally, there are reports of actual contact with the visitors, perhaps even of having been taken aboard the craft for a period and then released. Some people have even reported being sexually molested or seduced aboard a spacecraft (a very close encounter indeed).

I said that the basic data consist of *reports* of UFO sightings, not the existence of *what was reported*. This distinction is crucial because the fact that some people have reported such things has been verified by many investigators. There can be no doubt that people have made such reports. That the people in question actually saw or experienced what they say they did, however, is open to question.

How can we explain the fact that people report what they do in fact report? One explanation, of course, is they actually saw or experienced what they claimed and thus that ETV is a reality. But there are other possible explanations as well. For example, those making the report in question might be engaged in a deliberate hoax. Various motives for perpetrating such a hoax are possible: just for the fun of it, to attract attention, to get on television, to make money. A number of such cases have been discovered, some even involving deliberately contrived photographs.

Another possibility is that the person, or persons, making a report were suffering from hallucinations of some sort. Thinking that one sees bright lights is a common form of hallucination. It is even possible to generate a kind of mass hysteria in which many otherwise quite normal people think they see something that simply is not there.

It is also possible that the person or persons making the report are psychologically unbalanced to the extent that they cannot distinguish reality from fantasy. They are convinced that they experienced something that never happened. This kind of explanation is quite plausible for cases of reported kidnap by space visitors, especially those with sexual episodes. It is not implausible that a real, but traumatic sexual experience (e.g., rape) should be reinterpreted in the mind of a victim as an encounter with extraterrestrial visitors.

It may also be the case that the people in question really did see some-thing very much like what they described, only it turns out to be something not at all extraterrestrial. Several reported UFOs have turned out to be hunters using a helicopter equipped with a spotlight. Low-flying aircraft with their landing lights on have also generated numerous UFO reports. Once an old Russian booster rocket reentered the atmosphere and burned up along an arc extending across the United States from the Northeast to the Southwest. There were UFO reports all along the path, and those from the Northeast were earlier.

Even experienced observers can be fooled. Weather balloons, which are very large and shiny in the sun, have often been taken for "flying saucers" by experienced pilots. One pilot in a small plane chased a UFO to a very high altitude, without oxygen equipment. He blacked out and crashed. The UFO was later identified as a weather balloon.

Finally there is the possibility that what was experienced was a natural, terrestrial phenomenon, but one that has never been studied before. One cannot assume that we now know everything there is to know about phenomena here on earth. Such UFOs will of course be "unidentified," at least right now. One cannot, by definition, give examples of UFO reports that were due to something as yet undiscovered. But one can cite a phenomenon that was only recently investigated: ball lightning.

There are many manifestations of atmospheric electricity—lightning and the Northern Lights being among the best known. Ball lightning is among the least well known because it is fairly rare, short-lived, and has not been pro-duced in a laboratory. But it is now generally recognized that it is possible for something like lightning to concentrate itself in a ball perhaps a foot in diame-ter. Such balls typically last 5 to 10 seconds, glow, drift about relatively slowly, sometimes follow metal cables and sometimes melt them, and usually disappear with a bang, but sometimes silently. In addition to forming at ground level, there is some evidence that ball lightning may form in clouds near thun-derstorms. These balls are generally thought to be larger and carry more energy, but the evidence is skimpy. In any case, the verified existence of such phenomena helps to explain a number of otherwise very puzzling reports.

Now each of the other possible explanations of a UFO report is itself also a theoretical hypothesis. So we have a whole set of rival hypotheses, each of which might fit the case at hand. How are we to determine which in fact fits? Although popular writers rarely set out the structure of their argument, what they say can often plausibly be interpreted as an attempt to *eliminate* all the alternatives to the ETV hypothesis. Thus, in sympathetic accounts of UFO reports, the author is usually quite concerned to show that the subject—that is, the person making the report—is not the kind of person who would perpetrate a hoax or otherwise deliberately lie. Moreover, it is often argued that the subject was previously skeptical of claims of ETV, and even reluctant to report his/her experiences—mainly out of fear of ridicule. The reason for presenting such facts about the subject can easily be understood as an attempt to *eliminate* the hypothesis that the report is a hoax.

Similarly for hallucinations. If one can show that several people saw the same thing, or if one can produce a photograph of the phenomenon, that has the force of tending to eliminate the hypothesis that the report is the result of hallucinations. Joint hallucinations are fairly rare, and cameras never have them.

Again, it is generally regarded as strengthening the case for ETV if the subject is a pilot, police officer, or someone with similar training in making "cool, objective" observations. In addition, such people have at least to some extent been screened for psychological instabilities. One can fairly confidently eliminate the hypothesis that the report is the result of the subject being psychologically disturbed in some way. There is the added benefit that police and military people are not very likely to be involved in a deliberate hoax.

Popular books and articles on UFOs tend to focus on the people making the reports. There are two reasons for this. One is that by showing that the subjects are honest, stable people, one eliminates several possible terrestrial explanations of the reports. The other is that most popular writers on UFOs are unfamiliar with the wide variety of natural atmospheric phenomena, both optical and electrical, that might have been the basis of the subject's experience. So they are not in a good position to argue that what was experienced was not some natural phenomena that, although unusual and not well known or even well understood, has already been studied by various scientists. Indeed, since there are a great number of natural phenomena that might have led to the reported sightings, eliminating them all would be a very difficult task. On the other hand, unless one can, with some confidence, rule out all possible known natural causes of the observations, the argument by elimination fails to reach the desired conclusion: the ETV hypothesis. One would still be left with a number of other possible hypotheses, any of which might be true.

Let us suppose, however, that one did manage to eliminate all *known* possible natural causes of the reported phenomenon. One would still be left with one possible claim other than the ETV hypothesis. The phenomenon might be an *unknown*, but natural and terrestrial, occurrence. This claim is impossible to eliminate. By definition one does not know about what is unknown. So one cannot say that what happened was not the result of some such phenomenon. At this point the attempt to justify the ETV hypothesis by elimination breaks down completely. One will always be left with two possible hypotheses—unknown natural cause and ETV—and with no way to eliminate the first.

From the fact that one cannot justify the ETV hypothesis using an argument by elimination, it does not of course follow that this hypothesis cannot be justified; it simply cannot be justified in this manner. Nor does it follow that the hypothesis is false. It might be true. Even E. U. Condon, the physicist who headed the official government-sponsored study of UFOs in

1966–68, said that he would believe in ETV if a spaceship landed in front of his laboratory and its occupants came out to greet him. Lacking a well-formulated version of the ETV hypothesis that could be subjected to a *good test,* as we have defined it, this seems the only kind of evidence that could justify this hypothesis. In the meantime, it remains in the realm of speculation—fascinating, perhaps, but still speculation.

The Case of the Murdered Host

Having discussed an example in which the conclusion could not be justified using an argument by elimination, let us now look at a case in which it can.

Imagine an as-yet-undiscovered manuscript by Sir Arthur Conan Doyle. In this story, Holmes and Watson are invited for the weekend to the home of an eccentric nobleman far out in a moor in northern England. After a curious dinner (curious because none of the guests other than Holmes and Watson had ever previously met) and many glasses of fine port, all retire for the night. The next morning, the butler reports that the host is dead. He had been killed in the night with a dagger from his own collection, which was displayed prominently in the library.

Holmes immediately begins an investigation. There were only 13 people in the house that night: three servants (cook, butler, and gardener), Holmes, Watson, and seven other guests. The host was the thirteenth. There was absolutely no evidence of any other people near the house, and a bad storm on the moor made it impossible that someone could have come and gone during the night. After investigating the scene of the crime, Holmes and Watson carefully question each of the other ten people in the house. That evening, before dinner, Holmes calls everyone to the library.

He begins by assuming that neither he nor Watson committed the murder. Indeed, the reason Holmes was invited was because the host suspected someone wanted him dead. Holmes then considers each of the seven other guests in turn, explaining in each case why he thinks that person could not have been the murderer. He then explains why neither the cook nor the butler could have done it. Turning dramatically to the gardener he asks: "Why did you do it?" "How did you know?" the poor man blurts out. "Because you were the only one left," replies Holmes triumphantly.

In this case Holmes is obviously reasoning by elimination. Neglecting himself and Watson, there are ten hypotheses of the form "X did it," where X is either a servant or one of the other guests. The *disjunction* of these ten hypotheses—$H_1, H_2, \ldots, H_{10}$—is Holmes' first premise. The second premise consists of a *conjunction* of the negations of the first nine hypotheses. He concludes that H_{10} is true; that is, "The gardener did it." Symbolically:

H_1 or H_2, or . . . , or H_{10}.
Not H_1, and Not H_2, and . . . , and Not H_9.
Thus, H_{10}.

Now the reason elimination justifies this conclusion is that the premises are of a type that can themselves be justified. Holmes has good reasons for thinking that the murderer has to be one of these ten people. So the first premise is secure. He then offers reasons for each negation in the second premise. So the second premise is secure. The conclusion follows deductively. All the inductive reasoning, which Holmes himself always called "deductions," goes into establishing the two premises.

The general lesson is that if the original alternatives are definite enough, and few enough, and if one can be sure that they are all there, then one stands some chance of being able to eliminate all but one. This is almost never the case when the alternatives are theoretical hypotheses about some complex system. For a complex system there will be infinitely many different possible theories and thus infinitely many possible hypotheses, only one of which is true. Rarely would all the possible hypotheses be so neatly ordered that one could in some way eliminate all but one. Usually, no matter how many possibilities one succeeds in eliminating, there are still infinitely many left. It is impossible to get down to just one.

8.6   COMBINED FALLACIES

The four fallacies we have examined are to some extent ideal types. One rarely finds just one being used in its pure form. Most often they are used in combination. Thus people who rely on sifting out the successful predictions from among a large set also tend to make their predictions more or less vague so that more of them will come true. And when confronted with a failed prediction they are likely to come up with some *ad hoc* reason why that particular prediction failed. Moreover, a great many of the popular informal theories one meets were originally just patchworks based on known facts. And as time goes on, the patchwork keeps growing, maybe as a result of failed predictions from which the basic hypothesis has been rescued, but maybe just in light of new facts. And along with everything else there is likely to be an implicit argument by elimination that takes the form of attempts to show that everyone else is wrong.

In the midst of such a barrage of fallacious arguments, it is often difficult to keep one's logical footing. The only general advice is to stay calm and examine the situation step by step. Ask yourself:

1.   What is the system being discussed? What is the basic hypothesis being made about this system?
2.   Has there been a good test of the claims? Are Conditions 1 and 2, especially Condition 2, satisfied?

3. If there has not been a good test, what kind of argument is being offered? Is it an argument by elimination? What fallacy or fallacies are being committed?
4. Are the claimed facts really so? Can the author be trusted to report correctly what has already been well justified?

These, of course, are just general guidelines. But they can be helpful, and with practice you can come to approach many subjects with the confidence that you can judge whether some hypothesis is justified or not, even if you are not yourself an expert on the subject.

One final point. In discussions with people who hold various views you suspect of being unjustified, be careful not to let yourself be maneuvered into the position of having to prove *they* are wrong. It is not up to you to do that. It is up to them to *justify* their claims to you. All you have to do is *evaluate* their justifications. That is what is known as keeping the "burden of proof" where it belongs, on those who wish to convince you that their claims are correct. If you make the mistake of letting them shift the burden of proof to you, then you will almost certainly "lose" the argument. You will be in the position of not knowing as much about the subject as they do, which makes it difficult to argue that they are wrong.

Moreover, if you do come up with apparently false consequences of their hypotheses, they are likely to invoke an *ad hoc* rescue to avoid the conclusion that their hypothesis is false. They can surely change auxiliary assumptions faster than you can come up with false consequences, thus protecting their basic claim. Of course you may try to point out to them the fallacy of the *ad hoc* rescue, but it is much better not to let yourself get stuck with burden of proof in the first place.

CHAPTER EXERCISES

The following four problems have been written so that each provides a relatively clear-cut example of one of the four fallacies discussed in the text.

8.1 A recent article in the student newspaper of a Big Ten university was devoted to the claims of a student that he was able to reach a deep state of meditation in which he could levitate—that is, float above the ground (or floor) with literally no support. The reporter inquired whether he meant that he really was levitating or that it just felt like it. No, the student insisted, he was really doing it. At that point the reporter asked for a demonstration. Later the reporter watched while the student meditated, but observed no levitation. Informed that he had not levitated, the student explained that having somebody watching made him too nervous to achieve a sufficiently deep state of meditation. The reporter apparently accepted the explanation because he wrote up the story in a straightforward—not tongue-in-cheek—manner.

A. Identify the fallacy that apparently fooled the fledgling reporter.

B. Explain briefly in your own words the nature of the fallacy involved.

8.2 Many people believe that their daily horoscope provides clues to what will happen to them that day. Suppose you are having coffee and reading the morning paper before a class, and note that your horoscope for that day says that you should be on the lookout for unwelcome surprises. You become apprehensive because the instructor in your upcoming class gives 10-minute surprise quizzes about once every 2 weeks and you haven't read the assignment. Sure enough, when you arrive the instructor is writing the quiz question on the board—definitely an unwelcome surprise.

A. Identify the fallacy you are committing if you regard your experience as evidence that astrology, the source of the horoscopes, does work.

B. Explain briefly in your own words the nature of the fallacy involved.

8.3 When the stock market crashed in 1929, an economist with rather "far out" views claimed that the crash proved his theory was right. He had predicted the crash a year before. It turns out that every year since 1921 he had been predicting a crash for the following year. But this did not prevent him from continuing to cite the 1929 crash as evidence for his views.

A. Identify the fallacy committed by this economist.

B. Explain briefly in your own words the nature of the fallacy involved.

8.4 Many people believe that the gasoline shortage of 1973 was not a real shortage at all but was engineered by the large oil companies in conjunction with the oil-producing countries. In support of this belief they cite numerous facts. For example, there have been no lines at gas stations since 1973. The profits of the oil companies have risen dramatically. And the oil-producing countries are so rich that they are buying up coal mines in the United States. And so on.

A. If you think these facts support the belief in question, you are committing what fallacy?

B. Explain briefly in your own words the nature of the fallacy involved.

8.5 The following passages are taken from the opening paragraphs of Chapter 8 of Erich von Daniken's *Chariots of the Gods?* The chapter is titled "Easter Island—Land of the Bird Men." Read the passages, and then go on to the questions that follow.

The first European seafarers who landed on Easter Island at the beginning of the eighteenth century . . . saw hundreds of gigantic statues, some of which are between 33 and 66 feet high and weigh as much as 50 tons. Originally these colossuses also wore hats . . . which weighed more than ten tons apiece.

Easter Island lies far away from any continent or civilization. The islanders are more familiar with the moon and the stars than any other country. No trees grow on the island, which is a tiny speck of volcanic stone. The usual explanation, that the stone giants were moved to their present sites on wooden rollers, is not feasible in this case, either. In addition, the island can scarcely have provided food for more than 2,000 inhabitants. (A few hundred natives live on Easter Island today.) A shipping trade which brought food and clothing to the island for the stonemasons, is hardly credible in antiquity. Then who cut the statues out of the rock, who carved them and transported them to their sites? How were they moved across country for miles without rollers? How were they . . . polished and erected? How were the hats, the stone for which came from a different quarry from that of the statues, put in place?

A.   Nowhere in these passages does von Daniken explicitly refer to his ETV hypothesis—that the work was done by "ancient astronauts." But you may assume it is in the background as a possible explanation of the existence of the statues. Take the facts about the statues—their size, location, and so on—as given. Let F represent all these facts together. Then see if you can extract from the passages at least one other hypothesis, H, that might explain F. Also look for statements, G, that might be used to refute other hypotheses. Set out the refuting arguments in schematic form. Also set out the schematic form of the final argument by elimination.

B.   Discuss briefly how well justified is von Daniken's final conclusion, ETV. This involves considering both the first premise of the elimination argument (the one that sets out the hypotheses to be considered) and the second. The justification of the second premise, of course, depends on the justification of premises in the separate arguments used to refute alternative hypotheses.

8.6 The following short article comes from the pages of a weekly newspaper. The article is given in full. Read the article with an eye to analyzing it as an argument by elimination, with the unstated conclusion being a version of the ETV hypothesis—that the reported object was a vehicle piloted by extraterrestrial visitors. Take the various aspects of the report itself as the facts to be explained. Formulate alternative explanations that the report tries, implicitly, to refute. Evaluate both the refuting arguments and the overall argument by elimination to judge how much justification the report provides for the ETV hypothesis.

Cliff Robertson tells of Moment He Saw UFO

Actor Cliff Robertson this week described the startling moment he spotted a UFO "moving gently across a clear sky."

The Oscar-winning star said he saw the UFO 12 years ago, but never admitted it publicly before because he did not want to be thought of as a "kook."

He said: "I simply saw something in the sky. I'm a pilot and have flown for many years."

He described the object as metallic in color. "At first I thought it was a weather balloon. I was in my garden in California watching it through my field glasses for about 10 minutes.

"Suddenly it just darted and went up. It definitely went up from that already high altitude and it was gone."

The 50-year-old star said: "I know I'm sober and I certainly was sober when I saw whatever it was.

"I'm just sorry I saw it. I know from now on there will always be a certain fringe group who say: 'Well, you had one too many.'"

Robertson, who lives in New York City, is currently working on the adaptation of a Broadway play.

The following three exercises do not exactly exhibit any of the fallacies discussed in this chapter. They are cases in which some hypothesis is claimed to be justified, although the conditions for a good test of the hypothesis are not met. So the hypothesis is not really justified. That fallacy, then, is the general one of claiming justification when there is none. In each case your task is to explain why the stated hypothesis is not justified.

8.7 Astrology is a complicated and ancient practice. It rests on the basic theoretical hypothesis that the positions of the sun, moon, and stars at one's birth influence one's personality and one's actions in life. Along with the basic hypothesis goes specific hypotheses about particular "sun signs." Aquarians (those born between January 21 and February 18), for example, are said to be generally scientific but eccentric, that is, brilliant but unconventional. In support of this hypothesis it is often noted that a number of famous scientists (e.g., Copernicus, Galileo and Thomas Edison) were Aquarians.

8.8 The following "case" has been cited in support of the hypothesis that pyramids have special powers. A young woman who was having difficulty with her complexion was told to keep a pitcher of water under a pyramid and then wash her face in that water, with only the mildest possible soap, once in the morning and once in the evening. She was also told to put nothing else on her face, no creams, or medications of any kind, and no makeup. Although she has been in the habit of using quantities of makeup, she agreed to the experiment. Within 2 weeks there was a clearly noticeable improvement in her complexion.

8.9 The headline of a recent issue of a national weekly newspaper states that there have been hundreds of cases of reincarnation in the United States. These cases, the headline proclaims, provide convincing evidence that there is "life after death." Turning to the inside pages one finds accounts of patients who have been cured of various complaints by being hypnotized and then "regressed" to a "previous life." The following cases are typical of those presented.

A fifty-year-old woman claimed to have suffered from severe headaches, several a week for over 35 years. She claimed to have seen ten different doctors who had prescribed various pain killers and other drugs—none of which worked. Then in a single two-hour session under hypnosis she "discovered" the true cause of her headaches. In an earlier life she had been a young man in nineteenth-century New England. One day, while on the way to visit his fiancee, the young man fell into a gully, hit his head on a rock, and was killed. A year and a half later, the woman claimed she had suffered only one or two headaches since her session with the therapist.

A twenty-five-year-old real estate dealer complained of several serious allergies, including a very strong aversion to corn. Under hypnosis he was regressed back to an earlier life as a commander in a Mongolian army. In one campaign the commander refused to order his men to kill innocent women and children. Because of this disobedience, his superiors had him tortured by being force-fed corn and water which caused him to bloat up so much that he died. After "learning" of his earlier life, the realtor claimed to be rid of most of his allergies and to be able to eat corn with no ill effects.

Several of the psychologists engaged in this sort of therapy are quoted as being convinced that their work provides scientific evidence that reincarnation does occur and that there is indeed "life after death."

8.10 This is more a project than an exercise. Find a book on a popular scientific topic that interests you—not a topic that comes out of a formal science source. Read the book, keeping a lookout for indications of the basic theory and hypotheses and for specific arguments for the hypotheses. Are there any good tests discussed? If not, what types of arguments are there? Are there any obvious fallacies? Is the overall argument one of elimination? What are the alternative theories and hypotheses considered? Is their rejection justified? Is the author's conclusion justified?

You may turn this into a more extensive project by looking up other books and articles on the same subject. In particular, you may be able to find books or articles that explicitly set out to refute the views in the book you read first. In this case you can evaluate the counterarguments as well. Is the rejection of the original thesis justified? Why or why not?

Summary to **PART II**

What follows is a list, together with short definitions, of all the important new concepts introduced in Part II. If you find that you are now unsure of any of these, you should go back and review the indicated section.

Theory (5.1)

A statement that defines a type of natural system.

Theoretical Hypothesis (5.1)

A contingent statement asserting that some specified real system is in fact a system of the type defined by the corresponding theory.

State of a System (5.4)

Some complete set of characteristics that a system of that type might have. At any time, a system must have one and only one such complete set of characteristics.

Deterministic System (5.4)

A system whose state at any particular time completely determines its state at all other times.

Stochastic System (5.4)

A system whose state at any particular time determines only the probabilities of states at later times.

Basic Elements of a Test of a Theoretical Hypothesis (6.2)

These are the hypothesis H, a prediction P, initial conditions IC, and auxiliary assumptions AA.

A Good Test of a Theoretical Hypothesis (6.3)

An experiment or set of observations for which the following two statements are true:
If [H and IC and AA], then P.
If [Not H and IC and AA], then very probably Not P.

160

Linear Growth (7.2)

Growth by adding a constant amount at given intervals of time.

Exponential Growth (7.2)

Growth by adding a constant percentage of the existing total at given intervals of time.

Doubling Time (7.2)

The time it takes an exponentially growing quantity to double in size. Equal to 70 years divided by the growth rate in percent.

Superexponential Growth (7.2)

Growth by adding an increasing percentage at given intervals of time.

Justification by Elimination (8.5)

Justification using the valid argument of disjunctive syllogism (4.4). A good method of justification only if the first premise, which states the alternatives, and the second premise, which eliminates some of the alternatives, can themselves be justified.

We began Part II by distinguishing two broad categories of scientific hypotheses. Having learned "everything you always wanted to know about theoretical hypotheses," you are now ready to tackle the other category, which includes statistical, correlational, and simple causal hypotheses.

Unlike theoretical hypotheses, statistical and simple causal hypotheses tend to be the stock in trade of psychology and the social sciences—for example, economics and sociology—and such science-related fields as biomedicine, public health, and education. A great many of the examples and exercises in Part III involve public health and biomedical issues. For example, "Does smoking cause lung cancer?" or "Do birth control pills cause fatal blood clots?" Thus, the effort you put into learning to understand statistical reasoning will have an immediate and practical payoff. You will be learning to handle the kind of information that is likely to be directly relevant to what you do every day in your personal life and at work.

The reason the social and medical sciences rely so much on statistical reasoning is easy to understand. The systems they study (e.g., the human body) are very complex. It is very difficult to discover exactly why anything does what it does. It is not even known, for example, just how aspirin relieves the pain of arthritis. Moreover, because these systems are so complex, there is a great deal of variation from one to the next. Some light smokers get lung cancer and some heavy smokers never do. In many cases, then, the only way to discover what is going on is to study a large number of similar systems—for example, a large number of people—and see what happens "on the average." Thus the need for statistics.

Our focus, as always, will be on justification, that is, on what kind of statistical evidence justifies what kinds of hypotheses. But first we must get straight just what kinds of hypotheses are at issue. This is done in Chapter 9. There we learn, for example, the differences between a simple statistical hypothesis and a genuine causal hypothesis.

The justification of these hypotheses, like most scientific hypotheses, requires inductive arguments. And good inductive arguments are defined in terms of probability. So far we have been able to get by with only an intuitive understanding of probability. No

PART THREE

Causes, Correlations, and Statistical Reasoning

longer. To understand statistical reasoning, you will have to learn a little more about probability—but not very much, so don't worry. Chapter 10 is relatively short.

Chapter 11 deals with simple statistical hypotheses and correlations. We shall discuss survey sampling (e.g., Gallup polls) and the use of data from questionnaires. The main section of the chapter explains how statistical hypotheses are tested using statistical data. These ideas apply directly to correlations as well. The final section examines in some detail a study of the relationship between the use of marijuana and heroin, based on a survey of 3500 college students.

As you will learn, it is difficult to justify causal hypotheses using only correlations from survey data. But it is easy to justify a causal hypothesis if you have data from experiments utilizing a randomized design. In Chapter 12 we shall learn about such designs by studying the experiments that "proved" that saccharin causes bladder cancer—in rats. However, not all hypotheses can be investigated in this way. Studies of the effects of smoking and the use of oral contraceptives have used other designs. We shall learn how these studies differ from randomized tests and why they provide a "weaker" justification for causal claims.

By the end of Chapter 12 you should be able quickly to identify any claim based on statistical data and to decide for yourself whether the given data provide an appropriate justification for that claim.

9

Statistics, Correlations, and Causal Hypotheses

9.1 SIMPLE STATISTICAL HYPOTHESES

The following are all examples of *simple statistical hypotheses:*

51 percent of adult American men smoke cigarettes.
34 percent of adult American women smoke cigarettes.
One out of every 10 persons living and working in Los Angeles suffers from the disease of alcoholism.
Fewer than one-fifth of all households are "typical" American families, in which the husband is the sole breadwinner.
Nearly half of all adult women are employed.
In 1976, 53 percent of all graduating high school seniors had tried marijuana.
80 percent of homicides in the United States involve guns.

I have taken each of the above statements from a printed source, but I would not vouch for the truth of any. Nor should you believe them. The first rule in dealing with statistical hypotheses is to regard them as unjustified assertions until you know something about where they came from. In this chapter, however, we are solely concerned with the nature of the assertions themselves, not with their truth or falsity. We shall get to the problem of justifying such assertions in Chapter 11.

By now you know that one of the secrets of understanding scientific reasoning is learning to see the same basic structure in different statements and arguments. So it should not surprise you to learn that the above statements can all be read in a standard way. That is, they all identify a *population,* a *property* which each member of the population may or may not exhibit, and a *percentage.* The percentage tells you the *relative number* or proportion of members of the population that exhibit the property in question.

The second statement, for example, identifies the population of adult American women. It specifies the property of being a cigarette smoker. And it says that 34 percent of the members of the population have the property of being a smoker. Figure 9.1 shows a convenient way of picturing this simple

165

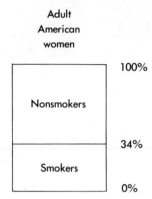

FIGURE 9.1   Simple statistical hypothesis: 34% of adult American women smoke cigarettes.

statistical hypothesis. The box represents the population. The area below the horizontal line represents all those members of the population that have the property. The percentage may be viewed as following a scale that goes from zero at the bottom of the box to 100 percent at the top.

You should note that the word "population" is not always to be taken literally as a group of people. The last statement in the list, for example, refers to the population of homicides in the United States. Homicides are not people, but *incidents* involving people—a killer and a victim, for example. Likewise, "households" are not people but rather small groups of people. But one can still formulate statistical hypotheses about the "population" of American households.

That's it, except for one general warning. When you see a statistical hypothesis, pay particular attention to the way both the population and the property are characterized. In particular, try to think what is being included and excluded. For example, the specification "families in which the husband is the sole breadwinner" rules out families in which the mother works part time, which is probably why the percentage is so low. Also, the meaning of "adult" is somewhat vague. Does it mean over 21? Over 18? Over 16? Which it is might make a lot of difference if, for example, one were concerned with the percentage of adult women that are married. Many women get married between the ages of 16 and 21.

In your daily reading, practice spotting simple statistical hypotheses and slipping them mentally into the standard form: *X* percent of (population) have/are (property). If you have a good grasp of simple statistical hypotheses, the "less simple" will be no problem.

9.2   DISTRIBUTIONS

Many of the statistical hypotheses you are likely to meet will not be presented singly but in conjunction with other related statistical hypotheses. Some of these groupings of statistical hypotheses are known as "distributions." As you will see, distributions of statistical hypotheses play an important role in statistical reasoning.

In talking about distributions it is very helpful to use the related notions of a "variable" and a "value of a variable." So we shall look first at these notions and then at distributions of statistical hypotheses.

Variables and Values of Variables

These notions are easy to grasp from examples, but difficult to define in the abstract. So let us stick with examples.

Very roughly, a variable represents any quantity that comes in different kinds. Thus, when we were discussing Mendelian genetics, we could have said that eye color is a *variable* which has two possible *values*, brown and blue. Or, height in sweet peas is a variable with two possible values, short and tall. Similarly, height in humans is a variable that could have any number of different values, depending on how fine you want to make the scale. For example, if you "round off" to the nearest whole inch, and run the scale from 1 foot to 9 feet, that makes 96 (i.e., 8 × 12) different possible values of the variable representing height in humans. Each human fits into one, and only one, of these 96 "slots."

Simple Distributions

The property referred to in any simple statistical hypothesis may be regarded as a variable with two values, e.g., smoker and nonsmoker. Once regarded as a variable, however, there is no reason why it should have only two values; it may have any number of values. Let's continue looking at the population of American women and let smoking, or rather, amount smoked, be the variable.

The standard measure of amount smoked is the number of cigarettes, or packs, smoked per day. Which scale one uses is purely a matter of convenience. Let's use half-pack, or ten-cigarette, intervals. The population of adult American women is thus divided into several categories according to the number of cigarettes, to the nearest half-pack, smoked per day. For nonsmokers, of course, the value is zero. Figure 9.2 shows a *statistical distribution* of the variable "amount smoked" for adult American women. You will note, of course, that all the percentages must add up to 100 percent. And the percentages for smokers add up to 34 percent.

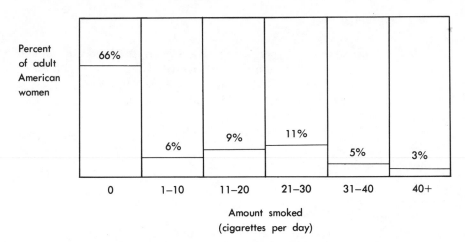

FIGURE 9.2

This way of representing a *distribution* makes it quite clear that a distribution is nothing more than a *conjunction* of simple statistical hypotheses. Regarding the population of adult American women it tells you: 66 percent are nonsmokers *and* 6 percent smoke from 1 to 10 cigarettes a day *and* 9 percent smoke from 11 to 20 a day, and so on. Each one of these components is itself just a simple statistical hypothesis. So, for example, the second "bar" in Figure 9.2 represents the simple statistical hypothesis that 6 percent of adult American women have the property of smoking from 1 to 10 cigarettes a day.

Figure 9.2 is an example of a *histogram*, or bar graph. This is the most common way of representing simple distributions. Another way of representing the same information is shown in Figure 9.3. The representation given in

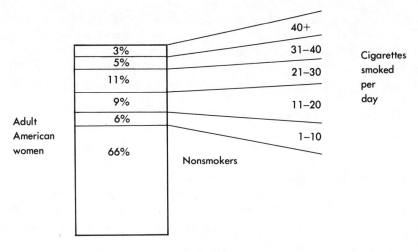

FIGURE 9.3

Figure 9.3 makes it obvious that the percentages must add up to 100 percent. But it makes less obvious the fact that the distribution is indeed just a set of simple statistical hypotheses. You should be aware that both of these ways of presentation give exactly the same information.

9.3 SIMPLE CORRELATIONS

Imagine a single population with two properties, each represented by a different variable. In any such case it is possible to ask whether there is a *correlation* between the two variables. This is true even if each variable has only two values. Indeed, this is the only case we will consider in any detail.

Adult Americans form a single population. Although you may not have thought of it this way before, sex is a variable with two possible values, male and female. And, as before, smoking may be taken as a variable with two possible values, smoker and nonsmoker. Now looking back to Section 9.1 we find the following two examples of simple statistical hypotheses:

51 percent of adult American men smoke cigarettes.
34 percent of adult American women smoke cigarettes.

Putting these two statements together yields the statement that sex and smoking are correlated in the population of adult Americans. This statement expresses a *correlational hypothesis*, or, more simply, asserts the existence of a *correlation*. Like simple statistical hypotheses or distributions, correlations can be easily pictured, as in Figure 9.4.

The specific hypothesis represented in Figure 9.4 may be written in the standard form for such hypotheses as follows:

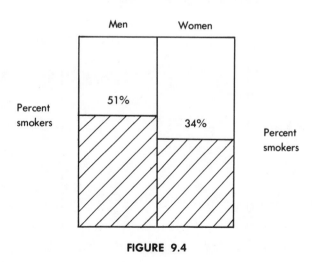

FIGURE 9.4

> Being a smoker *is positively correlated with*
> being male
> in the population of adult Americans.

The correlation is said to be "positive" because the percentage of smokers among men is *greater* than among women. Exactly the same correlation could be expressed as follows:

> Being a smoker *is negatively correlated with*
> being female
> in the population of adult Americans.

Here the correlation is said to be "negative" because the percentage of smokers among women is *less* than among men. Only in case the percentage of smokers were exactly the same for both men and women would we say that there is *no correlation* between smoking and sex.

Now let's generalize on this example. Take any population. Let A be a variable standing for any property that the members of the population might or might not have. So for each member of the population this variable has value A or Not A. The population is thus divided into two parts, the A's and the Not A's. Similarly, let B be another variable that likewise divides the population into two groups, the B's and the Not B's. Finally, there must be some percentage of the A's that are B and some percentage of the Not A's that are B. So our population with its two variables and corresponding percentages can be pictured as in Figure 9.5. We now define positive, negative, and zero correlation as follows:

> B is *positively correlated* with A if and only if the percent of B's among the A's is *greater than* the percent of B's among the Not A's.

> B is *negatively correlated* with A if and only if the percent of B's among the A's is *less than* the percent of B's among the Not A's.

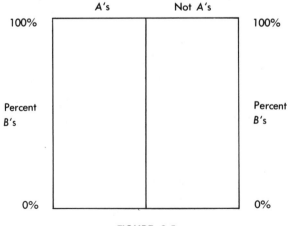

FIGURE 9.5

B is *not correlated* with *A* if and only if the percent of *B*'s among the *A*'s is *the same as* the percent of *B*'s among the Not *A*'s.

So there is little more to simple correlations than a conjunction of two simple statistical hypotheses. But it has to be the right two. Not just any two simple statistical hypotheses taken together express a correlation.

It is, of course, possible to define correlation for variables with more than two values, but we shall draw the line here. Most of the correlations one sees mentioned in literature for a general audience can be understood as either a single simple correlation or at most as a conjunction of several simple correlations. We shall deal with some of these later on. They should not cause us any difficulties.

Strength of Correlation

So far we have only specified when there is a correlation, and whether it is positive or negative. We have not said anything about *how strong* it is. But obviously some correlations are stronger than others. How should we define the strength of a correlation?

The standard measures for "strength of association" found in texts on statistics are more complicated than one really needs just to have a reasonably good idea of what is going on. So rather than use one of these measures, let us simply take the *difference* between the two percentages as our measure of the strength of the association. Following standard practice we will express the percentages as fractions or decimals, e.g., as ".34" rather than as "34 percent."

This measure at least has the right sort of intuitive properties. If the percentages are the same, the difference is zero, so no correlation corresponds to a strength of zero. Similarly, if all the *A*'s have property *B* and none of the Not *A*'s have *B*, then the strength of the correlation gets the value 1. If none of the *A*'s have *B* and all the Not *A*'s have *B*, that will be a strength of −1. So our measure goes from −1 to +1, with 0 being no strength at all. To make it official:

The *strength of the correlation* between *B* and *A* is the fraction of *A*'s that are *B* minus the fraction of Not *A*'s that are *B*.

Why Correlation Is Not the Same as Causation

Anyone who has ever worked with statistics will tell you that correlation is not causation. Textbooks are full of examples of the fallacy of inferring a causal connection from a positive correlation. A few of these examples are worth repeating.

Anthropologists studying a tribe in the South Seas found the natives believing that body lice promoted good health. It turns out that this was not just superstition. Almost every healthy person had some lice, but many sick people did not. So the percentage of healthy people with lice was clearly higher than the percentage of sick people with lice. Thus there was a clear positive correlation between having lice and being healthy.

The reason for the correlation, however, was not that having lice made you healthy. It was that being healthy caused you to have lice. Lice are not stupid. They prefer healthy bodies to sick ones, particularly to feverish ones. When a person's temperature gets much higher than normal, the lice start looking for cooler surroundings.

In a more serious vein, researchers at a state hospital connected with a major state university began comparing the recovery rates of patients. They discovered, among other things, that patients living within 50 miles of the university hospital had a much higher recovery rate than patients from farther away. That is, there was a positive correlation between recovering and living within 50 miles of the university hospital. Some people thought this showed that living near the university hospital was beneficial. Others thought it showed that quick treatment was important.

Both of these conclusions may be true, but the explanation for the correlation was simply that because the university hospital had the best facilities in the state, seriously ill patients from all over the state were brought there. Less seriously ill patients, of course, were not. They were treated at their local hospitals. So of course the recovery rate at the university hospital for the more distant patients was lower. They were on the average much more seriously ill to start with.

These examples are instructive, but they fail to reveal the underlying source of the difficulty. It is not just that attempting to justify causal hypotheses on the basis of positive correlations is risky. The very nature of correlations makes it impossible that a correlation could be the same thing as a causal connection. To see why this is so we shall return to the correlation between smoking and sex.

Rather than dealing with the whole population of adult Americans, we can just look at a population of 100 men and 100 women. You can think of each individual as representing a half-million people if you wish. Now according to the figures we have been using, there will be 51 smokers among the men and 34 among the women. This results in the positive correlation exhibited in Figure 9.4.

Now let us look at the correlation the other way around. That is, instead of asking whether smoking is positively correlated with sex, we shall ask whether sex is positively correlated with smoking. This means that we should divide the population of 200 men and women into smokers and nonsmokers and then compute the percentage of men (or women) in each group. There will be 51 + 34 = 85 smokers and 49 + 66 = 115 nonsmokers. Of the smokers, 51/85 = 60 percent will be men. Of the nonsmokers, 49/115 = 43 percent will be men. Thus you can see that being male is positively correlated with smoking. Both ways of looking at the correlation are shown in Figure 9.6.

Although we have worked it out just for this one example, the result is

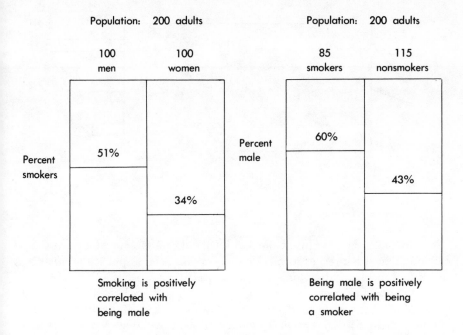

FIGURE 9.6

completely general. To put it abstractly, if *A* is positively correlated with *B*, then *B* will be positively correlated with *A*, and vice versa. This means that positive correlation is a *symmetrical* relationship. If *A* is related to *B* in this way, then *B* is necessarily also related to *A* in the same way.

Now whatever else may be true of causation, it is certainly *not* a symmetrical relationship. Speeding causes accidents, but obviously accidents do not cause speeding. Taking poison may cause death, but death certainly cannot cause the taking of poison. Abstractly, if being an *A* causes you to be a *B*, it does not follow necessarily that being a *B* would cause you to be an *A*. So causation and positive correlation are fundamentally different kinds of relationships. No wonder it is fallacious to argue directly from the one to the other.

So now we know that causal hypotheses must be different from hypotheses asserting the existence of a correlation. But this does not tell us much about what causal hypotheses do assert. That is another story.

9.4 CAUSATION IN INDIVIDUALS

To clarify the concept of causation we shall return briefly to looking at *individual* systems and then work our way back up to populations.

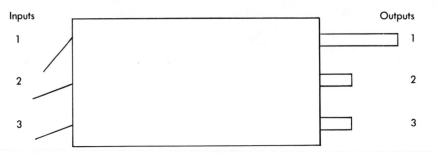

FIGURE 9.7

Causation in Individual Systems

No matter how confused one may be about causation in general, it should be clear that it is present in mechanical systems. Or, to put it the other way around, if mechanical systems are not causal, then what is?

As an example of a mechanical system, consider a big "black box" with several levers at one side and a matching number of rods at the other. The rods extend out from the box when levers are pushed. Although you do not know what is going on inside the box, you do know that it is a *deterministic* system. That is, pushing one of the levers uniquely determines which rod is pushed out. For each lever, there is one and only one rod that it controls. Suppose there are just three levers, marked 1, 2, and 3, and three rods marked similarly. Then it must be correct to say that pushing lever 1 *causes* rod 1 to extend. Similarly for the other two levers and rods.

In computer jargon, the levers are "inputs" to the system and the movements of the rods are "outputs." So each input *causes* one and only one output. The whole system is pictured in Figure 9.7.

Now we have a clear example of a causal system; moreover, we know what causes what. The only question is whether this is a good model for the kinds of systems we want to talk about, that is, humans. In other words, does it provide a useful analogy for dealing with more complex systems?

Is the Human Body a Deterministic System?

This question is important here because we want to be able to deal with a range of biomedical examples; for example, "Does smoking *cause* lung cancer?" Before you answer "No!" notice that I only said human "bodies." I am leaving questions about minds and free will out of the picture. Not that these aren't interesting and important questions. They are just too much for us to deal with now. If we can handle bodies, we shall be doing well.

Even with this restriction, you may think it rather "far out" to suppose that the human body is just a complex causal input-output machine like that in

Figure 9.7. Nevertheless, despite what you *thought* you thought about this question, I think I can convince you that you really do believe that the human body is a deterministic system.

Imagine two people as much alike as possible—say, identical twins. Suppose that both come down with the same fairly serious disease. Not having any other good options, the doctor in charge decides to try treating the disease with a new drug. Both twins get the same dose. However, the one twin recovers almost immediately while the second remains ill for a long time. Now, if you were the doctor, would you not wonder why the drug worked on the one twin and not on the other? And would you not immediately assume that there must be some difference in the two cases that explains this difference in result? If this is your reaction, then you are assuming that the twins (or at least their bodies) are deterministic systems. That is, you are assuming that there is some combination of input with internal makeup that leads deterministically to recovery and some other combination that leads to a failure to recover.

The alternative is to say, "No, there is no difference whatsoever in the two cases, except in the result." Most people find this response nearly incomprehensible. So most people are determinists, perhaps in spite of themselves. Certainly most doctors and medical researchers act as if they were determinists. Can you imagine a medical researcher saying, "Too bad. Some people get Legionnaires' disease and others don't. There is no other difference. There is nothing to research. That's just the way it is." Not likely.

One final point. A sophisticated reply might be that humans are not deterministic, but rather *stochastic* systems. That is, the inputs do not uniquely determine the outputs but only determine the *probability* of the various outputs. We cannot go into this here. (But see the references to this chapter at the end of the text.) It turns out, however, that what we shall say about causal hypotheses for populations would be little changed if we regard individuals as being stochastic rather than deterministic systems. And the arguments we shall use to justify such hypotheses would be exactly the same. So this bit of sophistication would not change anything important.

9.5 CAUSAL FACTORS IN POPULATIONS

Now we know what it means to say that some input *causes* some output in some *individual*. The individual is a deterministic system in which that input (cause) leads deterministically to that output (effect). Now let us see if we can use this idea to get at the meaning of simple causal hypotheses.

What Do Simple Causal Hypotheses Assert?

Let's begin by looking at some standard examples of causal hypotheses. At the moment we are only interested in learning what these hypotheses *mean*.

Later we will consider what evidence there is for thinking that they are *true.* It helps to imagine these statements being made by the surgeon general of the United States, or by some official of the Public Health Service.

The use of marijuana leads to heroin addiction.
Saccharin causes cancer of the bladder.
Cigarette smoking causes lung cancer.
Oral contraceptives cause fatal blood clots.
Vitamin C increases one's resistance to colds.

Note that in everyday communication the word "cause" may not appear in what are clearly causal hypotheses. For the moment let us stick to the example of smoking and lung cancer.

Now when such a statement is made, say by the surgeon general, it is not explicitly directed at any specific individual. The surgeon general is not saying, "Jones, if you smoke, you will get lung cancer." Nor is it being claimed that *each person* will get lung cancer if he or she smokes. The surgeon general knows as well as anyone that only a small percentage of smokers ever do get lung cancer. In fact, there are many cases of people who have smoked regularly for 50 years and did not get lung cancer. So just what is being claimed when it is asserted that smoking causes lung cancer?

The best way to understand such claims is as not referring to individuals directly but to a whole population of individuals. The claim is a claim about the population. In this case the population is presumably all Americans over the age of, say, 12. And the statement that smoking causes lung cancer says that there are *some* individuals in the population who will get lung cancer if they smoke. For these individuals, adding smoking as an input will cause lung cancer as an output, just as pushing a lever in the system in Figure 9.7 causes the corresponding rod to extend. However, neither the surgeon general nor anyone else knows which individuals these are. The hypothesis only says that there are *some.*

Why doesn't everyone who smokes get lung cancer? Because everyone is not exactly alike. We are assuming that individuals are deterministic systems. So if everyone were just like people for whom smoking does cause lung cancer, then smoking would indeed cause it in everyone. But we know it does not cause lung cancer in every individual. So, assuming determinism for individuals, we must conclude that not all individuals in the population are exactly alike. Indeed, the majority are not like those individuals for whom smoking does cause lung cancer.

A Standard Form for Simple Causal Hypotheses

Now that we know, in general, what simple causal hypotheses assert, let us develop a more detailed standard form for these statements. The idea is to develop a form that will make it obvious how the meaning of simple causal hypotheses connects up with standard methods for *testing* such hypotheses.

Consider a fixed, real population—for example, all Americans over the age of 12. Suppose this population has N members altogether. In the example, N is about 150 million. Now the simple causal hypothesis asserts that there are *some* members of this population for whom the stated cause will definitely produce the effect. So if we imagine *everyone* in the population as having the cause, but otherwise the same in all relevant respects, then we would presumably end up with *more* cases of the effect than there would be in the real population as it is. For example, if the hypothesis that smoking causes lung cancer is true, then presumably there would be more cases of lung cancer if everyone over 12 smoked than there will be as things are.

Now it might just happen that everyone who would get the effect from the stated cause already has the cause, although this is certainly unlikely. In this unlikely case, supposing that everyone in the whole population had the cause would not lead one to expect there to be more cases of the effect than in the real population as it is. For example, if it just happened to be the case that every American over 12 who would get lung cancer from smoking already smokes, then supposing everyone else smokes too will not lead to there being more cases of lung cancer. There would be no one left in the population who could get lung cancer from smoking.

This little difficulty is easily cleared up by simultaneously imagining the real population as it is except that *nobody* smokes. If there are indeed some individuals in the real population for whom smoking would make the difference, then there would be more cases of lung cancer if everyone smoked than if no one smoked. The reverse is true as well. If there would be more cases of lung cancer with everyone smoking than with no one smoking, there must be some individuals in the population for whom smoking makes the difference.

Figure 9.8 pictures a real population together with the two hypothetical populations we have just considered. For reasons that will be obvious in a few chapters, we will use the letter X to designate the hypothetical population in which everyone is imagined as having the cause in question. Similarly, the letter K will refer to the hypothetical population in which everyone is imagined as not having the cause. In each hypothetical population there would be some number of individuals who would get the effect. This number is called $\#E_X$ and E_K, respectively, in the two hypothetical populations. As you will learn, it is almost always better to work with percentages than with actual numbers. Dividing by the total number of members in the population, N, and multiplying by 100 gives the percentage of members of each population that get the effect.

We began with the statement that for *some* individuals in the population, smoking does cause lung cancer. We have now converted that statement into a statement about the relative numbers of cases of lung cancer that would occur in two completely hypothetical populations. Each is imagined to be exactly like the real population except that in one population everyone is imagined to smoke while in the other no one does. We can now define three related notions that provide a more precise way of expressing statements like "Smoking causes lung cancer."

Hypothetical
population, X.
Suppose all
members of popu-
lation have C.
(Population of
Americans over 12
supposing everyone
smokes.)

$$\frac{\#E_X}{N} \, (\times \, 100) = \%E_X$$

The real
population.
Some have C and
some do not.
(Population of
Americans over 12.
Some are smokers
and some are not.)

$$\frac{\#E}{N} \, (\times \, 100) = \%E$$

Hypothetical
population, K.
Suppose no
members of popu-
lation have C.
(Population of
Americans over 12
supposing no one
smokes.)

$$\frac{\#E_K}{N} \, (\times \, 100) = \%E_K$$

FIGURE 9.8

C is a *positive causal factor* for E in the real population if and only if $\%E_X$ is *greater than* $\%E_K$.

C is a *negative causal factor* for E in the real population if and only if $\%E_X$ is *less than* $\%E_K$.

C is *causally irrelevant* to E in the real population if and only if $\%E_X$ *equals* $\%E_K$.

Thus, instead of simply saying that smoking causes lung cancer, we now say: Smoking is a positive causal factor for lung cancer in the population of Americans over 12.

The notion of a negative causal factor gives us a precise way of saying that something prevents something else, as when it is claimed that taking vitamin C *prevents* colds. This hypothesis would be expressed by saying: Taking vitamin C is a negative causal factor for colds in the population of all humans (or all Americans, or whatever).

Talking about causal factors rather than just "causes" has the added advantage that it suggests the possibility of there being *other* causal factors, both positive and negative, for the same effect. Thus there is evidence that air pollution, as well as smoking, is a positive causal factor for lung cancer. It is well to keep this possibility in mind.

You may have wondered why we did not just put in zero for $\%E_K$. The reason is that $\%E_K$ is not in general equal to zero. The fact that there can be more than one positive causal factor for a single effect explains why. Thus, if you imagine the population of Americans over 12 without cigarettes, there would still be cases of lung cancer. Some of these would be caused by air pollution, and some by other substances, such as asbestos or radioactive materials. In claiming that smoking is *a* positive causal factor for lung cancer, you are saying that there would be *still more* cases of lung cancer if people smoke than if they do not.

The Effectiveness of a Causal Factor

For some purposes, like making a decision, it is often less important to know *that* something is a causal factor than to know *how much* of a factor it is. In the case of medical treatments, for example, it is more important to know *how many* people a given treatment is likely to help than simply to know that there are *some*. In the case of treatments, the term we would use to talk about relative numbers would be "effectiveness." We want to know how *effective* a given treatment would be if it were applied to a given population. Let us generalize this notion and use the word "effectiveness" to talk about any causal factor, whether it is good, like vaccinations, or bad, like air pollution.

We need a measure of the *effectiveness* of a causal factor. Once again there are various measures we could use. Let's again take the simplest, which is just the difference between $\%E_X$ and $\%E_K$ expressed, however, as a decimal rather than as a percentage. To get this into a simple "formula," let $\text{Ef}(C, E)$ stand for the effectiveness of C at producing E in the given population. Then

$$Ef(C, E) = \#E_X/N - \#E_K/N$$

Thus effectiveness can range anywhere from -1 to $+1$, with zero effectiveness corresponding to causal irrelevance. Be sure to note that this definition applies not to individuals but to populations. A given causal agent might be quite effective for one population and not at all for another.

How Causation Differs from Correlation

It may have occurred to you that there is a parallel between our definitions of positive correlation and of positive causal factors. Similarly, there is a parallel between our definitions of the strength of a correlation and the effectiveness of a causal factor. But causation is not the same as correlation, and now you can see better why this is so.

A correlation is a relationship between properties that exist in some *actual* population. Thus, the correlation between sex and smoking is defined by the relative numbers of smokers among men and among women in the population of adult Americans *as it now exists.*

Causal factors, on the other hand, are defined by relationships between two *hypothetical* versions of the real population. To say, for example, that smoking is a positive causal factor for lung cancer is not to say just that there are *in fact* more cases of lung cancer among smokers than among nonsmokers *in the existing population.* That is just a correlation. It is to say there *would be* more cases of lung cancer if everyone smoked than if no one smoked—everything else being the same. That is a very different statement.

To take an obvious example, being male is positively correlated with smoking. But this does not mean that there would be more men in the population if everyone smoked than if no one smoked. Contrary to what some people seem to think, taking over what used to be male prerogatives, like smoking, does not cause women to become men.

The difference between a correlation and a causal relationship in populations is pictured in Figure 9.9. (You may want to compare Figure 9.9 with Figures 9.5 and 9.8.) In particular, Figure 9.9 shows that a correlation is defined by relationships in the existing population, whereas a causal hypothesis is defined in terms of what would happen if the population were different in certain respects.

You should make sure you have a good grasp of the difference between correlations and causal relationships in populations. This difference will be crucial when we come to consider *justifying* these two types of hypotheses. In particular, there is a lot more to justifying a simple causal hypothesis than to justifying a claim of correlation.

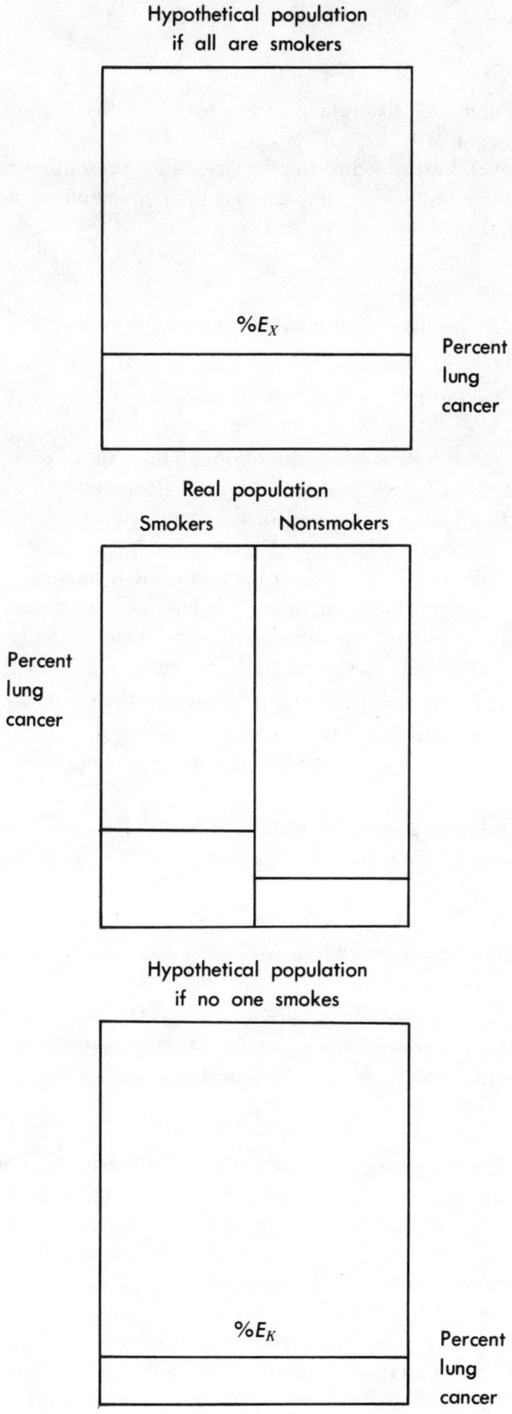

Hypothetical population
if all are smokers

Smoking is a *positive
causal factor* for lung
cancer in the real
population if and only if
$\%E_X$ is greater than $\%E_K$.

%E_X

Percent
lung
cancer

Real population

Smokers   Nonsmokers

Percent
lung
cancer

Lung cancer is *positively
correlated* with smoking in
the real population if and
only if the percentage of
lung cancer victims among
smokers is greater than
among nonsmokers.

Hypothetical population
if no one smokes

A *correlation* is defined by
a conjunction of statistical
hypotheses referring to the
real population.

A *causal relation* is
defined by a conjunction of
statistical hypotheses referring to two hypothetical
populations.

%E_K

Percent
lung
cancer

FIGURE 9.9

CHAPTER EXERCISES

9.1 Following the model of Figure 9.1, diagram any three of the last five examples that appear at the beginning of Section 9.1.

9.2 Using Figure 9.1, 9.5, or 9.8 as a model, draw a diagram representing the assertion made by each of the following statements. In each case the population consists of the inhabitants of a certain island in the South Pacific.

    A.   Most of the islanders have lice.

    B.   More of the healthy islanders have lice than do unhealthy ones.

    C.   Having lice promotes good health.

    D.   Being healthy leads to having lice.

9.3 Draw a diagram (or diagrams) representing the information given in each of the following statements. Use Figures 9.1, 9.2, 9.5, and 9.8 as models. Be sure your diagram contains all the relevant information contained in the statement.

    A.   According to a recent Gallup poll, 43 percent of Americans belonging to a church claim to have had a "religious experience" defined as "a particularly powerful religious insight or awakening." Among those not belonging to any church, 24 percent said they have had such an experience.

    B.   According to a private poll sponsored by a U. S. senator, 73 percent of the voters in his state favor a constitutional amendment abolishing the Electoral College, 23 percent oppose such an amendment, and 4 percent have no opinion.

    C.   A Washington public interest group has recently filed a class action suit based on the claim that DES—a synthetic hormone prescribed to prevent miscarriages—causes breast cancer.

    D.   According to the U.S. Bureau of the Census, "there are second marriages for five of every six men and three of every four women of ages 50 to 65 whose first unions ended in divorce."

    B.   According to a recent survey of 200,000 college freshmen, 58 percent of the entering class of 1978 answered yes to the question, "Do you want to be very well off financially?" In 1968, 40 percent of the entering freshmen answered yes to that question.

9.4 A recent book on loneliness as a cause of death cites data showing that the death rate among widows from heart attacks is greater than that of married women in general. Granting the assumption that widows are more lonely than married women, diagram both the conclusion and the information on which it is based. Can you think of any obvious explanation of the data different from the proposed hypothesis? Explain.

9.5 During World War I, the US Navy argued in its recruiting literature that being in the Navy during the Spanish American War was safer than living in New York City. The death rate in New York City during the war, they pointed out, was *higher* than that among sailors in the Navy. Diagram the information used in this argument. Try to explain why the information given did not provide any justification for a resident of New York City to think that he would have been safer if he had joined the Navy.

9.6 A careful study of college students came up with the conclusion that there is a positive correlation between having below average grades and smoking cigarettes. Opponents of cigarette smoking concluded that smoking causes students to get lower grades. Others concluded that getting low grades causes people to smoke. Diagram the correlation and the corresponding causal hypothesis. Then reverse the variables and again diagram the correlation and the corresponding causal hypothesis. Is there any reason to think that the correlation provides more justification for one causal hypothesis than the other? Explain briefly.

10

Probability

Before we get into probability, a word of encouragement for those who have in the past occasionally suffered from "math phobia" or "science anxiety." Recently educators have at last begun to recognize that many students develop a fear (phobia) of mathematics and science and thus suffer pangs of anxiety whenever they are faced with a mathematical or scientific problem. It is not hard to find the immediate source of the difficulty. It is the existence of similar fears and anxieties in parents and teachers. The attitudes and expectations of parents and teachers are passed on to their children and students. In addition, young children are very good at reinforcing the attitudes of adults in each other. This process seems to be particularly hard on women because of existing sex-role stereotypes. Little boys are expected to be good at "mechanical" (including "scientific" and "mathematical") things, and are encouraged in these directions. Little girls are not; in fact, they may even be actively discouraged.

It is, of course, easier to recognize the problem than to know what to do about it. A few schools have set up clinics or workshops to help students overcome math-science anxiety. Lacking such aids, the most important thing is for both students and instructors to be aware of the problem and to realize that it is widespread, not just something that happens to some students now and then. Instructors should try not to make things unnecessarily complicated, and students should try not to let themselves be overcome by what may be mainly irrational fears, generated in the past, that have little to do with what they are learning right now.

I hope that this digression is not really necessary for those of you who have been through Parts I and II of this text. Either you never had the problem or have by now made good progress at overcoming it. So my main message is for those who have had the problem and have been overcoming it. Don't get discouraged just because the title of this chapter is "Probability." We shall not go into the subject with the intention of learning how to do problems in probability itself. We need only learn enough about probability to *understand* standard methods of testing and justifying statistical hypotheses. And that turns out to be not very much at all.

10.1   WHAT IS PROBABILITY?

In addition to textbooks that deal with the mathematics of probability, there are many books dealing with the broader question of what more there is to probability besides just some mathematics. For our purposes, however, it is not so important to discover what probability "really" is. All we need is some easy way of understanding statements about probability so that we can evaluate arguments that rely on statistical data.

Probability and Statistical Hypotheses

The easiest way to understand simple statements about probabilities is to identify such statements with simple statistical hypotheses. Thus, the simple statistical hypothesis

34 percent of adult American women are smokers

may be expressed in terms of probability as follows:

The probability of an adult American woman being a smoker is 34 percent.

For our purposes these two statements may be taken as saying exactly the same thing. It is just that the second expresses it in terms of "probability" and the first does not. So far as the basic form of simple probability statements is concerned, you really have nothing new to learn.

Thus a hypothesis expressing a simple probability incorporates a reference to a population, a property, and a fraction or percentage. It tells you what percentage of the population has that property. That is just what a simple statistical hypothesis asserts. You may picture a simple probability statement in exactly the way you would picture a simple statistical hypothesis. For such a picture, see Figure 9.1 in the preceding chapter.

A similar analysis applies to *distributions* of statistical hypotheses. Just as distributions are conjunctions of statistical hypotheses, *probability distributions* are conjunctions of simple probability statements. For example, the distribution of percentages of American women according to the amount they smoke (Figure 9.2) is a probability distribution, i.e., a distribution of probabilities. As you will see, probability distributions play a central role in statistical reasoning.

A Simple Example

It is difficult to picture a population with millions of members, such as the population of adult American women. So let us take a population that is easy to picture—say, the population of students in a large lecture class. Figure 10.1. shows such a population. The numbers in the example have been chosen so that you can concentrate on the concepts involved and not be distracted by the

A class of 200

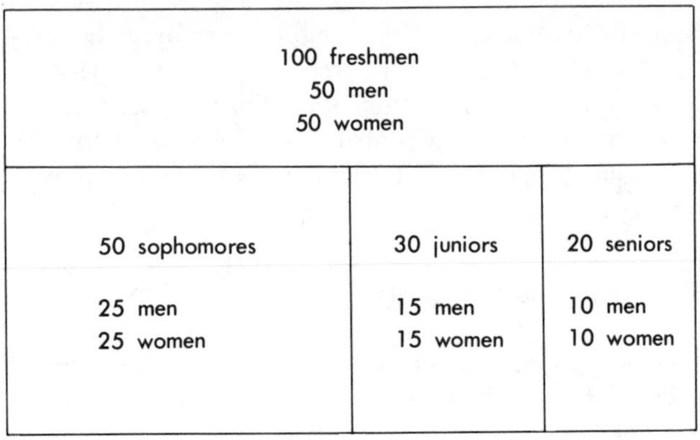

100 freshmen 50 men 50 women		
50 sophomores 25 men 25 women	30 juniors 15 men 15 women	20 seniors 10 men 10 women

FIGURE 10.1

arithmetic. A class of 200 with just this distribution of students would be a very unlikely occurrence.

In the example, there is just one *population* to consider: the class of 200 students. Both simple probability statements and distributions of probabilities always refer to a single, fixed population.

The example incorporates not one but two property variables: class and sex. The "class" variable is not socioeconomic, as in most studies, but simply class standing in a 4-year college or university. The sex variable is as always.

The distribution by sex is so simple as not to be worth picturing. It is simply 50-50, or, more precisely: The probability of a member of the class being male is .50. Since the sex variable has only two possible values, and since the percentages in the whole distribution must add up to 100%, it follows that the probability of a member of the class being female must also be .50, as indeed it is.

In picturing the distribution by class, it is useful to introduce the notation that is commonly used in talking about probability distributions. We shall let the letter V stand for the variable that represents class. This variable may have four different possible "values": F (freshman), S (sophomore), J (junior) or N (senior). The expression

$$\Pr(V = S) = .25$$

is to be read as: The probability that the variable V has value S is .25. This is a bit cumbersome, so one usually just writes $\Pr(S) = .25$, which is read as: The probability of a member of the class being a sophomore is .25. The expression $\Pr(V)$ by itself is used to refer to the whole distribution, which is shown in Figure 10.2. As required, the probabilities (percentages) all add up to 100%.

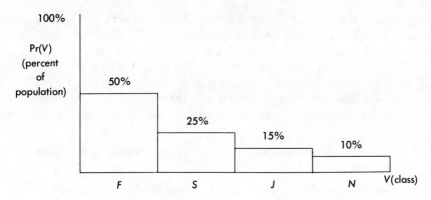

FIGURE 10.2 Probability distribution by class for the population shown in Figure 10.1

10.2 THE RULES OF PROBABILITY

The rules of probability are often called "the *laws* of probability," but the word "law" has so many overtones that it seems best to use another word. In general, since we are treating probabilities as just ratios (fractions or percentages), our rules of probability are just the rules of arithmetic that apply to ratios. But certain of these rules are especially important when talking about probabilities, so they get special billing.

We have already seen one rule of probability, namely, the rule that the probabilities in any single distribution must sum to one. We might write this rule as

$$\text{sum } \Pr(V) = 1$$

It is understood here that one takes each possible value of the variable V in turn and adds up the corresponding probabilities. Thus, in the example,

$$\begin{aligned}
\text{sum } \Pr(V) &= \Pr(F) + \Pr(S) + \Pr(J) + \Pr(N) \\
&= .50 + .25 + .15 + .10 \\
&= 1.00
\end{aligned}$$

In addition to this rule, which we might call "the total sum rule," there are two other rules that probabilities follow. A wide variety of calculations involving probabilities can be understood as being just a complex sequence of applications of these three rules. We are not going to do many such calculations. Mainly we just need to know what basic rules are involved.

The Addition Rule

Suppose you were interested not in the individual classes, but in combinations of classes. For example, you might want to know the probability that a member of the class is a "lower-division" student, where this is defined as meaning either a freshman or a sophomore. You should already be able to figure out what this probability is. Try it before reading on.

Looking back at Figure 10.1 you see that there is a total of 150 students that are either freshmen or sophomores. So the probability that a member of the class is a lower-division student is 150/200 = .75.

Now let us formulate a general rule that leads to the above result. The rule is this:

> *Addition Rule for Probabilities.* For two "mutually disjoint" values A and B of a probability variable V, $Pr(A \text{ or } B) = Pr(A) + Pr(B)$.

The restriction that A and B must be "mutually disjoint" means that the members of the population corresponding to the properties A and B *must not overlap*. You may also say that A and B must be *mutually exclusive*, meaning that no member of the population could be both an A and a B.

Now the categories of freshman, sophomore, and so on are generally understood as being mutually exclusive. So, letting A and B be F and S respectively, we calculate the desired probability as follows, where LD means lower-division student.

$$Pr(LD) = Pr(F \text{ or } S) = Pr(F) + Pr(S)$$
$$= .50 + .25 = .75$$

The need for the restriction to disjoint properties is easy to see. Just imagine a case in which it is not satisfied. For example, suppose you wanted to know the probability that a member of the class was either a freshman or a lower-division student. If you fail to restrict yourself to disjoint properties, you might be tempted to calculate this probability as follows:

$$Pr(LD \text{ or } F) = Pr(LD) + Pr(F)$$
$$= .75 + .50 = 1.25 \quad (\textit{Note: This}$$
example exhibits
a mistake.)

But this is impossible. The highest value any probability can have is 1.00. What has gone wrong, of course, is that you counted the freshmen twice, once as part of the group of lower-division students and once as themselves. So before applying the addition rule, make sure that you are talking about disjoint properties.

The Multiplication Rule

Suppose you wished to know the probability that a member of the class is *both* a sophomore and a woman. You could figure this out directly by going back to the population shown in Figure 10.1 and noting that it includes exactly

25 sophomore women. Letting M and W stand for the two values of the variable for sex, we thus conclude that $Pr(S$ and $W) = 25/200 = .125$.

The same result can be obtained from the individual probabilities, $Pr(S)$ and $Pr(W)$, using the following rule:

Multiplication Rule for Probabilities. If A and B, respectively, are values of two variables that are *not* correlated, then $Pr(A$ and $B) = Pr(A) \times Pr(B)$.

To apply the rule we must first determine that the two values are not correlated. Let A be S and B be W. To determine whether being a sophomore is correlated with being a woman we need only fill in the standard diagram for representing simple correlations. This is done in Figure 10.3.

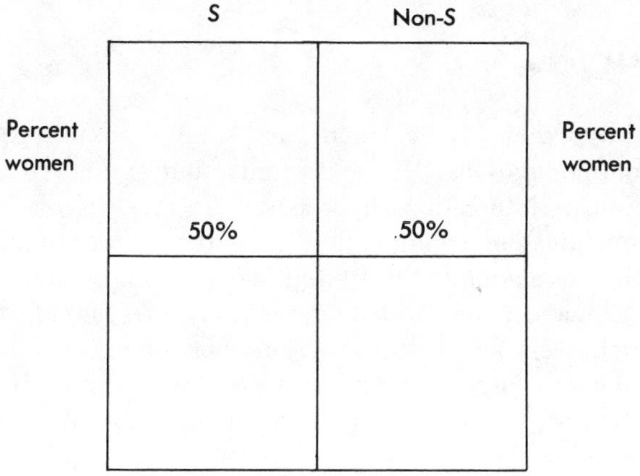

FIGURE 10.3

Since the percentage of women is exactly the same among both the sophomores and the nonsophomores, these two properties are not correlated.

We can now apply the rule. The distribution of probabilities by class (Figure 10.2) shows that $Pr(S) = .25$. We also determined that $Pr(W) = .50$. So $Pr(S$ and $W) = .25 \times .50 = .125$, which is what we got going directly back to the original population.

At this point you may wonder why we should bother with the rule if it is so much easier simply to go back to the population and compute the desired probability directly. The answer is that it is not always easier. You might not have such complete information about the population as we have in this example. In particular, you could be in the position of knowing only the probabilities needed to apply the rule, and not knowing the percentages needed for the direct computation.

Also, in thinking about probabilities in general, you need to have some feeling for the logic of probability relations. The multiplication rule is part of this logical structure. It is a crucial part of the nature of probability that for two uncorrelated properties, the probability that both occur together is just the product of the individual probabilities. This property will be very important in just a moment when we come to consider the probabilities involved in selecting a *sample* from a population.

By way of summary, you should note, if you have not already, that there is a correspondence between the two operations of conjunction and disjunction, and the two rules for probabilities. The *addition rule* applies when you wish to determine the probability of a *disjunction* of properties, Pr(A or B). The *multiplication rule* applies when you wish to determine the probability of a *conjunction* of properties, Pr(A and B). This is no accident, but a reflection of a deep similarity of structure between logic and probability. But that is a subject for a more advanced course.

10.3   SAMPLING

Statistical hypotheses, correlations, and simple causal hypotheses are all statements about *populations*. However, unlike our example, the populations considered by most "interesting" hypotheses are much too large to examine person by person. Imagine trying to determine the average number of calories consumed by the population of the United States or, more difficult still, by the population of China. In the United States, only the Bureau of the Census attempts to reach everyone, and that is done only once every 10 years. Most conclusions about any large population can be justified only by examining a small selection of individuals and using information about those individuals to argue for conclusions about the whole population.

Those members of the population that we do examine make up what is called a *sample* of the population. The procedure of selecting some members of the population to be examined is called *sampling*. And sampling, as you may already know, involves probabilities. That is why we need to know how probability functions in the specific context of sampling from a given population.

Random Sampling

The only kind of sampling we shall study is *random* sampling. This is because the most basic kinds of arguments for statistical and simple causal hypotheses require the assertion of random sampling as a premise. There are techniques for dealing with nonrandom sampling, but these are known only to specialists. You are unlikely to run across them in day-to-day life. For the most part, any suspicion that the data come from a nonrandom sample is reason to question the justifiability of the conclusion.

Sampling a population is a matter of selecting some members of the population in order to get information about the population. Let's begin by considering the selection of a single individual from our class of 200. The amount of information you could get from a sample of only one person would generally not be enough to justify any statistical hypothesis, so we shall want eventually to talk about larger samples. But let's start slowly.

If our selection of a single person is to be a *random* selection, every person in the population must have an equal chance—that is, an equal probability—of being selected. Since there are 200 people in the class, every individual should have a chance of 1/200 (or .005) of being selected. So, for example, if Jane Jones is a member of the class, the probability that our selection process picks out Jane Jones will be .005, if our selection procedure is a random one.

There are no fixed ways of doing the selecting randomly. Any procedure that ensures each individual an equal chance of being selected is acceptable. One simple procedure, for example, would be to take all the class cards, shuffle them, and pick out one without looking. Or we could assign each person a number. This would be easy if there were exactly 200 numbered seats in the lecture room. Each student would have a number ranging from 001 to 200. Then we might pick a three-digit number "at random." For example, we might flip through a telephone book, stop at an arbitrary page, and take the last three digits in the first telephone number on the page. If it happened to be over 200, we could go to the second number, and so on, until we got a number below 201.

An example of a nonrandom procedure, or at least one whose randomness is questionable, would be for the instructor just to look over the class and pick out one person. People in far corners or off to one side are less likely to be picked than people more near the middle. A person wearing a red jacket might be more likely to get picked than others. There is likely to be a sex bias. A woman instructor might notice other women more than men and be more likely to pick a woman than a man. Or there might be an age bias that would favor either freshmen or seniors. The possibilities for nonrandomness with this method of selection are practically limitless.

In scientific investigations, one is not usually concerned with the individuals, as such, but only with particular groups, or categories, of individuals. In terms of our example, one would not be so concerned which individuals were selected, but which class or which sex the selected individuals represent. Now if the selection process is random with respect to the individuals, it will automatically be random with respect to any category of individuals as well. For example, what is the probability of selecting a freshman on any random selection from the class? Well, since each individual is distinct, we can apply the addition rule. Each individual freshman has a probability of .005 of being selected and there are 100 freshmen. Adding up 100 probabilities of .005 gives us .50, which corresponds exactly with the probability of a member of the class being a freshman. The same reasoning leads to the conclusion that the probabilities of selecting a sophomore, a junior, and a senior are .25, .15, and .10, respectively.

The population in the example is so small that one would not think of determining the composition of the class by sampling. But if it were to be done this way, you see that it is not necessary that each *individual* have the same probability of being selected, so long as the probabilities for selecting members of the same *category* match the percentages of the categories in the population. For example, it would not matter if the selection process were biased toward people sitting in the middle of the room, so long as the percentages of freshmen, and so on, were the same in the middle as in the whole population.

The fact that most investigations focus on only a few categories of immediate interest thus means that random sampling is not as difficult as it might at first seem. There will only be a restricted range of biases to look out for. However, even eliminating biased selection among just a few categories can sometimes be quite difficult.

Multiple Selections

It is obvious that you cannot learn much about a reasonably large population by examining just one individual at random. You need to examine an appropriate number, which means there must be multiple selections.

In considering multiple selections, it is difficult to keep tabs on even just four categories. So let's lump all the nonfreshmen together in a category we will call "other." Let R be the variable representing this category. Applying the total sum rule, $\text{Pr}(R) = 1 - \text{Pr}(F) = 1 - .5 = .5$. You could also get this result by applying the addition rule.

Now there is just one more minor distinction to be made before we can talk about multiple selections. This is the distinction between sampling *with replacement* and *without replacement*. The difference is just whether or not an individual picked out by one selection is eligible to be picked out again on the next selection. That is, are the individuals "replaced"—that is, put back into the population—or are they not replaced? The distinction is only important if the whole sample is a noticeable fraction of the population. If we were to select 10 people from 200 *without* replacement, then by the time we got to making the tenth selection, anyone left in the population would have a probability of 1/191, not 1/200, of being selected. That is a noticeable difference. But if we were sampling from a population of 100 million, taking out a few thousand makes almost no difference to the probabilities on later selections. So you rarely have to worry about this difference. We shall for our example, however, because our population is small. We shall assume that each person is replaced—for example, by putting back the class card after making a selection. That way the probabilities stay the same for each selection. It is possible to do the calculations if the probabilities change each selection, but it is messy and definitely not worth the trouble here.

When making a single random selection from the population, the "probability of selecting a freshman" has the *same value* as "the probability of an individual in the population being a freshman." But strictly speaking these are not the same probability because they refer to different "populations." The probability of *being* a freshman in the class refers to the class of 200 people. But the probability of *selecting* a freshman refers to a "population" of *possible results* of selecting an individual from the population of 200 students. In our example there are only two possible results, either we get a freshman or we get a student in some other class. The result of getting a freshman is, logically speaking, not the same thing as a freshman. A result is an event or occurrence. A freshman is a person. The idea that there can be "populations" of things other than people should not be new to you. In Section 9.1 we considered populations of homicides and households, neither of which are people.

The "population" of possible results of a single random selection from the class of 200, is pictured in Figure 10.4. I shall follow the convention of using square brackets, [ ], to represent possible results of sampling and carry over this convention to probabilities as well. Thus, for example, the probability of a member of the class being a freshman, which we sometimes call a "population probability," will continue to be written Pr(F). The probability of selecting a freshman, which we will sometimes call a "sampling probability," will be written Pr[F]. It is easy to confuse these two probabilities because the requirement of random sampling means that they have the same value; that is, Pr(F) = .5 and Pr[F] = .5. If the sampling were biased, they would not have the same value.

The difference between the population probability and the sampling probability is easier to see as soon as you consider more than one selection, or "trial." For example, with two trials there are four different possible results. This set is pictured in Figure 10.5. The first possible result consists of picking a freshman on the first trial and a freshman on the second as well. The second possibility is getting a freshman on the first trial and a member of some other class on the second. And so on.

Now what are the values of these sampling probabilities? Perhaps you already know. But let us proceed systematically, from first principles. You might have obtained the right answers for the wrong reasons.

Selecting a freshman on the first trial *and also* on the second is a *conjunction* of results. This means we should be able to apply the multiplication rule to determine the value of Pr[F, F]. But before we can apply the rule, we must make sure that the two results satisfy the condition of not being correlated. The correlation in question is between getting a freshman on the first trial and getting a freshman on the second. This correlation is diagramed in Figure 10.6. The fact that we are sampling randomly, with replacement, ensures that the

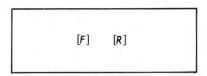

FIGURE 10.4 Possible results of a single selection from the class of 200.

$$[F, \ F]$$

$$[F, \ R]$$

$$[R, \ F]$$

$$[R, \ R]$$

FIGURE 10.5   Possible results of two selections from the class of 200.

probability of getting a freshman on the second trial is the same whether or not we got a freshman on the first. That tells us that these two properties are not correlated. So we can use the multiplication rule; that is;

$$\text{Pr}[F, \ F] \ = \ \text{Pr}[F] \ \times \ \text{Pr}[F] \ = \ .5 \ \times \ .5 \ = \ .25$$

Since $\text{Pr}[F] = \text{Pr}[R] = .5$, the other three probabilities all have the same value as well; that is, $\text{Pr}[F, \ R] = \text{Pr}[R, \ F] = \text{Pr}[R, \ R] = .25$.

The reason I suspect that you may have known the values of these probabilities all along is that this problem is analogous to tossing a fair coin. If you toss a coin once, there are two possible results, heads and tails, each with probability one-half. The probability of getting two heads in a row in two tosses is then $1/2 \times 1/2 = 1/4$. If you find coin tossing easier to understand on an intuitive level, then you can simply regard our selections as tosses of a coin. All the probabilities will be the same. I have chosen the sampling example mainly because sampling is more relevant to the justification of statistical hypotheses.

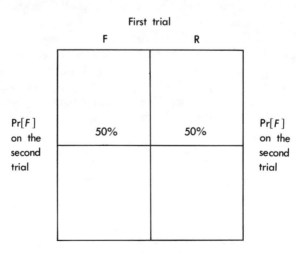

FIGURE 10.6

Another reason for thinking in terms of sampling is that, unlike fair coins, the different results of sampling *do not have to* have the same probability. They do in our example because we have half freshmen in our population. If we had considered sophomores instead, and called everyone else an "other," then the probabilities would not all be the same. For example, $Pr[S, S] = 1/4 \times 1/4 = 1/16$. But it is a little easier to picture the case in which the original population probabilities are 50–50, so we will continue to use that example.

In most applications of sampling, one is not interested in the *order* in which the results occur in a set of trials. For example, it usually makes no difference whether two trials had the result $[F, R]$ or $[R, F]$. The important thing is that these two results both have one freshman and one other. That is, the *number* of freshmen is the same in both of these results (i.e., one).

For two trials, the number of F's can have three different values: 0, 1, and 2. That is, the result might be no F's, one F, or two F's. Let us use the letter Y to represent the variable with these three values. Now what are the values of the variable $Pr[Y]$?

For no F's or two F's, the calculations are easy. $Pr[0 F's] = Pr[R, R] = 1/4$. Also, $Pr[2 F's] = Pr[F, F] = 1/4$. The probability of getting one F is the probability of getting the result $[F, R]$ *or* the result $[R, F]$. But these two sequences are *disjoint* possibilities. So we can use the addition rule:

$$Pr[1 F] = Pr\{[F, R] \text{ or } [R, F]\} = Pr[F, R] + Pr[R, F]$$
$$= 1/4 + 1/4 = 1/2$$

These three probabilities form a *probability distribution* for the number of F's in two random selections from the given population. This distribution is shown in Figure 10.7, which is our first example of a distribution of sampling probabilities, or, for short, a sampling distribution. You should study it carefully because most of the probability distributions that are important in understanding statistical reasoning are sampling distributions. If at a later stage you find your understanding of sampling distributions getting fuzzy, come back to Figure 10.7. You ought to be able to work out for yourself every detail of this distribution. Other sampling distributions are more complicated mainly because they involve larger numbers of trials. But they are not any more complex in principle.

Let us now try *three* selections in a row, with replacement. In this case there are *eight* different, mutually exclusive, possible arrangements of the results of the three trials. These form the "population" of sequences in Figure 10.8.

Once again, each of these sequences represents a *conjunction* of individual results. The first, for example, represents getting a freshman on the first trial *and* getting a freshman on the second trial *and* getting a freshman on the third. Also, the requirement that the sampling be random again ensures that selecting an F on any trial is *not correlated* with selecting an F on any other trial. That is, the first is not correlated with the second, nor the second with the

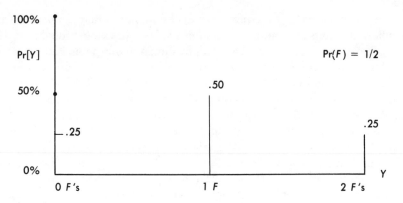

FIGURE 10.7   **Probability distribution for the number of F's in two random selections.**

third, nor the first with the third. There is no need this time actually to diagram these three zero correlations. We are assured that we can use the multiplication rule to compute the probabilities of the sequences of three trials. Thus

$$\text{Pr}[F, F, F] = \text{Pr}[F] \times \text{Pr}[F] \times \text{Pr}[F]$$
$$= 1/2 \times 1/2 \times 1/2$$
$$= 1/8 = .125$$

Again, since $\text{Pr}[F] = \text{Pr}[R] = 1/2$, each of the other seven sequences will have exactly the same probability.

For a sample consisting of three trails, the variable Y has *four* different

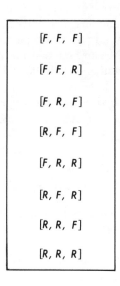

FIGURE 10.8   **Population of possible sequences in three trials.**

possible values. That is, the *number* of *F*'s obtained in three trials could be none, one, two, or three. We have just determined that Pr[3 *F*'s] = .125. The probability of *no F*'s is the same. The probability of obtaining 1 *F* or 2 *F*'s is determined, as before, by using the addition rule. This time there are three mutually exclusive sequences that have only one *F*. These are [*F, R, R*], [*R, F, R*], and [*R, R, F*]. The probability of getting one *or* another of these three is, by the addition rule, just the sum of the three probabilities—that is, 3/8 (or .375). As you can easily verify for yourself, the probability of getting 2 *F*'s is also 3/8. So the *probability distribution*, Pr[*Y*] for three trials is as shown in Figure 10.9.

To work out the distributions of Pr[*Y*] for four and for five trials would require looking at 16 and 32 different sequences respectively. That is best left as an exercise for the student. The general procedure is just as above. Use the multiplication rule to calculate the probabilities of the individual sequences and then use the addition rule to calculate the probabilities for the various possible numbers of *F*'s.

It is, however, interesting to compare the distributions of Pr[*Y*] for four and five trials with those for three and two trials. The end result of going through the calculations for both four and for five trials is shown in Figure 10.10. You may verify these results at your leisure.

As you will soon learn, a sample of four or five is still much too small to justify any interesting hypotheses. But you should look closely at the sampling distributions in Figures 10.7 through 10.10 and make sure that you understand where they come from. Sampling distributions for more trials are very tedious to calculate by hand. For only 10 trials there are 1024 different sequences of results to consider, and for 25 trials over 30 million. So to work out these sampling distributions from first principles requires a good calculator or a small computer. But in principle nothing is involved except repeated applications of the multiplication rule and the addition rule.

Even with only a few trials you can begin to see why larger samples are

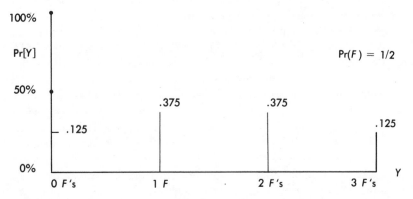

FIGURE 10.9 **Probability distribution for the number of F's in three random selections.**

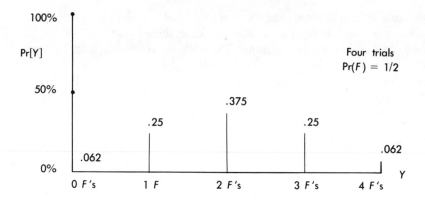

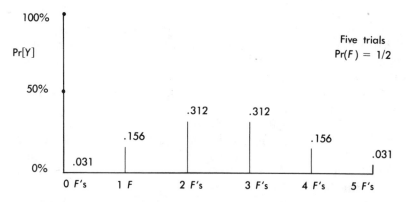

FIGURE 10.10 **Probability distributions for the number of F's in four and five random selections.**

desirable. For larger samples, the probability of observing a relative number of F's "near" the ratio of F's in the population (i.e., .50) increases. The relative number of F's in the sample, of course, is the number of F's divided by the size of the whole sample. In Figures 10.7 through 10.10 this is best seen by looking at the ends, or "tails," of the distributions. For larger samples, the probability of selecting a relative number of F's that is *not* near to half *decreases*. Obviously, selecting either no F's or all F's is as "far" from getting half F's as possible. With only two trials there is a 50% probability that you will select either no F's or all F's. (Use the addition rule.) For three trials, the probability of selecting no F's or all F's drops to 25%. With four trials it drops to about 12% and with five to about 6%. So if selecting "near" to half F's means just not selecting some number as far as possible from half, the probability of getting near to half F's increases from 50% to roughly 94% as the size of the sample increases from two to five. This trend becomes more pronounced as the number of trials is increased still further.

10.4 LARGE SAMPLES

Figure 10.11 exhibits the sampling distributions for 10, 20, and 50 trials. Figure 10.12 exhibits the sampling distributions for 100, 250, and 500 trials. Figure 10.13 is a "blow-up" of Figure 10.12. That is, it exhibits the same distributions, but on an enlarged scale so that you can see the individual probabilities a bit better.

Remember that these distributions are not *in principle* different from the sampling distributions for small numbers of trials. It is just that for large numbers of trials there are many more possible numbers of freshmen in the sample to be considered. So the distribution has considerably more probability values to calculate (or look up in a table).

Sample Frequency

The distributions we computed for just a few trials were done in terms of the *number* of freshmen in the sample. We found, for example, that the probability of getting two freshmen in four trials was .375 (see Figure 10.10). In general, however, it is more useful to think in terms of the *relative number* (or relative frequency, or just plain "frequency"), rather than actual number. This is because we will be wanting to compare two or more samples, and the relative frequency provides a better means of comparison since the samples may not all be the same size.

Thus, looking at Figures 10.11 and 10.12, you will see that the variable Y represents not the possible numbers of freshmen, but the possible *relative frequencies* of freshmen, that is, the number of freshmen divided by the total size of the sample. Thus, instead of recording the probability of getting five freshmen in a sample of ten, we record the probability of getting a relative frequency of .5, that is 5/10. If you want to see the effect of having a larger sample (e.g., 50), on the distribution of probabilities, the probability to look for is not the probability of getting 5 freshmen out of 50, but the probability of getting a *relative frequency* of .5, which is 25 out of 50. That will be of more help to you in seeing what it means to have a sample of 50 rather than only 10.

Expected Frequency

One obvious feature of all these sampling distributions is that the *most probable* sample frequency agrees with the ratio in the population. In our example, the ratio of freshmen in the population is .50 and the sample frequency with the greatest probability is always .50 (or as near to .50 as possible). In a sample of four from this population the most probable number of freshmen is two, for a relative frequency of .50. Similarly, in a sample of 250 the most probable number of freshmen is 125, again for a relative frequency of .50.

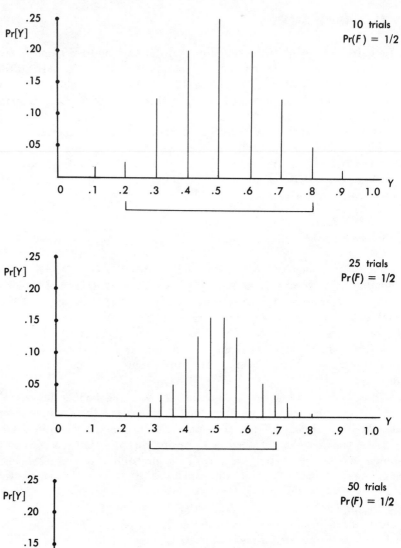

FIGURE 10.11   Probability distributions for relative frequencies of F's with 10, 25, and 50 trials. (Note: Pr[Y] goes only to 25%.)

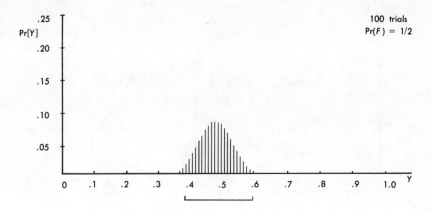

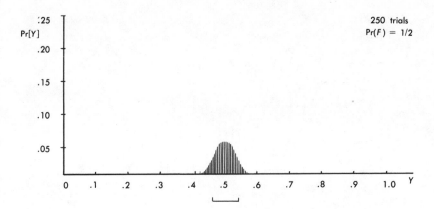

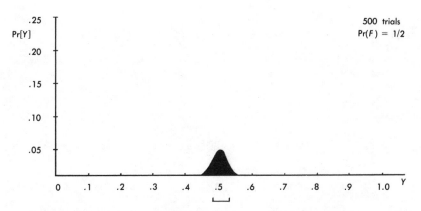

FIGURE 10.12 Probability distribution for relative frequencies of F's for 100, 250, and 500 trials.

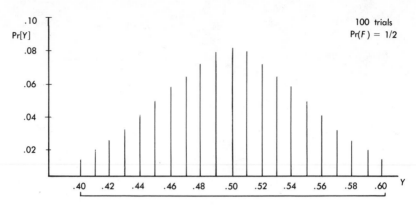

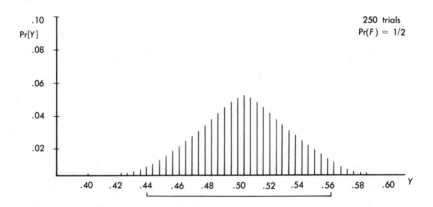

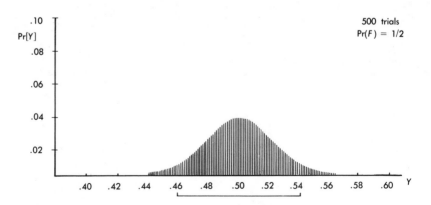

FIGURE 10.13   Probability distributions for relative frequencies of F's with 100, 250, and 500 trials. (Note: Pr[Y] goes only to 10% and [Y] from .40 to .60.)

Not only is .50 the most probable frequency, it is also, in a clearly definable sense, the "average" or "expected" frequency. This notion of expected frequency is easiest to see for small samples, but applies no matter how big the sample.

Consider, for example, a sample of only two. The sampling distribution for two random selections from our population is shown in Figure 10.7. Now imagine selecting two members of the class on each of four different occasions. Since the probability of getting no freshmen in two trials, is one fourth, you could expect to get no freshmen on one of the four occasions. Likewise, since the probability of getting one freshman in two trials is one half, you could expect to get one freshman on two of the four occasions. Finally, since the probability of getting two freshmen in two trials is again one fourth, you could expect to get two freshmen on just one occasion. Thus on all four occasions you could expect a *total* of four freshmen ($1 \times 0 + 2 \times 1 + 1 \times 2 = 4$). That amounts to an average of 4/4, or one per occasion. So on the average a sample of two yields one freshman, which is a relative frequency of .50.

In random sampling, then, both the most probable relative frequency and the expected relative frequency match the ratio in the population sampled. And this holds for all sample sizes. In textbooks on probability and statistics, the expected frequency is often called the *mean value* of the frequency. We will occasionally use this terminology.

In looking at the distributions in Figures 10.11 and 10.12 you may have noticed that the probability of the most probable frequency is *lower* for large samples. The probability of getting half freshmen is 50 percent for a sample of two, 25 percent for a sample of 10, 8 percent for a sample of 100, and only 5 percent for a sample of 250. Thus, the larger the sample the *less* likely you are actually to get the expected frequency. This may strike you as strange, even contradictory. But it makes sense if you think about it.

If you toss a fair coin just a few times, say twice or four times, there is a pretty good chance that you will get half heads. (In fact the probability is .375 for four tosses.) But if you toss it 100 times, you would not really expect to get *exactly* 50 heads. You would, however, expect to get pretty "near" to 50 heads. For probability distributions, "nearness" to the expected value is defined in terms of *deviation* from the mean.

Standard Deviation

We will approach the concept of deviation indirectly. That way we can avoid technicalities and extract just what is useful for understanding statistical reasoning.

It may have occurred to you that the sampling distributions for large numbers of trials look something like the "bell-shaped" curve known as the "normal" distribution. This is correct. For 100 or more trials the difference between our sampling distributions and the normal distribution is slight. Even for a relatively small sample of 25, the difference is not great. So what is true of normal distributions is roughly true of our distributions as well.

The rough correspondence between our sampling distributions and the normal distribution is important because the usual measure of deviation, the *standard deviation* (SD), is most commonly associated with normal distributions. For our purposes all we need to know about standard deviation is the following: In a normal distribution, the combined probability of all values within *one* standard deviation of the mean value is about 2/3, i.e., 67 percent. The combined probability of all values within *two* SD's of the mean is about 95 percent. And the combined probability of all values within *three* SD's of the mean is about 99 percent. These relationships are shown in Figure 10.14.

Look again at the sampling distributions in Figures 10.11 and 10.12. Below the axis in each graph there is a bracket indicating all the sample frequencies corresponding to two standard deviations from the expected frequency. These frequencies have a combined probability of roughly 95 percent. In a sample of 50, for example, this set includes all possible sample frequencies from .36 to .64. Now as the size of the sample gets larger, the frequencies in this set get closer and closer to the expected frequency. This means that the *difference* between the mean and the frequencies within two SD's of the mean gets smaller and smaller. That is to say, the likely *deviation* of the sample frequency from the mean gets smaller and smaller.

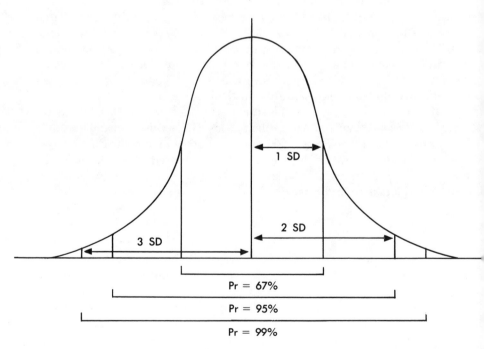

FIGURE 10.14

We have just seen that if you pick a fixed high probability, say 95 percent, then the set of possible sample frequencies with that high probability gets closer to the mean value for larger samples. But we might turn things around by fixing the range of frequencies and letting the probability change. For example, let's say that any frequency in the interval .40 to .60 is "near" to .50. Then for a sample of ten the probability that the sample frequency is "near" to .50 is only 65 percent. (By the addition rule, Pr[.4 or .5 or .6] = Pr[.4] + Pr[.5] + Pr[.6] = .20 + .25 + .20 = .65.) However, for a sample of 100, the probability that the sample frequency will be similarly "near" to .50 is 95 percent. Thus the larger the sample the more probable it is that the sample frequency will be "near" to the expected frequency.

It should now be fairly clear to you why scientists in general prefer large samples to small ones. The expected frequency of any property in a random sample is the same as the ratio of that property in the population. Thus, the larger the sample, the more probable it is that the frequency observed in the sample is "close" to the actual ratio in the population. So, in general, large samples should be better indicators of the actual ratios in populations. We will see how this works in the following two chapters.

Sampling with Unequal Probabilities

The example we have used for most of this chapter involved sampling with *equal* probabilities. That is, the two possible results of any single selection, "freshman" or "other," both had a probability of one-half. For most people this case is easiest to grasp because it is like tossing a coin.

Just so that you can see that there is no fundamental difference when sampling with unequal probabilities, Figure 10.15 shows two sampling distributions for the frequency of *sophomores* from our population of 200 students. In the example, $Pr(S) = .25$. So in a random selection of one individual, $Pr[S] = .25$ as well. In Figure 10.15, the number of trials is 25 and 50 respectively. You should compare these sampling distributions with those shown in Figure 10.11. The main difference is that the distribution shifts "downward" so that its mean value is .25 rather than .50. It is also a little asymmetrical. The asymmetry, however, is less pronounced for larger samples. And, as before, the larger the sample, the greater the probability that the sample frequency will be near the population ratio, which in this case is .25.

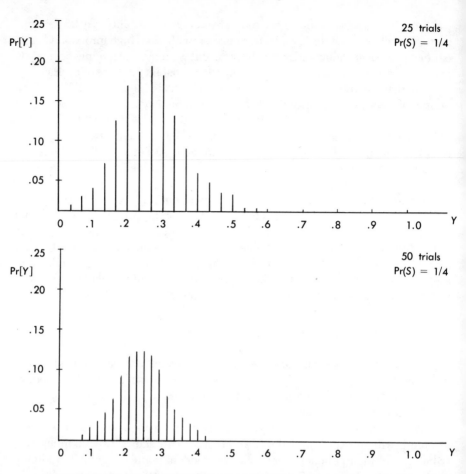

FIGURE 10.15 **Probability distributions for relative frequencies of S's for 25 and 50 trials.**

CHAPTER EXERCISES

10.1 Use the addition rule to calculate the following probabilities for the population pictured in Figure 10.1. You may check your answer by going directly to the ratios in the population.

A. Pr(*F* or *J*) B. Pr(*F* or *N*)
C. Pr(*S* or *J*) D. Pr(*S* or *N*)

10.2 Use the multiplication rule to calculate the following probabilities for the population pictured in Figure 10.1. You may check your answer by going directly to the ratios in the population.

A. Pr(*F* and *W*)   B. Pr(*S* and *M*)

C. Pr(*J* and *W*)   D. Pr(*N* and *M*)

10.3 Use the multiplication and addition rules, together with your answers to Exercises 10.1 and 10.2 to calculate the following probabilities for the population pictured in Figure 10.1. You may check your answer by going directly to the ratios in the populations.

A. Pr((*F* or *J*) and *M*)   B. Pr((*S* or *N*) and *W*)

C. Pr((*F* and *W*) or *S*)   D. Pr((*S* and *M*) or *F*)

10.4 Following the method used to determine the sampling distributions in Figures 10.7 and 10.9:

A. Work out the details leading to the sampling distribution for four trials, as shown in Figure 10.10. You will have to consider 16 different possible results of the four trials.

B. Do the same for the five trials. This means considering 32 different possible results of the five trials.

10.5 Following the method used to determine the sampling distribution in Figure 10.7, work out the sampling distribution for the number of *sophomores* on three trials. In this case Pr(*S*) = 1/4 and Pr(*R*) = 3/4.

10.6 The following are some probabilities for the *number* of freshmen on ten trials when Pr(*F*) = 1/2. Pr[0] = .001; Pr[1] = .01; Pr[2] = .04; Pr[3] = .12; Pr[4] = .20; Pr[5] = .25. Using these values and the fact that the distribution is perfectly symmetrical (i.e., Pr[6] = Pr[4], etc.) determine the following probabilities:

A. Pr[0 or 10].

B. Pr[0 or 1 or 9 or 10].

C. Pr[3 or 4 or 5 or 6 or 7].

10.7 The following are some probabilities for the *number* of freshmen on 100 trials when Pr(*F*) = 1/2. Pr[35] = .001; Pr[40] = .011; Pr[45] = .049; Pr[46] = .058; Pr[47] = .067; Pr[48] = .075; Pr[49] = .078; Pr[50] = .080. Using these probabilities and the fact that the distribution is perfectly symmetrical, determine the following probabilities:

A. Pr[35 or 40 or 60 or 65].

B. Pr[48 or 49 or 50 or 51 or 52].

C. Pr[0 through 44 or 56 through 100].

10.8 Listed below are three possible *sample frequencies*. For which of the following six sampling distributions are these frequencies within two standard deviations of the mean? 10 trials? 25? 50? 100? 250? 500? You will have to look at Figures 10.11, 10.12, and 10.13 to answer this question.

A.   .7.   B.   .6.   C.   .55.

10.9 Following the method used in Section 10.4, show that the *expected frequency* of freshmen in a sample of *four* is indeed 1/2. (Hint: Refer to Figure 10.10 and imagine selecting a sample of four students on *sixteen* different occasions.)

11

Testing Statistical Hypotheses

By now we should be familiar with statistical hypotheses, correlations, and simple causal hypotheses. And we know something about the main tool used in justifying or refuting such hypotheses, namely, probability. So now we can put these things together and see how such hypotheses are in practice justified or refuted. In this chapter we shall concentrate on statistical hypotheses and correlations. The following chapter deals with causal hypotheses.

We shall begin with simple statistical hypotheses and the methods of survey sampling in order to learn what lies behind polls that supposedly justify conclusions like: "Sixty percent of the voters favor the president's energy program." We then consider tests of specific statistical hypotheses and correlations. At the end of the chapter we use what we have learned to evaluate a study of the relationship between the use of marijuana and of heroin based on a large survey of college students.

11.1 ESTIMATING A SIMPLE PROBABILITY

Earlier we discussed the simple statistical hypothesis that 34 percent of adult American women are smokers. That conclusion, however, dates from 1965. There are indications that things have changed somewhat in the past 15 years, so one might wonder, just what is that percentage now? We can put the question this way: What is the probability today that an adult American woman is a smoker? It is easy to outline the general procedure one would use to answer that question. Get a large random sample of adult women, interview them, and find the percentage that are smokers. Then use this percentage to estimate the percentage in the whole population of adult American women. This procedure has two parts: getting the information and then using it to construct an estimate. We shall take up both parts in turn. Remember, however, that our aim is not to learn how to do such studies ourselves. It is just to learn enough about how they are done to be able to understand the statistical claims we see every day.

Getting the Sample

You have only to give the question a few minutes of serious thought to realize how difficult it would be to get a *random* sample of American women. There are women living in cities, suburbs, and small towns. There are women who work and those who do not; women who raise children and those who do not. There are women in Boston and in Tulsa. The smoking habits of all these different groups might be different. Your sample must take into account all these possible differences. We shall discuss these general problems a bit more when we look at "survey sampling" in the next section of this chapter. Our main concern in this section is to learn how information about a sample can justify "estimates" of probabilities.

To simplify things, let us narrow the question to: "What is the probability that a woman student at your school this year is a smoker?" This is easier to think about, but it is still difficult. How would you get a random sample of this population? Setting up a table to interview people coming into the student union is probably the first method that comes to mind. But many people come to the union just for coffee, and people who drink coffee are more likely to be smokers than people who do not. Better to get a list of all students and go through picking every tenth woman, or every fiftieth. Then telephone them.

There is also a question of how many women to sample. How big should your sample be? We will know better how to answer that when we know how estimates are made. If you were part of a group doing this as a class project, you would probably think 250 a big sample. Let's use that number and see how it works out. (I am assuming that your school has many more than 250 women. If not, imagine you are in a larger school.)

Estimating the Probability

Assume now that you have taken a sample and, to make things easy, suppose you found 100 smokers in a sample of 250. That is 40 percent. What do you say now? That the probability is 40 percent? If you did, it is very likely that your statement is false!

Looking back at Figure 10.13, you see that the probability of getting half freshmen in a sample of 250 from a population with half freshmen is only about 5 percent. That is, the (sampling) probability is only 5 percent that the (population) probability *exactly* matches the frequency in the sample. So whatever frequency you got in your sample, it is not very likely that the ratio in the whole population is exactly the same.

So you want to say something like, The probability is "near" 40 percent that a woman at my school is a smoker. But how near is "near"?

Your situation is pictured in Figure 11.1. $Pr(S)$ refers to the probability that a woman at your school is a smoker. The population in question is women at your school. Note that the size of this population, N, doesn't matter (unless it is relatively small and you sampled without replacement). The size of your sample, n, does matter. The symbol $f(S)$ represents the *frequency* of smokers in your sample. We are supposing that $f(S)$ is 40 percent.

Logically speaking, your situation is this. You want to justify some statement about $\Pr(S)$. As a premise for an argument you have the statement that $f(S)$ in a random sample of 250 is 40 percent. Now what exactly will the conclusion be? And will the argument be deductive or inductive?

The difficult parts of the reasoning from the frequency in a random sample to the ratio in the population are all *deductive*. This reasoning uses the mathematical principles of probability, the addition rule, multiplication rule, and so on. Because this reasoning is mathematically difficult, we will not worry about the details. We'll concentrate just on the conclusion.

Let us return for a moment to the example of sampling (with replacement) from the class with half freshmen. Remember the probability distribution for possible frequencies of freshmen in a sample of 250 (Figure 10.13). It showed that the probability is 95 percent that a random sample of 250 from that population would show a frequency of freshmen between .44 and .56. This means that 95 percent of the time, on the average, taking samples of 250 from that population would yield a frequency of freshmen between .44 and .56 (i.e., between 110 and 140 freshmen in the sample of 250).

Now it can be proved that, roughly speaking, something like the reverse is true as well. That is, if you observe a frequency of 50 percent freshmen in a random sample of 250 from a population with an *unknown* ratio of freshmen, the probability is 95 percent that the interval from .44 to .56 includes that unknown ratio. This means that 95 percent of the time, on the average, if you observe a frequency of 50 percent in a random *sample* of 250, the unknown *ratio* of freshmen in the *population* will be in that interval.

This leads us to a two step argument. The first step is purely deductive. The conclusion of the deductive argument is then used as the premise for a simple inductive argument.

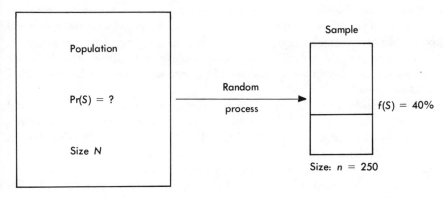

FIGURE 11.1

Premise:	A random sample of 250 shows 50 percent freshmen.
Deductive Conclusion:	The probability is 95 percent that the interval from .44 to .56 includes the true ratio of freshmen in the population.
Inductive Conclusion:	The ratio of freshmen in the population is in the interval from .44 to .56.

The second step of this argument is a *good inductive argument* because there is a high probability (i.e., 95 percent) that the conclusion is true if the premise is true.

Now a similar argument can be constructed for any frequency you might observe—for example, 40 percent. For a sample of 250, the interval roughly six percentage points either side of the observed frequency will include the true ratio in the population 95 percent of the time. The exact boundaries of the interval may change slightly for observed frequencies closer to .10 or .90. But that need not concern us here. We are not trying to learn how to make estimates, just to understand what people who do make them are doing. Returning, then, to our sample of women with 40 percent smokers, our conclusion (i.e., our estimate) should be:

> The probability that a woman at this school this year is a smoker is between .34 and .46.

In textbooks on statistics, the problem we have just discussed is treated under the title "Interval Estimation." The name is well chosen, because the conclusion of the process is always a statement about an *interval* of probabilities or population ratios. These intervals are often called "confidence intervals" and the 95 percent probability a "confidence level." Now you may have wondered what would happen if we kept the sample size the same, but picked a different "confidence level." Why not 90 percent? Or 85 percent? Or why not 99 percent? Or 99.9 percent?

In general what would happen is this. If you decide on a *lower* confidence level, you get a narrower (i.e., more precise) confidence interval. If you decide on a *higher* confidence level, you get a wider (i.e., less precise) interval. It follows that if you want a more precise estimate of the population ratio, you will not be able to have such a high level of confidence that your estimate is correct. The more precise the estimate, the more doubtful it becomes. On the other hand, of course, if you are happy with *less* precision, you can have a higher confidence level.

The standard practice is to set 95 percent for the confidence level and then take whatever interval that yields. The reason for this practice has something to do with the purely historical fact that the first probability distribution that was studied extensively was the normal distribution, back in the nineteenth century. And as we have seen, 2 SD either side of the mean encompasses 95 percent of the whole distribution. In the cases when scientists do want to have a higher confidence level, they generally go to 99 percent, which corresponds to plus or minus 3 SD on the normal distribution.

Besides the historical precedent, 95 percent is a comfortably high probability to take as the standard for a good inductive argument. Most scientists seem to think that science can get along with one mistake in 20, but not with too many more. Higher confidence levels provide less precision or require much larger samples, and both of these situations are undesirable, though for different reasons. Let us take a closer look at this relationship between sample size and precision.

Sample Size and Precision

The conclusion of our make-believe survey of women at your school was that the percentage in the whole population is between .34 and .46. What if you are not happy with an answer that has a total spread of 12 percentage points? Couldn't we get a more precise answer than that and still be 95 percent sure it is correct? Yes, if you would be willing to take a much larger sample. A sample of about 1500, for example, would give you a total spread of only four points, two on either side of your observed frequency. But would you be willing to put out the effort for such a big sample in order to get that much precision? The American Cancer Society would, and so would most tobacco companies—though probably not just to survey one school.

As a general guide to reading results of various polls, it may be helpful to set out the intervals corresponding to various sample sizes. That way you can develop a rough feeling for how much spread goes with various sizes of the sample. This is done in Table 11.1. As you will note, Table 11.1 assumes a 95 percent confidence level. Most of the time that is what is used. If you do not see it stated, 95 percent is usually a safe assumption—if the information comes from a respected source. Also, as noted above, the size of the interval does not vary much for frequencies other than .50, unless they get pretty close to 0 or 1.

TABLE 11.1

APPROXIMATE 95 PERCENT CONFIDENCE INTERVALS FOR AN OBSERVED SAMPLE
FREQUENCY OF .50 IN SAMPLES OF VARIOUS SIZES

Sample Size	Confidence Interval
10	.20 to .80
25	.28 to .72
50	.36 to .64
100	.40 to .60
250	.44 to .56
500	.46 to .54
1000	.47 to .53
1500	.48 to .52

11.2 PROBLEMS WITH SURVEY SAMPLING

Much of the information available to the public is based on surveys, especially public opinion surveys. It is thus worthwhile to note some of the difficulties peculiar to survey sampling. One problem is getting a random sample from the right population. The other is that the information one obtains is not always reliable. People do not always tell the truth, or they may not remember clearly, or they may simply not care. Looking at a few examples of these sorts of difficulties will make you more aware of what might have influenced the sample results you see.

Nonrandom Sampling

In survey sampling, especially public opinion sampling, the population of interest (the "target population") is a fairly large subgroup of all Americans—for example, all adults of voting age, all registered voters, all women. It would be very difficult and costly to set up a system of sampling that gave every person in the whole target population an equal chance of being selected—which is what a truly random sampling system would have to do. The population of people actually sampled—that is, those who do have an equal chance of being selected—is already only a subpopulation of the target population. So there is always the danger that the subgroup actually sampled is not sufficiently representative of the target population. If it is not, then no matter how large the sample, the result will not be correct for the original population of interest.

The classic example of such a mistake was the political poll conducted by the *Literary Digest* in 1936. Ten million sample ballots were mailed to people randomly selected from telephone directories around the country. More than 2 million were returned and showed a clear majority for Republican Alfred M. Landon over Democrat Franklin Roosevelt. In fact, Roosevelt got 60 percent of the vote on election day. Stop and try to guess the mistake before reading any further.

The mistake was that 1936 was the middle of the Great Depression. People with telephones tended to be people in the upper economic brackets. People in the lower brackets, especially people out of work, tended not to have telephones. But these people were strongly for Roosevelt. So due to strong correlations involving telephone service, economic status, and party preference, the poll showed a majority for Landon when in fact the majority of voters preferred Roosevelt. Largely due to this fiasco, the *Literary Digest* ceased publication in 1937.

The moral of this story is that you should pay particular attention to the method of sampling used—if you are fortunate enough to be given this information. Try for yourself to think of ways the sampled population might differ from the target population and then consider whether these differences might be relevant to the conclusions reached.

Of the various methods of conducting surveys, sending questionnaires by mail is particularly bad because the randomness of the sampling process is strongly influenced by the "return rate"—the percentage of questionnaires returned. Usually this rate is fairly low simply because most people are not willing to take the time to fill out a questionnaire and send it back. Those who are willing tend to have very strong opinions on the questions at issue. If the sponsor of the poll is known to support a particular view, the people agreeing with this view are generally more likely to respond.

To see how this works, consider the example of a former congresswoman who sent 200,000 questionnaires to her constituents asking their opinion on the issue of school busing for purposes of integration. The congresswoman was widely known for her opposition to busing. Her poll showed 80 percent of the respondents also against busing. It turns out, however, that only about 23,000 of the questionnaires were returned, or roughly 12 percent. What about the other 177,000 people? Well, suppose that they were evenly split. Half of 177,000 plus 80 percent of 23,000 gives 106,900 people against busing, or only 53 percent of the whole 200,000. And given that supporters are more likely to respond than opponents, probably less than half of the nonresponders agreed with her stand. Certainly there was no justification for her claim of 80 percent support.

The moral of this story is always to look for the return rate when you read the results of a poll done by handing out questionnaires, by mail or otherwise. If the return rate is low, ask yourself whether there is likely to be any strong correlation between having the motivation to return the questionnaire and answering the questions one way or the other. If so, the percentages in the returned questionnaires are bound to misrepresent the whole population. And even if you cannot think of any obvious correlations between answers and motivation to respond, there probably are some. You should remain skeptical.

In fairness to the major polling organizations—Gallup, Harris, Roper, and so on—it must be said that they devote much effort to getting random samples, and generally use personal or telephone interviews to get their information. You may have read that the typical sample in standard polls is between 1200 and 1500. How could this relatively small sample accurately reflect the true facts about more than 100 million American adults? What they do is to *stratify* the whole population into a number of groups known from previous polls to be relatively uniform. The groupings are done by economic status, occupation, place of residence, region of country, race, sex, age, and so forth. They then make sure that their sample reflects the correct percentages of these groups. For example, if half of all adults live in cities with populations over 100,000, then half the sample is selected from this subpopulation. By being careful in this way, one can ensure that differences of 4 or 5 percent will be meaningful. Thus, if on the basis of such a sample it is reported that 36 percent of those interviewed said "yes" to a given question, you can be fairly confident that overall percentage would be between 34 percent and 38 percent.

Unreliability of Information

Now let us turn to the second type of problem with survey data. People do not always give true answers. Political opinion polling again provides a good example. In 1972 the ex-governor of Alabama, George Wallace, ran in the primaries in a number of northern states, including Indiana, Wisconsin, and Massachusetts. In all of these states Wallace got a much higher percentage of the vote than indicated by the polls prior to primary day. In Indiana, for example, the polls said Wallace would get about 20 percent of the vote, but on election day he got over 40 percent. The generally accepted explanation is that because of his segregationist reputation, people in the north were reluctant to admit to an interviewer that they preferred Wallace to the other candidates. So they gave an "acceptable" reply, but voted for Wallace.

Even when people are not deliberately lying, they may often be mistaken about simple facts. For example, imagine trying to test the hypothesis that smoking causes health problems simply by a survey. In the first place, anyone currently hospitalized because of lung cancer or heart disease probably would not be interviewed. They certainly would not be at home to answer their door or telephone, and they would probably not answer their mail either. But this is a problem with the randomness of the sample, and not our major concern here. The real problem is that if you just ask once about smoking habits (e.g., when they started smoking, how much they smoke a day, whether they inhale, etc.) you will not get very accurate answers. People forget when they started, probably thinking it was later than it was. Someone in poor health might see smoking as a convenient scapegoat and overestimate the amount they had smoked. People in good health, but worried about possible bad effects, might well underestimate how much they smoke, not wanting to admit even to themselves that it is not just a pack a day, but well over a pack and a half. Similarly with questions about one's state of health. One might simply forget having had a mild case of pneumonia or strep throat. People often do not realize how much they actually cough. And so on.

It is because of this type of problem that most serious studies of health issues are not based just on interviews but employ actual medical examinations. Keep this in mind, so that when we come to discuss "prospective" studies in Chapter 12, you will realize how such studies can be more reliable than survey sampling.

11.3 TESTING A STATISTICAL HYPOTHESIS

There are many situations in which one wants to do more than merely *estimate* a probability, that is, the percentage of individuals that have some property. When making an estimate, the only justifiable conclusion is always an *interval* of possible values of the probability. But sometimes one would like to consider a particular simple statistical hypothesis and ask whether it is true or not. In such cases one is really after a *test* of that simple statistical hypothesis.

Imagine that you are at a basketball game and get into an argument over the question whether there are more men than women at the game. You think that there are more men than women and your companion thinks that the opposite is true. With between 5 and 10 thousand people all moving about, it is really difficult to tell just by looking. Then a man behind you butts in saying that he thinks you are both wrong. He claims that the audience is evenly split, 50 percent men and 50 percent women. Of course both you and your companion think that he is wrong.

In this situation it would be natural for you and your companion to focus on the third hypothesis, namely, that the ratio is 50–50. If you can show that hypothesis to be mistaken, you will have refuted the other man and justified one of your other two alternatives.

You decide to try to test the third hypothesis empirically. There is obviously no way you could count all the people in the audience, but you could get a pretty good sample by standing at the end of one of the main aisles at the end of the game and counting men and women as they come by. Now you must face the purely logical question of how you would use the data you got to test the hypothesis at issue.

In the language of statisticians, the hypothesis that the ratio is 50–50 would be called a "null hypothesis." It says that there is "no difference" between the percentage of men and the percentage of women. Null hypotheses are usually abbreviated by the symbol H_0. In this particular example, there is no real need to use this terminology. You are just trying to test the hypothesis that $\Pr(M) = .50$ (or that $\Pr(W) = .50$). And what we say about testing this hypothesis could be said about any other; for example, $\Pr(M) = .60$. But when we come to testing for correlations and testing causal hypotheses, the hypotheses at issue will be genuine null hypotheses, so it will smooth the way if we introduce this terminology right from the beginning.

For those who studied Part II of this text, the way to proceed should be obvious. We want to set up a *good test* of H_0. A good test is an experiment or set of observations that is arranged so that the results can be used to justify either that H_0 is close to true, or that it is not. And since the justification will involve inductive arguments, we want it to be very probable that our conclusion is true if our premises are true. Now obviously one of our premises will be the *observed frequency* of men in a random sample of people in the audience. The only question is what the *other* premises will be.

There were two conditions that defined a good test of a theoretical hypothesis, each being a conditional statement. In testing statistical hypotheses there will also be two conditions. These will be very similar to the conditions that defined tests of theoretical hypotheses, and they will play a very similar role in the justification of statistical hypotheses.

First Condition for a Good Test of H_0

Once you have grasped some basic facts about probability, understanding tests of statistical hypotheses is actually easier than understanding tests of theoretical hypotheses. There is really only one kind of prediction that you can make using a statistical hypothesis. You have to predict something about a relative frequency in a random sample. The main difference when dealing with statistical hypotheses is that the prediction itself cannot be deduced from the statistical hypothesis. All that follows deductively from any statistical hypothesis is a probability distribution for all possible relative frequencies in some sample.

In the present example, what could you predict with high probability on the assumption that H_0 is true, that is, on the hypothesis that there are equal numbers of men and women in the audience? Given our discussion of sampling in Chapter 10, the answer should be easy. If H_0 is true, then with a probability of 95%, the frequency of men in a random sample, $f(M)$, should be within two standard deviations of the expected value, which is .50. Just how close to .50 that is depends on how big a sample you have.

Now let us work this idea into a conditional statement that can serve as the first condition for a good test of H_0. What we have just said can be expressed by saying that the following argument is valid:

H_0; that is, $Pr(M) = .50$.
A sample of size n is selected.
The sampling is done randomly.
Thus (deductively), the probability is 95 percent that $f(M)$ is within 2 SD of .50.

Before stating the true conditional statement that corresponds to this valid argument, there is one bit of terminology to be learned. This new terminology won't allow us to say anything we can't say perfectly well already, but since this is the language generally used in reporting results of statistical tests, you should know what it means.

If H_0 is true, then the expected value of $f(M)$ is .50. Values of $f(M)$ that differ from this expected value by more than two standard deviations are called *statistically significant at the .05 level*. By definition, then, all statistically significant values of $f(M)$ taken together have a joint probability of 5 percent. All the possible values of $f(M)$ that are *not* statistically significant at the .05 level have a joint probability of 95 percent.

It gets rather tiresome writing, and saying, "statistically significant at the .05 level." We shall abbreviate this as "SS (.05)." Figure 11.2 shows a sketch of the probability distribution of possible values of $f(M)$ if H_0 is true. You will see that there is nothing new here except the vocabulary of "statistical significance."

Note that these relationships hold for any sample size. The size of the sample just determines which *particular* values of the sample frequency count as statistically significant and which do not.

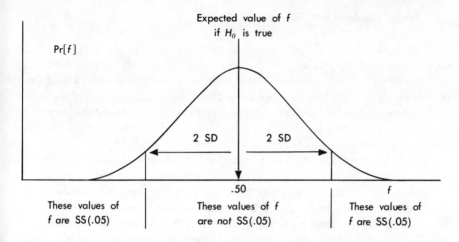

FIGURE 11.2   Statistically significant frequencies.

Now we can formulate Condition 1 for a good test of H_0. The conclusion of
our valid deductive argument was:

The probability is 95 percent that $f(M)$ is within 2 SD of .50.

In our new vocabulary, this is the same as saying:

With probability 95 percent, $f(M)$ will not be SS(.05).

This will be the consequent of our first condition.

Finally, let us call the statement that gives the *size* of the sample the
initial condition (IC) of the test. And let us call the statement that says that the
sampling was done *randomly* the auxiliary assumption (AA). Then we can state
the first condition for a good test of H_0 in a form that is very much like
Condition 1 for a good test of a theoretical hypothesis.

Condition 1.   If $[H_0$ and IC and $AA]$, then, with Pr = 95 percent, $f(M)$ will not be
SS(.05).

The truth of this conditional statement is guaranteed by the fact that its
consequent follows deductively from its antecedent. (This point was first made
way back in Section 4.5).

Those of you who have read (and remembered) what we did in Part II can
probably anticipate how Condition 1 will be used in an argument to justify a
conclusion. But let us first develop the second condition for a good test of H_0.
Remember that the notion of a good test is defined solely in terms of the design
of the experiment. One can, and should, be sure that the design yields a good
test before even beginning to collect any data and thus before it is known what
conclusion will in the end be justified.

Second Condition for a Good Test of H_0

When we finally do get around to considering a sample of people at the basketball game, we can be sure that one of two things will happen. Either we will observe a frequency of men that *is* SS(.05), or we will observe a frequency of men that *is not* SS(.05). Now we know that it follows from H_0 (with probability 95 percent) that we *will not* observe a statistically significant frequency. So we need to discover what hypothesis leads to the prediction (with 95 percent probability) that we *will* observe a frequency that is SS(.05).

At first glance you might be tempted to say that the negation of H_0, that is, Not H_0, implies that there is a high probability that the observed frequency will be statistically significant. But this is not correct. The negation of H_0 says that the true ratio of men in the audience is something other than .5, which means that it could be any ratio from 0 to 1, *except* .5. So the hypothesis, Not H_0, is really a *disjunction* of many simple statistical hypotheses. Let us call this hypothesis H.

Now suppose that, unknown to anybody, the true ratio of men in the audience is .51. That means that H_0 is false and H is true. The hypothesis that the ratio is .51 is one of the hypotheses that together make up H. We might call this simple hypothesis $H_{.51}$.

The question now is whether it follows from $H_{.51}$ (with high probability) that the frequency in the sample *will* be SS(.05). The answer is, "In general, no." This is because $H_{.51}$ is so "close" to H_0 that the corresponding sampling distributions will generally overlap to a great extent. And this means that assuming $H_{.51}$ to be true would not lead us to predict (with probability 95 percent) that $f(M)$ will be SS(.05). This is illustrated in Figure 11.3. As you can see, most of the frequencies we can expect to find in a sample even if $H_{.51}$ is true are *inside* the set of values of $f(M)$ that are *not* statistically significant. So, in this case, even if H_0 is false, we cannot predict (with high probability) that $f(M)$ *will* be SS(.05).

One minor qualification. What we have said about the hypotheses H_0 and $H_{.51}$ is true for all reasonable samples of, say, less than 2000. If the sample were really large, say 10,000, then the standard deviations of the distributions would be so small that H_0 and $H_{.51}$ would no longer overlap very much at all. $H_{.51}$ would then lead us to predict that $f(M)$ would be SS(.05). However, we could then consider an even nearer alternative to H_0 (say, $H_{.501}$) and what we said above about $H_{.51}$ would be true for $H_{.501}$. So the general point is correct. No matter what the sample size, the negation of H_0 does not logically imply a high probability that $f(M)$ will be SS(.05).

It should now be pretty clear what hypotheses would lead to the prediction that $f(M)$ *will* be SS(.05), namely hypotheses that are "further away" from H_0. These will include any hypothesis that says that the frequency of men in the audience is "quite different" from .5. Actually to determine what hypotheses these are requires lots of calculations or some good tables giving the relevant

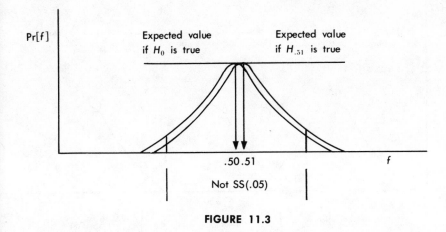

FIGURE 11.3

probabilities. We need not know how to do that. All we need to know is that there must be some hypotheses that do imply a high probability for observing a statistically significant frequency, and *roughly* what they are.

You can move "away" from H_0 in two directions. Moving up the scale from .50 corresponds to the hypothesis that there are more men than women in the audience. Moving down corresponds to the hypothesis that there are fewer men than women.

Let us begin by considering hypotheses asserting values of the ratio greater than .50. You should be able to see from Figure 11.2 that we will have to consider hypotheses that correspond to an expected value of $f(M)$ that is *greater* than the value at the upper boundary of the "not statistically significant" region. If the hypothesis with exactly that expected value were true, we would have about a 50–50 chance of getting a statistically significant frequency. About half the sampling distribution would be on either side of the boundary.

But if we consider still higher values of the ratio of men in the audience, we must eventually find one that has 95 percent of its distribution over values of $f(M)$ that are SS(.05). Call this hypothesis H_+. Now we can do the same thing moving down from .50. Eventually we must find another hypothesis that has 95 percent of its distribution over values of $f(M)$ that are SS(.05). Call this hypothesis H_-. Now, finally, we need only consider one more hypothesis. This is the hypothesis that the true ratio of men in the audience is somewhere *between* the ratio postulated by H_- and the ratio postulated by H_+. Call this hypothesis ΔH_0. ΔH_0 is thus a disjunction of simple statistical hypotheses that includes H_0 and some components of H that are "close" to H_0. These are all pictured in Figure 11.4.

Now we have it. If one of the component hypotheses of the *negation* of ΔH_0 (i.e., Not ΔH_0) is true, then our random sample should, with probability of at least 95 percent, show a frequency of men that is SS(.05). The hypothesis ΔH_0 was defined in such a way that this must be true.

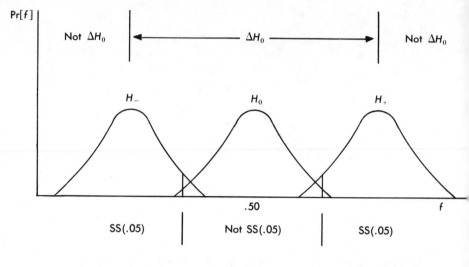

FIGURE 11.4

Putting this all together, from the hypothesis Not ΔH_0, together with the initial condition giving the sample size and the auxiliary assumption stating that the sample was selected randomly, it follows deductively that with probability 95 percent, the frequency of men, $f(M)$, should be statistically significant at the .05 level. And just because this is a valid argument, the following conditional statement must be true.

> *Condition 2.*   If [Not ΔH_0 and IC and AA], then, with $Pr = 95$ percent, $f(M)$ will be SS(.05).

Now we have, at last, the two conditions that define a good test of the "null hypothesis" H_0.

What Is a Good Test of H_0?

This has been a lot to keep straight. You should not expect it all to be clear the first time through. Also, we have not yet carried our example through to the end. That is, we have not yet said what specific hypothesis would be justified if we took some definite sample and actually observed a particular frequency of men. We shall do that in a moment.

Probably the most difficult part of the whole idea of testing a statistical hypothesis is keeping straight what refers to the *population* and what refers to the *sample*. Figure 11.5 may help you in this. Examining Figure 11.5, you see that the *hypotheses* always refer to the *population*. In terms of our example, the hypotheses always assert some ratio, or set of ratios, to be or include the true ratio of men in the population of people in the audience. The *frequencies* of men that might be observed are frequencies in the *sample*. And all the proba-

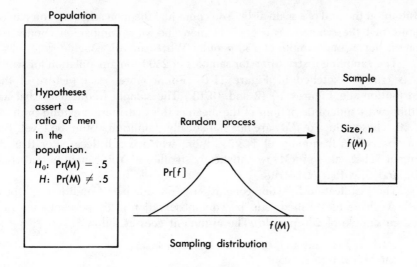

FIGURE 11.5

bility *distributions* describe the likelihood of obtaining various possible frequencies in the sample. The *connection* between the possible ratios of men in the population and the possible relative frequencies of men in the sample consists of the sampling distributions themselves. Each simple statistical hypothesis—that is, a hypothesis asserting some one particular value of the ratio in the population—determines a unique probability distribution over possible observed frequencies in a sample. And the mean, or expected, value of this distribution always has the same value as the ratio that the hypothesis asserts to exist in the population. Finally, the set of sample frequencies that are statistically significant is determined solely by that particular probability distribution that would exist if the hypothesis H_0 is true.

For easy reference, the two conditions for a good test are stated below. Try reading through these conditions in conjunction with Figures 11.4 and 11.5. If you are still a bit confused, go on to the end of this section (Section 11.3) before turning back to earlier pages. The examples may clear up your difficulties. If not, then you may have to go back to pick up specific details.

Condition 1. If $[H_0$ and IC and $AA]$, then with Pr = 95 percent, $f(M)$ will not be SS(.05).

Condition 2. If [Not ΔH_0 and IC and AA], then with Pr = 95%, $f(M)$ will be SS(.05).

Rejecting the Null Hypothesis

Now let's suppose that you decide actually to carry out the test by taking a sample of people as they leave the game. You and your companion take up positions at the end of a main aisle. You count just the men and your companion counts just the women. You get 143 men and your companion counts 107 women, for a total sample of 250 people. What can you conclude?

The sampling distribution for samples of 250 from a population for which H_0 is true is sketched in Figure 11.6. (For a more exact picture of this distribution see Figures 10.12 and 10.13.) The sample frequencies that are within two standard deviations of the mean of this distribution extend from .44 to .56. These values of $f(M)$ are *not* statistically significant at the .05 level. But you observed a frequency of 143/250 men, which is a little greater than 57 percent. That value of $f(M)$ *is* statistically significant! You conclude that H_0 is *false* and thus that H is true.

The justification for your conclusion is provided by Condition 1, whose truth we have established, and by your observation of 57 percent men in a random sample of 250 people. The argument goes as follows:

> If [H_0 and IC and AA], then, with Pr = 95 percent,
> $f(M)$ will not be SS(.05).
> $f(M)$ was SS(.05) and IC and AA.
> Thus (inductively), H (i.e., Not H_0).

The hypothesis H, of course, says only that the ratio of men in the audience is *not equal* to the ratio of women. Actually, since the frequency you observed was statistically significant in the direction of *higher* ratios for men, you are justified in making the stronger claim that the ratio of men is higher. Strictly speaking, this stronger conclusion is a little less well justified, but the difference is not important.

I have given only the "short form" of the argument. If you want to see all the steps, they are given in Section 6.4. This is a good inductive argument because there is at most only a 5 percent probability that you have reached a false conclusion from true premises. Your knowledge of the sampling distribution for 250 trials tells you that there was a 5 percent probability that you should have obtained a statistically significant frequency of men even if the ratio of men to women was equal.

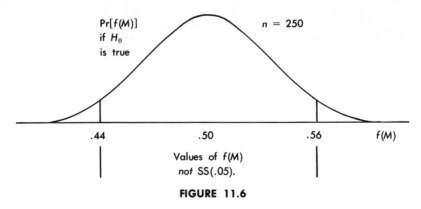

FIGURE 11.6

Failing To Reject the Null Hypothesis

Let us now suppose that the man behind you also decided to test his hypothesis. But he was less thorough than you and managed to count only 100 people: 57 men and 43 women. So he also got a count of 57 percent men. What can he conclude?

The sampling distribution for samples of 100 from a population for which H_0 is true is sketched in Figure 11.7 (see also Figures 10.12 and 10.13). For this sampling distribution, 2 SD from the expected value assuming H_0 includes all frequencies from .40 to .60. So his frequency of 57 percent is *not* statistically significant. He concludes that the ratio of men to women was not noticeably different from .50 (i.e., ΔH_0), which is just what he thought all along.

The argument that justifies his conclusion uses Condition 2 and his observed frequency of men. The "short form" of the argument is:

If [Not ΔH_0 and IC and AA], then, with Pr = 95%,
$f(M)$ will be SS(.05).
$f(M)$ was not SS(.05) and IC and AA.
Thus (inductively), ΔH_0.

In this argument, of course, IC says that the sample contained 100 people, not 250. The argument is a good inductive argument because ΔH_0 is defined so as to ensure that there is only a 5 percent probability that the conclusion is false if the premises are true.

By this time you may be wondering how it can be that both experiments obtained the "same" result (i.e., 57% men) and yet you were able to conclude that H_0 is false while the other experiment concluded that something around H_0 is true. The answer lies in the fact that 57 percent in a sample of 250 is *not* the "same" as 57 percent in a sample of only 100. The bigger sample gives a better indication of the true ratio in the population.

Also, the two conclusions are not contradictory. H_0 can be false (as you

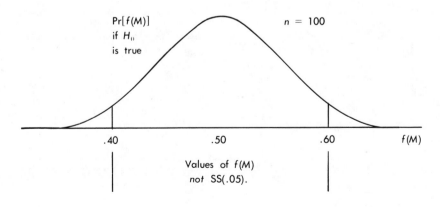

FIGURE 11.7

concluded) and ΔH_0 can be true (as he concluded). ΔH_0 only says that the true ratio is not enough different from H_0 to be reliably detected in that experiment. That is compatible with H_0 being false and with its falsity being reliably detected by a different experiment.

There would be a conflict, of course, if the other conclusion was H_0 itself. Indeed, that seems to be what the man originally claimed. But no experiment could justify that conclusion. Using Condition 1 and the fact that $f(M)$ was not SS(.05) to infer that H_0 is true would be an example of affirming the consequent. But this type of argument is no better an inductive argument than it is a deductive argument. It is fallacious in either case (see Section 6.4).

Now let's look more closely at the other conclusion. Just what simple statistical hypotheses does ΔH_0 include? You already know enough to make a close guess at this, although in general you could not be expected to come up with a precise answer. Imagine just sliding the whole distribution in Figure 11.7 up to the right. You know that when centered at .50 this distribution has 95 percent of its probability within the interval from .40 to .60. This means that it has roughly 2½ percent in either tail. So if the curve keeps its shape as you slide it up, 97½ percent of the distribution will be in the "statistically significant" region when the mean gets to .70. This means that the hypothesis $H_{.70}$ has a somewhat better than 95 percent probability of yielding a statistically significant frequency in a sample of 100. (See Figure 11.8.) The same is true of $H_{.30}$ at the lower side. So you cannot go far wrong if you take these two hypotheses to be H_+ and H_- respectively. Thus ΔH_0 asserts that the true ratio is roughly between .30 and .70. (It is .31 to .69 if you do it more precisely.) Thus the other man is claiming merely that as far as he can tell (with 95% probability of being right), the true ratio of men at the game was somewhere between .31 and .69. That is hardly a very informative conclusion.

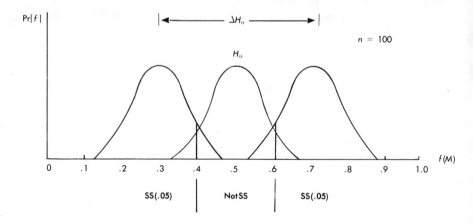

FIGURE 11.8

What "Statistically Significant" Really Means

For us the main payoff from learning what lies behind tests of statistical hypotheses is learning to understand reports of tests conducted by someone else.

The first thing to realize is that when the word "significant" appears in a discussion of statistical data, it probably means "statistically significant." Now the word "significant" is a significant word. Its regular meaning is something like "highly interesting" or "very important." But *statistically* significant results may be neither interesting nor important. This misleading use of the word "significant" was introduced into the language of statisticians in the 1920s. Since there is little chance of changing it now, you just have to learn to watch out for it.

An observation that is statistically significant (at the .05 level) is merely a sample frequency that is more than two standard deviations from the expected value if the null hypothesis is true. That is all it is. There is nothing more to it. Statistical significance is thus solely a property of the *data* in a *sample*. In our example, what is statistically significant or not is the relative frequency of men observed in the sample. And exactly which relative frequencies are statistically significant depends only on the size of the sample and the null hypothesis. So the only connection between the *population* and the statistically significant sample frequencies is through the sampling distribution determined by the null hypothesis.

So it is sample frequencies that are statistically significant or not. But sample frequencies are never in or of themselves "important" or "interesting" in a broader sense. They are only important if they reveal something important about the *population*. It is always the population ratios that we care about. For example, you may care enough about the relative numbers of men and women at a basketball game to argue about it, but you wouldn't care at all about the relative number of men in a sample unless the sample tells you something about the population. So the fact that a sample frequency was significant (read statistically significant) is only of interest if it tells you that the population ratio is significant (read important or interesting).

The main source of data that are statistically significant but may not reveal anything interesting about the population is *large samples.* You can show almost any null hypothesis to be false if you have a large enough sample. For example, if the true ratio of men in the audience at the basketball game had been 52 percent, that hardly seems enough different from 50 percent to be worth arguing about. But if you took a large enough sample, say 2500, you would stand a good chance of observing a statistically significant frequency. This is because the sampling distribution for 2500 trials has such a small deviation that an observed frequency of 52 percent would be 2 SD from the expected value implied by the null hypothesis.

On the other hand, a statistically significant observed frequency with a *small* sample is a good indication of an "important" difference between the null hypothesis value and the true ratio in the population. For small samples, the sampling distribution has a very large deviation. Thus, only if the true ratio in the population differs greatly from that postulated by the null hypothesis is there much chance that the observed frequency will be statistically significant. For example, with a sample of 25, the ratio of men in the population would have to be over .80 before there would be a 95 percent probability of selecting a "statistically significant" number of men. Even a population ratio of .70 would only give about a 50–50 chance of obtaining a value of $f(M)$ that was SS(.05). Even with a sample of 100, as we have seen, the true ratio of men in the audience would have to be .60 for there to be a 50 percent chance of getting a statistically significant value of $f(M)$.

What we have said about the expression "statistically significant" applies equally well to the phrase "*not* statistically significant." Just the details are more or less reversed. The expression "not statistically significant" applies only to observed frequencies in a sample. Strictly speaking, all it means is that the observed frequency was *not* more than two standard deviations from the expected value postulated by the null hypothesis. Whether this implies that the ratio in the population is not "importantly" different from that stated by the null hypothesis depends primarily on the size of the sample.

In this case the main source of data that are not statistically significant when the population ratio may be quite different from the null value is *small samples*. As we have just discussed, the difference between the population ratio stated by the null hypothesis and the real ratio has to be very large before there is much chance that a small sample will register a statistically significant frequency. Even fairly large differences from the null hypothesis are quite likely to produce sample frequencies that are not statistically significant.

The reverse is true for large samples. If a large sample shows a frequency that is not statistically significant, it is a good bet that the population ratio really is not much different from the null value. Because large samples yield sampling distributions with small deviations, the difference between the true ratio and the null value cannot be too big if the sample showed a frequency that was *not* SS(.05).

Figure 11.8 showed all the relevant sampling distributions for a test based on a sample of 100. Figure 11.9 shows the corresponding distributions for samples of 25 and 250 respectively. These should help you to judge the "importance" of data that are and are not "statistically significant."

In summary, you may get by reasonably well using the following rule of thumb. If the report says "the data were statistically significant" and the sample is *small*, you may be fairly sure that the real population is quite different from what the null hypothesis says. But if the sample is quite large, then you cannot tell. The real population might or might not be much different from what the null hypothesis says. On the other hand, if the report says "the data were *not* statistically significant" and the sample is *large*, then you can be pretty sure that

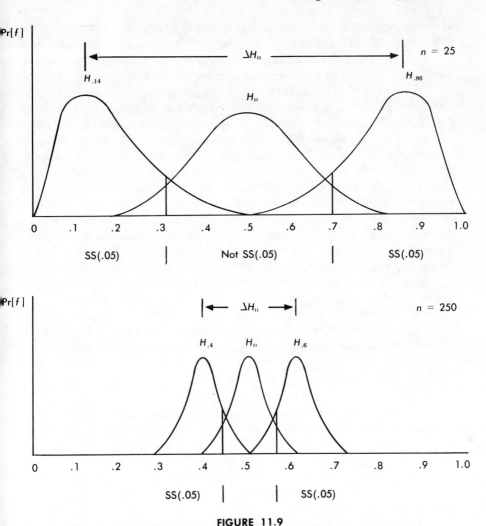

FIGURE 11.9

the real population is not much different from what the null hypothesis says. But if the sample was small, then you cannot tell. The real population might have a ratio quite different from the value given by the null hypothesis. And if you are not told the sample size, then you can't tell much of anything at all.

Can a Good Test Be Not Good Enough?

There is a fairly clear sense in which a test of the hypothesis about the ratio of men at the basketball game with a sample of 100 is not good enough. That test has only about a 50 percent chance of yielding a statistically significant value of $f(M)$ if the real ratio of men is as high as .6. Yet to know that ratio of men to women is 60-40 might be interesting if it were true. So testing the hypothesis with a sample of only 100 is not very likely to give you information that is worth having. You are likely to end up not rejecting the null hypothesis, which leaves you with a not very informative conclusion.

Yet the test with a sample of 100 is a "good test" by the definition of a good test. It will give you a justified conclusion whether or not you observe a value of $f(M)$ that is SS(.05). So there is a need for another notion, something like an "adequate" or "appropriate" test. This would be a good test which is designed to yield "interesting" or "important" conclusions. Mainly this is a matter of choosing the right sample size. And anyone with the proper technical training should be able to figure out what sample size is appropriate to have a good chance of yielding conclusions worth knowing.

Of course knowing what the sample size ought to be and getting a sample that size are two different things. If the appropriate sample size is very large, the investigators simply may not have the time or resources to get so large a sample. In that case they presumably do as well as they can. Unfortunately, most of these considerations never show up in the kind of published reports you are likely to see. So the best you can do is "read between the lines" to see if you can discover whether the actual test was appropriate to the question at issue. Knowing a little about the characteristics of probability distributions for a few different-size samples can be a big help in making this sort of judgment.

Can a Good Test Be Too Good?

This happens less often, but it does happen. If someone went through the trouble to get a sample of 2500 people at a sporting event just to be able to say that the observed frequency of men was significant (read "statistically significant") that would seem a waste of time if the ratio was really 52 to 48. Who cares? Unless there is some good reason why that value is "significant" (read "important"), having a test that good was a waste of time and effort—and maybe the taxpayers' money as well.

Unfortunately there is still some tendency among scientists themselves, to regard any significant (read "statistically significant") result as necessarily important and worth publishing. But some real differences in real populations are not important to anyone. The fact that they were discovered is merely a reflection of the fact that someone got a very big sample. It is nice sometimes to be able to tell when you are being given information that is not worth having. You can turn your attention to more "important" things.

Testing and Estimation

If you have an exceptionally good eye for figures, you may have noted that there seems to be a discrepancy between what one can conclude from the same sample depending on whether one uses the sample to *test* some hypothesis or simply to *estimate* a population probability. Thus, in our example of the failure to reject the null hypothesis about the ratio of men at the basketball game, the justified conclusion was that the ratio could be anywhere between .3 and .7. And that was with a sample of 100. If we had used that same sample to *estimate* the ratio of men, we would have concluded (see Section 11.1) that the ratio in the population was roughly between .47 and .67, given that the observed ratio was 57/100. This latter is a much narrower interval and thus a much more informative conclusion, although it includes the null hypothesis value of .50.

This discrepancy lies solely in a difference in the aims of testing a hypothesis versus estimating a simple population ratio. When testing a hypothesis, one focuses on a particular value of the ratio and asks: Is that the real ratio in the population? A test is designed to answer this question and can give a quite precise answer—if the conclusion is that the null hypothesis is false. But one pays for this possibility by a loss of precision if the observed frequency is not a statistically significant result.

This helps to explain why hypotheses that are tested tend to be genuinely "null hypotheses." They assert a ratio that is thought, or at least hoped, to be mistaken, and the point of the experiment is to prove that it is mistaken. Thus one is usually hoping to reject the null hypothesis. But of course that does not always happen.

The smart thing to do when the observed frequency is not SS(.05) is to abandon the testing framework and use the sample in the most informative way—which is simply to give the 95 percent confidence interval. This is not the way results of tests are usually reported, but you can interpret them that way yourself *if you are given the necessary information*. This means being given *both* the *sample size* and the actual *sample frequency*. Unfortunately this much information is not usually given in reports generally available to the public, and the more "popular" the source, the less information you are likely to get. You should generally count yourself lucky if you are told the sample size. But if you are given more information, by all means use it in the most efficient manner.

11.4 TESTING FOR CORRELATIONS

As you know (recall Section 9.4), a hypothesis that asserts a simple correlation is just a conjunction of two simple statistical hypotheses. So what we have just learned about testing simple statistical hypotheses can be applied with only minor modifications to testing for the existence of correlations as well.

How To Test for a Correlation

In the following section (11.5) we shall examine a study done on the connection between the use of marijuana and of heroin among college students in the early 1970s. In preparation for dealing with that example, let us imagine how one would go about testing for the existence of a similar correlation among students at your own school this year. The question we shall ask is: Is there a correlation between smoking marijuana and smoking tobacco?

A simple correlation is a relationship between two variables, each of which has just two values—roughly, "yes" and "no." Let's let M stand for a marijuana smoker and T for a tobacco smoker. Testing for the existence of a correlation will, of course, involve obtaining a random sample from the population. The population and the sample are represented in Figure 11.10.

We have already discussed the difficulties in obtaining random samples, so we shall not dwell on that problem here. The only new wrinkle in this example is that marijuana is an illegal drug. If you were to try to interview people by phone, indicating you knew their name, or standing by the entrance to a large classroom building, you might have a hard time convincing some people you were not working for the police. This might bias your sample in favor of people who say they do not smoke marijuana.

Testing a hypothesis requires that there be a specific simple hypothesis that is the object of the test. This makes it pretty natural to focus on the null hypothesis that there is *no correlation*. So the hypothesis to be tested is H_0: $Pr(M)$ is the same for both smokers and nonsmokers.

You will note that the null hypothesis does not assert a ratio of marijuana smokers in either part of the *population*. It just says that whatever the ratios are, they are the *same*. This means that H_0 cannot be used to calculate a sampling distribution over possible *frequencies* of marijuana smokers in a *sample* of (tobacco) smokers and nonsmokers. Rather the sampling distribution gives the probabilities of the possible *differences* between these two frequencies. Let us use the letter d to stand for the *difference* between the relative frequencies of marijuana users among the smokers and nonsmokers in the sample. So the sampling distribution gives values of $Pr[d]$.

We know enough about distributions in general that we do not have to worry how one determines the sampling distribution for d. We don't even need to know much about the shape of this distribution. Like most sampling distributions, it will look something like a normal distribution when the sample is

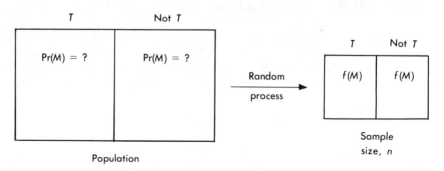

FIGURE 11.10

moderately large. The expected value of the distribution, of course, will be zero, since the null hypothesis asserts that the difference in ratios in the population is zero. Finally, the standard deviation of the distribution will be fairly great for small samples and relatively small for large samples.

Figure 11.11 shows a sketch of a typical sampling distribution for d. As always, all those values of d that lie within two standard deviations of the mean, $d = 0$, have a joint probability of roughly 95 percent. By definition, these values of d are not statistically significant at the .05 level. Values of d more than 2 SD from the mean are statistically significant at the .05 level. Let us introduce the abbreviation SSD(.05) to mean "statistically significant difference at the .05 level." This will be useful as we proceed since in tests of causal hypotheses it is always *differences* between two sample frequencies, not the frequencies themselves, that are statistically significant or not.

The two conditions that define a *good test* for the existence of a correlation look very much like those for tests of simple statistical hypotheses. They are:

Condition 1.   If $[H_0$ and IC and $AA]$, then with Pr = 95 percent there will be no SSD(.05).

Condition 2.   If [Not ΔH_0 and IC and AA], then, with Pr = 95 percent, there will be an SSD(.05).

To justify a conclusion, these two conditions are used just like the analo-

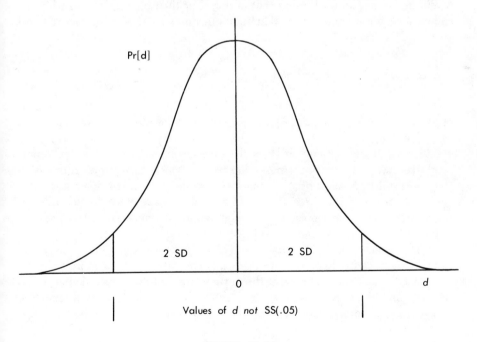

FIGURE 11.11

gous conditions for a simple statistical test. If the observed difference in sample frequencies *is* SSD(.05) you use Condition 1 as your first premise and conclude that H_0 is false; that is to say, you conclude that there is a correlation. And as before, you may safely take the "direction" (i.e., positive or negative) from the direction of the difference in the observed frequencies.

If the observed difference is *not* SSD(.05), you use Condition 2 to conclude ΔH_0. The hypothesis ΔH_0 asserts that any correlation that may exist in the population is too weak to be reliably detected by the test in question. And, as always, how strong the correlation would have to be to have a high probability (e.g., 95 percent) of producing a SSD (.05) depends primarily on the size of the sample. With small samples, even fairly strong correlations will not have a high probability of producing a statistically significant difference in the sample frequencies. And with large samples, even weak correlations may have a high probability of producing an SSD in the sample. In short, everything we have learned about tests of simple statistical hypotheses applies to tests of correlations as well.

From the standpoint of someone actually designing an experiment to test for the existence of a correlation, careful attention should be given to deciding how big a sample is needed. This depends on how big a divergence from the null hypothesis would be worth discovering—if it exists. Similarly, if there is not much difference in the two ratios in the population, how tight an interval around zero should one be justified in claiming with a 95 percent confidence level? But our position is more that of onlooker than participant. So what we should look for are indications that such questions have been considered and answered satisfactorily.

Evaluating Tests for Correlations

Let us now consider some different possible results of our imagined study—or, rather, some different possible *reports* of the results of this study.

You might be told only that the percentage of marijuana users was significantly higher among smokers than nonsmokers. The rest of the report might be given over to background and personal-interest matters. If this is all the "scientifically relevant" information you are given, you may of course justifiably conclude Not H_0—that is, H, the percentage of marijuana users, is higher among tobacco smokers in the population. We need not go through the argument. You should know it well enough by now.

The conclusion, H, by itself, is not very informative. Presumably the phrase "significantly higher" just means that the value of d was more than two standard deviations to the right of the mean value. But how much higher that is depends on the size of the sample. It might be that the SSD is more a reflection of a very large sample than of any appreciable difference between the two ratios in the population.

Now suppose that you are also told that the sample included only 100 people—say, 40 smokers and 60 nonsmokers. That is more impressive, because this sample is relatively small. If it showed an SSD, then presumably there is an appreciable spread in the population, too—at least 10 and maybe as many as 20 or 30 percentage points.

On the other hand, you might be given the percentages in the sample, but nothing else. For example, you might be told that 40 percent of the cigarette smokers, but only 20 percent of the nonsmokers, smoke marijuana occasionally, but no statement is made that this difference is indeed at least "statistically significant." And suppose no sample size is given, so you cannot make an intelligent guess as to whether the reported difference is indeed statistically significant. This is a frustrating situation. If the sample only included 10 or 20 smokers and nonsmokers, the observed difference is probably not even statistically significant, let alone "significant" in any more serious sense. On the other hand, if the sample included several hundred each of smokers and nonsmokers, that begins to look like the reported difference is statistically significant—and rather interesting, too. But if you are not told the sample size and not assured that the difference was statistically significant, you do not know whether to take the report seriously or not.

Another frustrating situation is to be told, for example, that the frequency among cigarette smokers sampled was 60 percent *higher* than the frequency among the nonsmokers. Again, you at least want to be assured that this difference is statistically significant. And failing that, you hope to be given a sample size. But even so, you are left pretty much in the dark. After all, 8 percent is 60 percent higher than 5 percent. But this much difference in the population would not be very interesting, even if the sample difference were statistically significant. On the other hand, 48 percent is also 60 percent higher than 30 percent. That would be interesting—if the sample difference is statistically significant.

In short, there are many possible ways you can be left not knowing what conclusion is justified, if any. And even if you know what conclusion is justified, you may not be told enough to know whether it is "worth knowing." It may sound informative to say that there were "significantly more" marijuana users among the smokers, but if this only means that the sample difference was "statistically significant," the conclusion may not be so interesting after all. In any case, knowing what you ought to be told will help you to decide whether the report tells you anything new and interesting.

11.5 MARIJUANA AND HEROIN ADDICTION

The question of whether using marijuana leads to the use of more dangerous drugs, particularly heroin, was sharply debated in the 1920s and 1930s and again in the 1960s and 1970s. In both cases the debate suffered from a lack of well-designed scientific studies. Many of the classic studies, dating from the

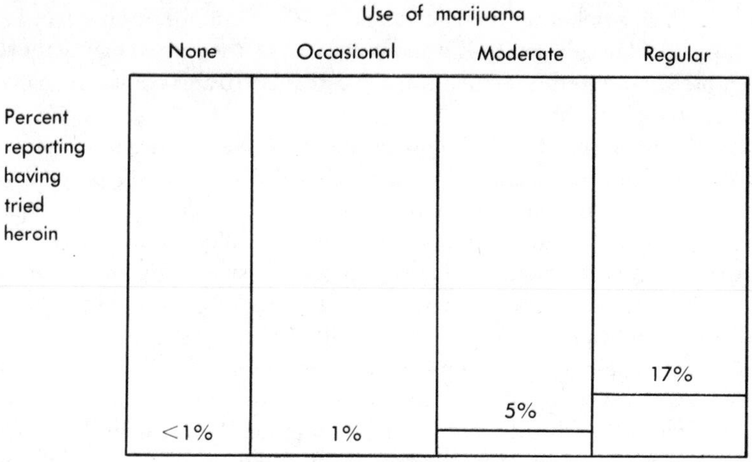

Use of marijuana

None	Occasional	Moderate	Regular

Percent
reporting
having
tried
heroin

17%

5%

<1%   1%

FIGURE 11.12

period between World War I and World War II, used data on prisoners in federal and state prisons or subjects in other countries (e.g., India), where the use of marijuana was legal. It is difficult to argue the relevance of such data to the population of middle-class American college students in the 1970s. And it is the widespread use of marijuana by this population that caused the issue to be raised again.

The Basic Correlation

One of the most widely cited studies in recent years was done by a sociologist in 1970. It was based on 3500 questionnaires that had been filled out anonymously by college students in and around New York City. The questionnaire asked for information on a variety of drugs, both legal and illegal. (*Note:* The report on which this discussion is based does not explain how the questionnaires were distributed, nor does it give the return rate. This is a possible source of bias, but other professionals who have commented on the study seem not to be worried about this aspect of the work.)

Another drawback of the study is that it only records whether heroin was tried—not actual addiction. This is understandable since there would be a tendency for actual addicts to drop out of school and thus not show up on a sample taken from the population of active students. Because people who did become addicted to heroin would tend to select themselves out of the population sampled, the study probably *underrepresents* the use of heroin in the "target population." But some unknown proportion of people who try heroin do become addicts and surely no one becomes an addict who does not try it, so recording percentages of people trying it is relevant to the larger question of actual addiction.

The basic correlations between use of marijuana and at least trying heroin are shown in Figure 11.12. In this study, "occasional use" is defined as less than once per month; "moderate" means more than once a month but less than once a week; "regular" means at least once a week.

Figure 11.12 shows several sets of data at once, all evidence of the same correlation. The difference in heroin use between nonusers and occasional users is not statistically significant. But all the other differences are. Thus it seems a justified conclusion that trying heroin is positively correlated with regular use of marijuana in the target population. Moreover, the actual percentage difference is hardly trivial.

The Causal Hypothesis

Insofar as there is an "official" view of the connection between the use of marijuana and heroin, it would be expressed by the simple causal hypothesis that the use of marijuana causes the use of heroin and thus, in some cases, heroin addiction (see Section 9.5). It is certainly believed, as our account of causal hypotheses requires, that the more people who use marijuana the more cases of heroin addiction there will be. If this were not the case, it would be difficult to understand why it is thought necessary to keep marijuana illegal, to intercept shipments into and within the country, and to destroy plants at the source—for example, in Mexico.

So let us say that the official view is that marijuana is a *positive causal factor* for heroin addiction in just the way that taking aspirin is a positive causal factor for getting ulcers or taking poison a positive causal factor for dying.

Now it is surely correct that the causal hypothesis *would explain* the existence of the basic positive correlation (Figure 11.12) between the use of marijuana and trying heroin. Indeed, one could put this in the form of Condition 1 for a good test of a *theoretical* hypothesis (Section 6.3). Let H be the causal hypothesis, IC the statement that certain numbers of people do use marijuana, AA whatever else needs to be assumed, and C the basic correlation. Then C follows logically from H, IC and AA, so,

If [H and IC and AA], then C

is true. Looking at the issue this way, however, reminds us that the correlation does not provide any justification for the causal hypothesis. What are needed are some "predictions," perhaps of other correlations, *that would be unlikely to be true if the causal hypothesis were false*. What might these be?

Now we can begin to see a difficulty with the causal hypothesis. The action of marijuana on the individual to produce heroin addiction is not *just like* the action of a regular drug or poison that produces some internal effect once it enters the system. The individual has to take some additional steps to go from using marijuana to using heroin. It does not come automatically as drunkenness follows too much alcohol. Some mechanism must be suggested that explains how using marijuana causes individuals to start using heroin.

Several mechanisms have been proposed, or rather, assumed. One is that marijuana somehow either chemically or psychologically produces a desire or "craving" for other drugs, such as heroin. Now there is considerable evidence that using marijuana produces in some people, but hardly all or even a majority, a desire for more and better *marijuana*. But the only evidence of a desire for drugs like heroin that is ever cited is the existence of the correlation between marijuana and heroin use. But the desire is supposed to be part of the explanation of that correlation. The justification for believing the desire exists must be given independently. Unless it is, some other mechanism must be found.

The other mechanism leading from marijuana to heroin that is often suggested is a kind of "tolerance-disillusionment" syndrome. The more one uses marijuana the more one builds up a tolerance for it. It stops being so enjoyable and so the person looks for something "stronger," such as heroin. But there is considerable evidence that this mechanism is not operative. Experienced marijuana users regularly report a reverse tolerance. They require less, not more, of the drug to get the desired effect. Moreover, the questionnaire showed a clear positive correlation between numerous measures of enjoyment and regular use. The percentages vary, but they are generally strongly in the direction of high enjoyment being positively correlated with regular use. This is the opposite of what would be predicted if a tolerance-disillusionment mechanism were operative. Indeed the explanation seems to be that people use it regularly because they like it. The people who don't like it don't use it regularly.

So depending on the particular mechanism suggested, the causal hypothesis is at best unproved. In any case, the simple causal model seems not to work.

The Social Hypothesis

The other major hypothesis is that the correlation between marijuana and heroin use is produced primarily by social interactions, not particularly by anything the drug does to the user's body or mind. People who use marijuana regularly tend to get drawn into a drug subculture. This is partly because the drug is illegal. Being criminals together tends to make the drug using group more cohesive. But people who deal in one illegal drug tend, for economic reasons, to deal in others as well. So the regular marijuana user comes into association with people dealing in and sometimes using other illegal drugs, such as heroin. So marijuana users end up trying heroin primarily because some of their friends use it. It is part of the subculture's way of life.

Not only does this hypothesis imply the existence of the basic correlation (Figure 11.12), it predicts *other* correlations. For example, it predicts that the more involved one is with marijuana, the more likely one is to have a friend who uses heroin. And it predicts that people having friends who use heroin are more likely to have tried it themselves. More precisely, the social hypothesis predicts a positive correlation between drug involvement and having heroin-using friends, and a positive correlation between having a heroin-using friend and trying it oneself.

The 1970 study was designed, in part, to see if these predictions are true. Before looking at the results of these tests, we should ask whether these predictions constitute a *good test* of the social hypothesis. That is, are these correlations unlikely to exist if the social hypothesis is false. It is hard to argue that they would be *very* unlikely if the social hypothesis were wrong. The hypothesis is not specific enough to predict the exact amount of the correlation, only that it will show up as statistically significant in the positive direction. But that is still something, and maybe about as much as one can expect in contemporary sociology. In making these predictions, the sociologists were taking some risk. If the correlations had not been there, they would have had a difficult time explaining why not.

As it turns out, the correlations were found. The first "prediction" involves an explicitly stated auxiliary assumption, namely, that "dealing" in drugs (i.e., buying and selling) indicates a high degree of involvement in the drug culture and in particular a higher degree of involvement than indicated by just buying marijuana. The data, for both moderate and regular marijuana users, are given in Figure 11.13. For both moderate and regular marijuana users, the correlation between "dealing" and having a friend that has used heroin is statistically significant, and quite strong.

The success of the second prediction is equally dramatic. The data are shown in Figure 11.14. (HUF means "having a heroin-using friend.") The strong positive correlation between having a heroin-using friend and trying it oneself is especially striking in the case of regular marijuana users.

Probably it is a little hasty to regard the social hypothesis as "established,"

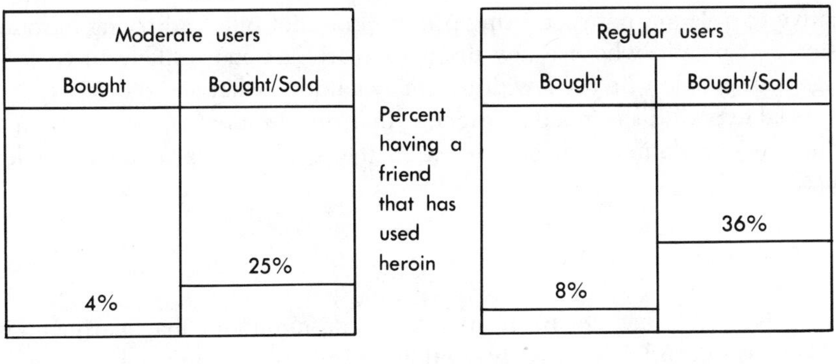

Moderate users			Regular users	
Bought	Bought/Sold		Bought	Bought/Sold
		Percent having a friend that has used heroin		
				36%
	25%			
4%			8%	

FIGURE 11.13

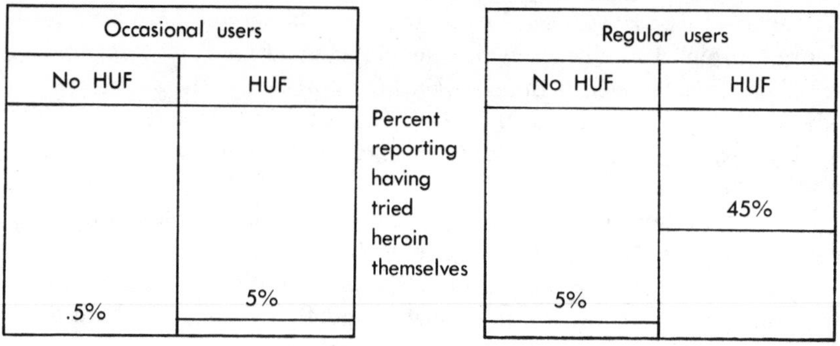

Occasional users	
No HUF	HUF
.5%	5%

Regular users	
No HUF	HUF
	45%
5%	

Percent
reporting
having
tried
heroin
themselves

FIGURE 11.14

but it is the best-supported hypothesis that explains why there should be a positive correlation between trying heroin and using marijuana.

It is very tempting to conclude that using marijuana is not causally related to using heroin. But this is a mistake. If the social hypothesis is correct, there is a causal relation. Indeed, the social hypothesis is that marijuana and heroin use are part of a complex social system that causes the correlation to exist. The social hypothesis is based on a social theory that defines a type of complex causal system.

What the social theory denies is that the causal relationship is correctly represented by the simple model developed in Section 9.5. Indeed, on the social hypothesis, if everyone used marijuana, the number of heroin addicts should go *down*. This is because if everyone used it, it could no longer be illegal. Marijuana use would then no longer be associated with a subculture in part held together by the fact that its members use and deal in illegal drugs. Breaking the social contacts would break the correlation between using marijuana and trying heroin.

Actually, we need to be a little more careful here. One of the findings of the 1970 study was that there is a positive correlation between almost any legal drug (e.g., tobacco and alcohol) and heroin use. Indeed there was even a positive correlation between using prescription diet pills and trying heroin. It seems that people who use any drugs regularly are more likely to try other drugs than people who use few drugs of any kind. So what one should expect, if the social hypothesis is indeed correct, is that "legalization" of marijuana would reduce its correlation with heroin use to the level that exists for other legal drugs.

Conclusion

This section has not been intended as an argument that marijuana is a perfectly good and harmless drug. It is merely an example of the kind of conclusions one can reach using correlations in conjunction with what are really theoretical hypotheses. There are many *other* aspects to the use of marijuana, including some serious medical ones. If there is an underlying message in this section, it is that each of these issues should be dealt with as carefully and scientifically as possible.

CHAPTER EXERCISES

Each of the following exercises begins with a report adapted from popular sources, mostly newspapers and magazines. The conclusion in each case is either an estimate of one or more simple probabilities, a statistical hypothesis, i.e., either H or ΔH_0, or a statement of a correlation, again either H or ΔH_0. In some of the problems, more than one type of conclusion may be involved. Your task, as always, is to analyze the report to determine whether the stated conclusion (or conclusions) has been adequately justified.

Your general strategy in approaching each of these exercises should be roughly as follows:

First, identify the conclusion (or conclusions). Is the conclusion an estimate of a simple probability, a statistical hypothesis, or a claim of correlation? Identify the target population. Also, be on the lookout for causal hypotheses or theoretical hypotheses that may be asserted using statistical data for justification.

Second, determine whether the data are being used to *estimate* some probabilities in a population or to *test* a specific null hypothesis. If the latter, try to formulate the null hypothesis for yourself. Again be on the lookout for cases in which a causal hypothesis or a theoretical hypothesis is being tested against statistical data.

Third, picture the study following the model of Figure 11.1, 11.5, or 11.10. You must do this for problems to be handed in, but it is helpful in any case. Put in as many details as you can from the information given. Missing facts, like the size of the sample, should get question marks (?) in your diagram. Note in particular whether the population actually sampled is importantly different from the target population and look for indications that the sampling was done randomly. It is helpful to try yourself to think of ways the sampling might have been biased. If the study involved testing a statistical hypothesis or testing for a correlation, look for some indication that the data does or does not exhibit statistical significance. Note the significance level, if that should happen to be given.

Finally, decide whether the conclusion is justified or not. You need not in general actually write out the appropriate argument, although some exercises will ask you to do that. In general, try to put the conclusion in one of the following three categories. (a) Conclusion very well justified. The sample size was "appropriate," the sampling was carefully done, and, (if applicable) the results were clearly statistically significant. (b) Results fairly well justified, but some mistakes or omissions, for example, the sample size was not given even though a "significant" result was obtained. (c) Conclusion not well justified. For example, no information about the sample size or the method of sampling was given. The conclusion is ΔH_0, but the sample was so small that this conclusion is not very informative. The conclusion is a causal hypothesis but the data only justify a correlation.

The first few exercises contain specific questions and hints. The later ones leave you more on your own. As always, some of the exercises are worked out for you at the back of the book. But don't look until you have tried to do them by yourself. The only way to learn is by trying.

11.1

Young, Western Women in Volvos Most Likely to Buckle Up, Study Indicates

In 1977 the National Highway Traffic Safety Administration commissioned a study of the use of seat belts by American drivers. Investigators in 16 cities observed 84,682 drivers during 1977 and 1978. The observations were made at intersections while traffic was stopped by red lights. The observers, who worked in pairs, were trained to record the type of belt systems in each vehicle, whether the driver was using it, the sex and approximate age of each driver, and the license number and vehicle model. The license numbers were used at motor vehicle bureaus to verify the manufacturer and model of the car. No observations were made in rural areas because it was too difficult to make the observations.

The researchers found that only 18.5 percent of the motorists in cars equipped with seat belts had fastened them.

Volvo owners were found to use their seat belts more than drivers of any other cars—44.6 percent. Only 14 percent of Cadillac people were found to have fastened their seat belts.

Out of 30,819 women drivers observed, 20.6 percent had fastened seat belts, but only 17.3 percent of 53,769 men did so.

Among drivers between ages 16 and 24, 20.4 had fastened seat belts, but among drivers over 50 only 15.4 percent did.

Drivers in the East fastened their belts at a rate of 12 percent compared with 27.3 on the West Coast.

In part the study was designed to test the effect of various systems to get people to use seat belts. The model year with the highest rate of use, roughly 25 percent, was 1974—the one year there were mandatory interlocks that would not allow the engine to start until the seat belts were fastened. The director of the study said that neither a buzzer nor a warning light had any noticeable effect on buckling habits.

You should find sufficient information here to consider one simple statistical hypothesis, four different correlations, and two causal hypotheses. Note that the report does not give any direct information on young, western women in Volvos. That was the creation of the headline writer.

11.2

Newsmen Aren't Always in Touch with Their Readers' Interests

In a recent Harris poll, 60 percent of the adults interviewed said they were "very interested" in national news, a subject only 34 percent of the journalists surveyed said they thought interested their audiences. Similarly, 75 percent of the newsmen said they thought the public was very interested in sports news; only 35 percent of the public indicated a high interest in sports.

It is probably best to regard this report as providing you with data for four different statistical hypotheses, which are then used to justify a more general conclusion. It is a little difficult to put the data in a form that would test correlations, but you might try viewing it this way. Could it be that the "adults interviewed" were not all telling the truth? Why might they not?

11.3

True or False?

Women are better than men when it comes to remembering people's names and connecting them with the right faces.
True. In studies conducted by California State University and San Diego State University behavioral scientists, who tested men and women undergraduates on their ability to match people's names with their faces, women made significantly better scores than men.

Since this ability often can assume crucial social importance—the easiest way to slight a person you've met is to forget who he is—it is suggested that "females may be more socially oriented than males, probably because of society's differential role expectations."

The data here apply to a correlation, but a theoretical hypothesis is suggested as well.

11.4

True or False?

The pleasure you derive from your work depends on how well you like yourself.
True. In studies at Ohio University, more than 100 men and women subjects were given personality tests measuring self-assurance and self-esteem. They were then asked to rate their job or occupation on a single nine-point scale, ranking from "Like very much" to "Do not like at all." The results showed that self-assurance and self-esteem tended to go hand in hand with a similar liking for the work in which they were engaged.

In evaluating this study you should consider the possibility that how well you like yourself "depends" on the pleasure you derive from your work.

11.5 The following paragraphs are adapted from a report of a four year study of the effects of television on adults. A large number of adults—men and women, old and young, well educated and not—were interviewed. Subjects were classified as "light" viewers if they watched 2 hours or less of TV per day and as "heavy" viewers if they watched 4 or more hours per day.

Heavy TV Viewer Sees Scary World

Violence on television leads viewers to perceive the world as more dangerous than it really is. When asked, "Can most people be trusted?" the heavy viewers were 35 percent more likely to check "Can't be too careful." (48 percent of light viewers and 65 percent of heavy viewers checked this response.)

When we asked viewers to estimate their own chances of being involved in some type of violence during any given week, they provided further evidence that television can induce fear. The heavy viewers were 33 percent more likely than the light viewers to pick such fearful estimates as 50-50 or 1 in 10, instead of a more plausible 1 in 100. (39 percent of light viewers and 52 percent of heavy viewers picked the higher estimates.)

The data here are sufficient for us to consider two correlations. There is also an underlying causal hypothesis to consider.

11.6

Study Links Heavy Drinking and High Blood Pressure

Regular consumption of three or more alcoholic drinks each day has been linked to the development of high blood pressure, according to results of a large study in California. This link was reported by epidemiologists at the Kaiser-Permanente Medical Center in Oakland on the basis of their statistical analysis of responses to health checkup questionnaires and the medical records of 83,947 men and women of three races.

The findings, published in the current issue of the *New England Journal of Medicine*, showed a solid statistical association between consumption of alcohol and high blood pressure. The study found that the blood pressures of men taking two or fewer drinks each day was similar to those of nondrinkers. Women who took two or fewer drinks each day had slightly lower blood pressures. Men and women who took three or more drinks each day had higher blood pressures.

There is evidence from other studies that people who take alcohol seem to be less likely to have heart attacks. Yet high blood pressure is a well-established risk factor for heart attacks. The researchers urged further studies to explain this seeming paradox.

You should consider one correlation involving men only, one involving women only, and one involving both men and women. You don't have to worry about the "paradox" for the purposes of the exercise, but you might want to think about it.

11.7

Poll Finds More Liberal Beliefs on Marriage and Sex Roles

Americans are more likely to believe that marriages in which the partners share the tasks of breadwinner and homemaker are a more "satisfying way of life" than they are to prefer the traditional marriage in which the husband is exclusively a provider and the wife exclusively a homemaker and mother, according to a New York Times—CBS poll.

Of those interviewed, 53 percent said they preferred the idea of shared marriage roles and 47 percent said they preferred the "traditional" marriage. But among those under 40 years of age, 61 percent preferred shared marriage roles, while 59 percent of those over 40 preferred the traditional marriage.

The survey also detected sharp differences of opinion as to whether working women make better mothers than women who do not work. Of women who work, 43 percent said working women make better mothers, whereas only 14 percent of women who do not work shared this opinion.

The survey was conducted between October 23 and 25, 1977, in telephone interviews with 1603 adult Americans from all parts of the nation and representing different races, religions, ages, and occupations.

11.8

Studies Show Men Better at Quitting Cigarettes

Men, the evidence shows, are better at giving up cigarettes. This was one of the conclusions discussed at a recent International Conference on Smoking Cessation.

Referring to her review of 33 different studies spanning a 19-year period, Dr. Ellen Fritz said that at first she was not convinced that men are better quitters. Now she is convinced that the percentage of men who succeed in giving up smoking is definitely greater than for women.

Most of the discussion at the conference centered on various possible explanations for this difference. The most widely held theory is that women use smoking more as a "coping mechanism" than men, and using smoking this way makes it more difficult to give it up.

11.9

Biorhythm Is Out of Synch

According to the "science of biorhythm," everyone's life moves in cycles of 23, 28, and 33 days that govern physical, emotional, and intellectual performance. The phases include critical and favorable periods in which things can be expected to go poorly or well. No matter that scientific tests have failed to prove the theory, athletes, horse breeders, gamblers and many among the health-concerned populace at large take it seriously.

The latest test of biorhythms, reported in the *Archives of General Psychiatry*, is the most rigorous yet. Scientists at Johns Hopkins University and other institutions in Maryland studied 205 automobile accidents for which the drivers had been judged legally responsible. The researchers figured out at what point in each driver's "cycles" the accident had occurred, with the idea that if biorhythm theory was right, more crashes would have occurred in critical periods than at other times. It turned out that the frequency of accidents in "down" periods was no greater than could have been expected by chance.

11.10

What Men Are Really Like

The typical American man believes that a faithful marriage is the ideal relationship. He wants sincerity, affection, and companionship in women above everything else. That is the conclusion of the recently published book *Beyond the Male Myth*.

The book is based on the results of a 40-item multiple-choice questionnaire designed to answer the questions women most wanted to know about men. The questionnaire was distributed by a public-opinion polling organization to men in communities around the country. The men, who were approached in such places as shopping malls and office buildings, were asked to fill out the questionnaire and to deposit it into a sealed box. Four thousand sixty-six men responded.

More than half of the men surveyed regarded a faithful marriage as their ideal. And almost three-quarters of the men said that, besides love, their primary reason for getting married is the desire for companionship and a home life.

11.11

Union Hotel Not Source of Legionnaires' Disease

Three more cases of Legionnaires' disease traced to the hotel in the University Student Union have been confirmed, but there still is nothing to suggest that the union is the source of the outbreak, state health officials said yesterday. Of 17 confirmed cases of the disease linked with the university, all but one person were guests at the campus hotel in the union building.

"Any association of Legionnaires' disease with the union does not constitute a cause-and-effect relationship," the officials said.

The state health board and the Center for Disease Control in Atlanta have determined that the risk of exposure to Legionnaires' disease is "not significantly greater" than at other, similar facilities in the area.

The investigation also showed that during the first week in May, which was chosen at random, the ratio of ill to well guests at the union was not significantly greater than at other hotels in the area.

11.12

Tying Abortion to the Death Penalty

Do women who have had an abortion have less regard for human life than those who haven't? Psychologist Paul Cameron thinks the answer may be yes, based on a survey he conducted on attitudes toward capital punishment earlier this year.

Shortly after a highly publicized execution in Utah, researchers asked more than 300 residents of Pasadena, California, whether they were totally against the death penalty, favored it for heinous crimes, or thought it should be used more frequently.

Overall, about 75 percent of the respondents in this conservative community favored capital punishment in some form. About half as many men as women were opposed to the death penalty under any circumstances.

Among women who had undergone an abortion, only 6 percent were unqualifiedly against the death penalty; 11 percent thought it should be used more frequently; and the rest favored it for some crimes. Twenty-six percent of the women who had not had an abortion were against execution under any circumstances; 20 percent favored more frequent use; and the rest wanted it for heinous crimes.

Cameron suggests that two factors combine to produce this difference: Women who regard human life less highly are more apt to have abortions and endorse capital punishment; and the abortion itself led them to depreciate the value of life in general as a way of vindicating the abortion to themselves.

12

Testing Causal Hypotheses

Does saccharin cause bladder cancer? Does smoking cause heart attacks? Does "the pill" cause fatal blood clots? In this chapter we shall use these three questions to illustrate the three basic methods used to justify simple causal hypotheses. The first method provides the best justification, and we shall use it as the standard by which to judge the shortcomings of the other two methods. The reason for taking the other methods seriously is that there are many questions that, for scientific, practical, and moral reasons, cannot be investigated using the "best" method.

Immediately following the discussion of each method there are several exercises that require you to analyze reports of studies of that type. These are to help you latch on to the essential features of that method. The exercises at the end of the chapter include reports based on all three types of study. Part of each of these exercises is recognizing which method of investigation has been used in that particular study. Popular reports don't usually come right out and tell you exactly what type of study was involved. But it is important to learn to recognize the differences because the amount of justification provided by these three methods is quite different.

12.1 SACCHARIN AND CANCER

When the U.S. Food and Drug Administration first announced its intention to ban the use of saccharin in food and soft drinks, their main scientific evidence came from a Canadian study of the effects of saccharin on rats. The U.S. officials did not have to justify the conclusion that it was harmful to humans because it is written into U.S. law that substances known to produce cancer in humans *or animals* cannot be added to any food products sold to the public. Let us postpone the problem of drawing inferences from rodents to humans and concentrate on the justification of the claim that saccharin causes cancer *in rats*. At the time, the Food and Drug Administration regarded this hypothesis as being proved beyond reasonable doubt. Why?

In the Canadian study, which took several years, 100 carefully bred laboratory rats, both males and females, were maintained on a diet that was 7 percent saccharin. A hundred offspring of these rats were continued on the same diet. Meanwhile, a second group of 100 similar rats were kept in the same place on the same diet—minus the saccharin.

In the first generation, 3 of the rats on saccharin and none of the others developed cancer of the bladder. In the second generation, 14 of the saccharin-fed rats developed bladder cancer. There were only two cancers in the other group.

You will notice that the second-generation saccharin-fed rats were exposed to saccharin literally from conception until death. This is another aspect of the study that we shall take up later. For the moment our question is simply whether this procedure justifies the conclusion that exposure to saccharin causes cancer. Or, to put the question more precisely, does this experiment justify the hypothesis that saccharin is a positive causal factor for the occurrence of bladder cancer in rats? If so, why?

12.2   RANDOMIZED EXPERIMENTAL DESIGNS

The Canadian study is a perfect example of a randomized experimental design. We shall first see why such designs produce good tests of causal hypotheses, and then see how the causal hypothesis is justified. Then we shall come back to some of the broader issues concerning the use of animal studies to study human diseases.

Causal Hypotheses and Randomized Design

According to our account of simple causal hypotheses (Section 9.5), the hypothesis at issue is:

> Exposure to saccharin is a positive causal factor for the development of bladder cancer in laboratory rats.

This means, on our analysis, that there should be more cases of bladder cancer if all rats are exposed to saccharin than if none are. Figure 12.1 shows the populations, both real and hypothetical, that express this hypothesis. Note that in Figure 12.1 the percentages of the effect in each population are written as $\Pr(E)$—rather than simply as $\%E$, as in Figure 9.8. This is because we now know that percentages in populations can be regarded as simple statements of probabilities. Otherwise the two figures are the same. With this one change, our causal hypothesis may be expressed by saying that $\Pr_X(E)$ *is greater than* $\Pr_K(E)$.

As we already know, it is rarely feasible to examine the whole of any population. We generally settle for a sample and then infer from the sample

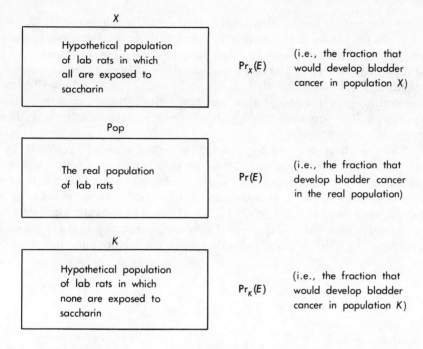

FIGURE 12.1

back to the population. But it is obviously impossible even to sample a hypothetical population, let alone two hypothetical populations with incompatible characteristics. So how are we to test the causal hypothesis?

The answer is that we *create* samples that play the role of samples from the two hypothetical populations. We do this by taking a sample from the *real* population and then dividing it into two parts. All the individuals in one part are experimentally manipulated so that they have the condition that defines the hypothetical population we have called X (for "experimental"). The individuals in the other part are manipulated so that they have the condition that defines the hypothetical population that we call K (for "kontrol"). These two groups, then, are just *as if* they had been sampled from the two hypothetical populations. These two groups are called the experimental group (x group) and the control group (k group) respectively.

In the saccharin experiment, the researchers began with 200 lab rats, which, given the careful conditions under which rats are bred, may be regarded as a random sample of all such rats. Next they divided this sample, randomly, into two samples of 100 each. The rats in the one sample (the x group) were fed the 7 percent saccharin diet and the rats in the other sample (the k group) received the same diet minus the saccharin. For the moment let us ignore the fact that there were actually two generations of rats in each group. We shall just take the rats in the second generation as those studied.

In the end, of course, there was a certain relative number of rats in each group that developed bladder cancers. We shall designate these two frequencies $f_x(E)$ and $f_k(E)$ respectively. The "logical skeleton" of the experiment is pictured in Figure 12.2. The two hypothetical populations are represented by broken lines indicating that they do not really exist. In general our diagrams representing experimental studies will not include these two hypothetical populations. The first time through, however, it is convenient to have them handy for easy reference.

Now we have reduced the problem of testing a causal hypothesis to a problem that has exactly the same structure as the problem of testing a simple correlation. We have, in effect, a sample from each of two different populations. The null hypothesis, H_0, is that the incidence of bladder cancer is the *same* in both populations; that is, $\Pr_X(E) = \Pr_K(E)$. The null hypothesis, together with the statement that 100 individuals (IC) were sampled randomly (AA) from each population, logically implies a distribution of probabilities for the possible

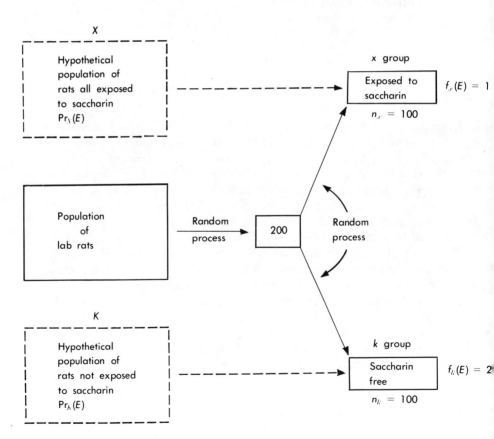

FIGURE 12.2

differences, *d*, between $f_x(E)$ and $f_k(E)$. The mean value of this distribution is, of course, zero.

The data from a randomized experimental test can be pictured just like the data from a test of a correlation. The data for the Canadian study is shown in Figure 12.3.

The set of all values of *d* within two standard deviations of the mean has a probability of 95 percent if H_0 is true. These differences are thus, by definition, not statistically significant at the .05 level. Differences greater than two standard deviations are statistically significant at the .05 level.

We have then satisfied the first condition for a good test of the hypothesis H_0, that saccharin is *causally irrelevant* to bladder cancer in the population of laboratory rats. That is, the conditional statement,

If [H_0 and *IC* and *AA*], then with Pr = 95 percent, there will be no SSD(.05)

is true.

Similarly, there must be some minimum value of the *effectiveness*, Ef(*C*, *E*) = $Pr_X(E) - Pr_K(E)$, such that if saccharin were that effective, the probability would be 95 percent that the observed value of *d* would be statistically significant. In principle we should consider both positive and negative values of the effectiveness of saccharin as a causal agent. It is in principle possible that saccharin should be a *negative* causal factor for bladder cancer in laboratory rats. That is, it is possible that saccharin acts to prevent bladder cancer and thus lowers the incidence of such cancers in the population exposed to it. So we should consider the minimum effectiveness, in both positive and negative directions, that would have a 95 percent chance of producing an SSD(.05). These two values of Ef(*C*, *E*) determine the hypothesis ΔH_0, which asserts that the effectiveness of the causal factor is below what would be detected with 95 percent reliability. These relationships are all shown in Figure 12.4. (These relationships are completely analogous to those shown in Figures 11.4, 11.8, and 11.9.)

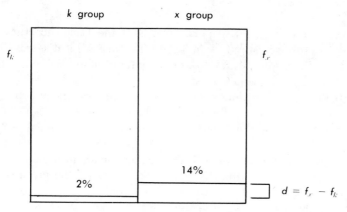

FIGURE 12.3

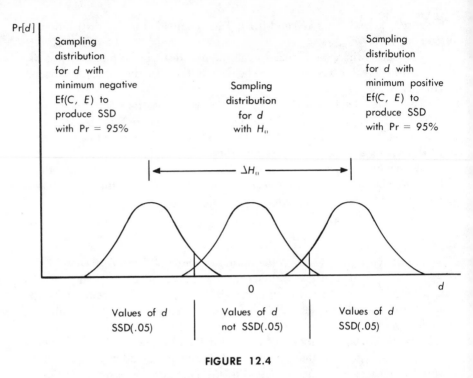

Pr[d]

Sampling
distribution
for *d* with
minimum negative
Ef(C, E) to
produce SSD
with Pr = 95%

Sampling
distribution
for *d*
with *H*₀

Sampling
distribution
for *d* with
minimum positive
Ef(C, E) to
produce SSD
with Pr = 95%

◄─────── Δ*H*₀ ───────►

0   *d*

Values of *d*
SSD(.05)

Values of *d*
not SSD(.05)

Values of *d*
SSD(.05)

FIGURE 12.4

The second condition for a good test, then, is satisfied. It is, as usual;

If [Not ΔH_0 and *IC* and *AA*], then with Pr = 95 percent, there will be an SSD(05).

Now all we need to do is determine whether the observed value of *d* was statistically significant or not, and we can justify either *H* or ΔH_0.

Justifying a Causal Hypothesis

Even in the best newspapers, reports of the Canadian study did not contain the words "statistically significant difference." But it seems likely that the Food and Drug Administration would not let itself in for so much trouble if that difference had not been at least SSD(.05). (*Note:* You may remember from Figure 10.13 and our previous discussions that, for a sample of 100, 2 SD was about 10 percentage points—for example, the difference between .60 and .50. The actual difference in the Canadian experiment was 14 percent − 2 percent = 12 percent).

So there was an SSD(.05). This means that you are justified in concluding that H_0 is false, and thus that *H* is true. Saccharin (in the given doses) is a positive causal factor for bladder cancer in laboratory rats. At this point there is no need actually to go through the argument, so long as you know what it is. It is repeated here for future reference only.

If [H_0 and *IC* and *AA*], then, with Pr = 95 percent, there will be no SSD(.05).
There was an SSD(.05) and *IC* and *AA*.
Thus (inductively), *H* (i.e., Not H_0).

Justifying a causal hypothesis is therefore not as difficult as it is often alleged to be. You just need the right kind of an experiment and an SSD.

Failing To Justify a Causal Hypothesis

As a bonus, the Canadian study provides us with a convenient example of the failure to justify a causal hypothesis. You will remember that the experiment involved two generations of rats. The SSD was observed in the second generation, that is, in those rats which had been exposed to saccharin from conception. The first generation, which was exposed only from birth, yielded only three cancers (in 100 rats) compared to none in the control group. This 3 percent difference was not SSD(.05). Thus, if the experiment had only gone that far, the correct conclusion would have been ΔH_0. That is, saccharin is not a sufficiently effective causal factor for bladder cancer to yield an SSD (.05) 95 percent of the time in a sample of 100. That is literally what ΔH_0 means in this experiment. The argument, again just for future reference, is

If [Not ΔH_0 and *IC* and *AA*], then, with Pr = 95 percent, there will be an SSD(.05).
There was no SSD(.05) and *IC* and *AA*.
Thus (inductively), ΔH_0.

A somewhat more informative conclusion could be obtained by using the observed difference in the frequency of the effect to give an *interval estimate* of the difference in the populations, (i.e., the effectiveness). But, as you should know just on general principles, if there is no SSD(.05), the 95 percent confidence interval is going to include an effectiveness of zero. So the causal hypothesis remains unproved.

As a matter of fact, based on previous research, the Canadian researchers did not really expect to get an SSD in the first generation. But if they had, you can bet that they would have said so, and fast.

Effectiveness and Statistical Significance

In Chapter 11 we learned that a "statistically significant" result may or may not indicate something "important" about the population sampled. The same lesson applies here. A *statistically significant difference* between an experimental and control group may or may not indicate a high degree of *effectiveness* of the causal factor in question. Effectiveness, remember, is defined in terms of the relative number of members of the *population* that would or would not develop the effect if they were or were not subjected to the cause.

TABLE 12.1

Sample Size	SSD	Effectiveness
Small	Yes	High
Large	No	Low
Small	No	Can't tell
Large	Yes	Can't tell

The link between effectiveness in the *population* and a statistically significant difference between two *samples* is provided by the *size of the sample*. Before reading further, you should try to figure out for yourself when the sample size and the existence (or nonexistence) of an SSD indicate a high (or low) degree of effectiveness. Figures 12.4 and 11.9, together with the discussion surrounding Figure 11.9, should provide all the information you need.

As you should now have figured out for yourself, an *SSD* with a *small* sample indicates a *high* degree of effectiveness of the causal factor. *No SSD* with *large* sample indicates a *low* degree of effectiveness. If the sample is small and there is no SSD, or if the sample is large and there is an SSD, you cannot tell whether the cause is very effective or not. It could be either. These relationships are summarized in Table 12.1. If you did not first figure out these relationships for yourself, you should now go back and make sure that you understand why they hold. The way scientific tests are standardly reported in the popular press, you will often be forced to make a judgment of effectiveness knowing at most only the sample size and whether or not there was an SSD. Knowing when such a judgment is possible and what it should be is thus a matter of considerable practical importance.

Ideally, of course, you would like to be provided with an estimate of the effectiveness of a causal factor. Even just the actual sample frequencies would be helpful. You could then make your own rough estimate of the effectiveness based on these frequencies. Lacking such information, you are left with the rules of thumb summarized in Table 12.1.

Of Rats and Humans

In all the furor over the proposed ban on saccharin, no responsible critics questioned that the Canadian study proved that saccharin caused cancer in laboratory rats. But lots of people, particularly representatives of companies that manufacture saccharin and diet soft drinks, questioned the relevance of the study to humans. And on two counts. First, humans are different from rats. Or, to put it in our language, the population studied was not the human population. Second, the amount of saccharin used (7% of the rats' diets) was quite large. It was easily calculated that 7 percent of a human diet corresponds to the amount of saccharin in 800 bottles of diet soda. Who drinks 800 bottles of diet soda a day? There is something to these criticisms, but not nearly so much as many people thought.

Most environmental hazards are not highly effective. If they were, lots more of us would be dead—or never have been born. The typical cancer producing agent strikes only one person in thousands. But that is a lot. In a population of 200 million people, one in a thousand is 200,000 people. Even just one in 50 thousand is 4,000 people.

One would not expect saccharin, or most substances, in normal doses, to cause more than one cancer in a thousand cases. But to detect such a low degree of effectiveness using normal doses would require an experiment using many thousands of rats. To breed and maintain so many experimental animals over a several-year period is almost impossible. So, instead, typical animal studies use many fewer animals and larger doses.

The assumption behind this strategy is that the number of cases of the effect is roughly proportional to the dose. Double the dose and you roughly double the number of cancers, or whatever. So knowing what large doses do, one just "extrapolates" back to what "normal" doses would do. Now the assumption that the effectiveness is proportional to the dose can be legitimately questioned. In particular, there could be a threshold for the effect. That is, it might be that below a certain dose there is no effect at all in anyone. The number increases with increasing dose only after the threshold is reached.

The problem is that almost no biological agent has been shown, in fact, to have a threshold. Saccharin, of course, might be an exception, but there is no positive justification for believing that it is. Nor, as we shall see in Part IV, is it a wise strategy to base decisions on the assumption that it is an exception when there is no specific knowledge one way or the other.

It is often said by people who know nothing about medical research that any substance will cause cancer if given in big enough doses. That simply is not true. Many chemical substances have been tested on animals using large doses, and only a relatively few have been found to cause cancer. It is true that large doses of anything usually have some effect, but the effect is by and large not cancer.

As for the statement that humans are not rats, that is obviously true. But of the roughly 30 agents known definitely to cause cancer in humans, all of them cause cancer in laboratory rats—in high doses. From this fact it does not follow, with logical necessity, that anything causing cancer in rats will also do so in humans. But again, it is difficult to justify basing practical decisions on the assumption that saccharin is an exception. And taking account of differences in dose and body weight, those 14 cancers in 100 rats translate into 12,000 cases of bladder cancer in a population of 200 million people drinking less than one can of diet soda a day.

You can easily calculate that 12,000 cases of bladder cancer in a population of 200 million means that any individual is facing a chance of 6/100,000 of getting bladder cancer. So it may be argued that the risk is small and people ought to be allowed to decide for themselves whether or not to take this risk. That, however, is an entirely different sort of question than the "scientific" question of whether there is any risk, and if so, how much. The answer to the scientific question seems to be that there is some risk, with 6/100,000 being a very rough guess as to how great the risk might be.

PRACTICE EXERCISES

The following exercises concern experiments designed to test the hypothesis that taking vitamin C prevents colds. They are taken from Linus Pauling's famous book *Vitamin C and the Common Cold*. Answers to all questions in these exercises are given in the back of the book.

12.1   THE RITZEL STUDY

The following report is adapted from Pauling's book:

A Swiss investigator, Dr. G. Ritzel, reported the results of a double-blind study. He studied 279 skiers, half of whom received 1 gm of ascorbic acid [vitamin C] per day and the other half an identical inert placebo. He reported a reduction of 61 percent in the number of days of illness from upper respiratory infections in the vitamin C group as compared with the placebo group. These results have high statistical significance (probability in two samples of a uniform population only 0.1 percent).

This experiment introduces a notion we have not seen before, namely, a "double-blind" experiment using a "placebo." Since this is a common technique in experiments using human subjects, you should know what it involves. To understand what "double-blind" means, let us consider first what it means to do a study "blind." By definition, a study is blind if the *subjects* themselves cannot tell whether they are in the experimental or control groups. Now in order for this to be so, the control subjects must be given something that they cannot detect as not being vitamin C. In this case they are simply given pills made of a known inactive substance that looks and tastes like vitamin C. Any such inactive substance that *is* given to control subjects as a substitute for the suspected cause is known as a "placebo."

The reason for going through all this trouble, particularly for human subjects, is to avoid a placebo effect. That is, it can happen that just knowing one is in the experimental group and believing that the suspected causal agent does work is enough to produce some effect. For example, patients suffering from headaches or other pain often report a lessening of the pain when given pills that they believe to be pain killers but which in fact contain no pain-relieving drugs at all. So one can produce an effect with a placebo. In the present example the worry is that people might be able psychologically to suppress a mild incipient cold simply because they believe they are taking something that prevents colds.

A study is double-blind if the experimenters are also kept in the dark about which subjects are in which group. The reason for this is that an experimenter may diagnose borderline cases in the direction that favors personal bias if it is known to which group a particular subject belongs. For example, there may be a question whether to diagnose a particular slight sniffle as a cold or not. If the experimenter wants the experiment to show that vitamin C does prevent colds and knows this particular subject is taking vitamin C, there may be an unconscious bias in favor of not calling it a cold. Such biases, even if completely unconscious, might produce an apparent difference when none really exists.

Now we can begin analyzing the experiment.

A. State the causal hypothesis at issue in the form: *C* is a positive/negative causal factor for *E*, in the population of _____. The population referred to in the causal hypothesis is the "target population"—that is, the population to which one wants to apply the hypothesis. The target population is usually more inclusive than the population actually sampled.

B. State the null hypothesis. Note the population actually sampled.

C. Draw a diagram of the experiment following the model of Figure 12.2. (You may omit the two hypothetical populations in your diagram.) You should fill in as much specific information as you are given, such as size of samples, treatment given the "experimental group," explicit mention of random sampling, frequencies of the effect, and so forth.

D. Diagram the data following the model of Figure 12.3. In this case you are not given the actual sample frequencies, but you can still indicate the *relative positions* of the percentages on your diagram.

E. On your diagram of the data indicate whether the difference in sample frequencies is reported as statistically significant. Note the significance level if given. (*Note:* The phrase "probability in two samples of a uniform population of only 0.1 percent" is just another way of saying SSD (.001).)

F. What conclusion is justified by the data? Sketch the "short form" of the argument that justifies this conclusion.

G. What, if anything, can you conclude about the *effectiveness* of the causal factor?

H. Summarize your evaluation by noting the main strengths and weaknesses of the study as it is reported. What is your overall assessment of how well the conclusion has been justified?

12.2 THE SALISBURY STUDY

The following report on a study done at the Common Cold Research Unit in Salisbury, England, is also adapted from Pauling's book.

These investigators reported observations on human volunteers and concluded that "there is no evidence that the administration of ascorbic acid has any value in the prevention or treatment of colds produced by five known viruses." Of the 91 human volunteers, 47 received 3 gm of ascorbic acid per day for 9 days and 44 received a placebo. They were all inoculated with cold viruses on the third day. In each of the two groups, 18 developed colds. The incidence of colds observed in the subjects receiving ascorbic acid (18/47) was 6 percent less than that in the control group (18/44). This difference is not statistically significant.

Answer questions A through H of Exercise 12.1 as they apply to the Salisbury study.

12.3 In discussing the Salisbury study, Pauling comments as follows:

> The number of subjects, 91 in the two groups, was not great enough to permit a statistically significant test of a difference as large as 30 percent in the incidence of colds in the two groups to be made, although a difference of 40 percent, if it had been observed, would have been reported as statistically significant (probability of observation in a uniform population equal to 5 percent).

A. State in your own words (including words you have learned in studying this text) what Pauling's complaint is all about.

B. What do you think about this issue? Be brief.

12.3 TOBACCO AND HEALTH

Ever since Columbus took tobacco from the New World back to the Old, people have been saying that it is an "evil weed" that causes all manner of illness and disease. Of course, others claimed it to be a blessing and cure for diverse ailments. It was not until the 1930s that there was any serious scientific evidence that smoking tobacco might cause lung cancer. And deliberate, well-planned studies of the question did not occur until the 1950s.

You can easily appreciate the difficulty of studying this question. Laboratory experiments of the type used to study the effects of saccharin (and many other drugs) are out of the question. It is impossible to subject enough animals (e.g., laboratory rats) to high enough doses of tobacco smoke to produce statistically significant differences. At normal doses you could not expect more than a few cases in a thousand over the lifetime of the subjects. But at above normal doses, rats and other experimental animals tend to die of smoke inhalation.

Cancer can, however, be produced in experimental animals by painting their skin with "tars" derived from cigarette smoke. Many experiments have shown that this process does cause cancers. And these tars are similar to those that collect on the lips, tongue, throat, and lungs of smokers. Still, that does not prove that *smoking* causes lung cancer, heart failure, or other diseases.

Direct experiments on humans are, of course, out of the question. One would have randomly to select several thousand nonsmokers, say age 12, randomly divide them into two groups, have everyone in the "experimental" group smoke two packs a day, and make sure that no one in the other group smokes. In 20 years there might be enough data to conclude either that smoking is a causal factor for lung cancer (i.e., H) or that its effectiveness, if any, is very low (i.e., ΔH_0).

But such experiments cannot be made on humans. It would be immoral, if not actually illegal. And it is not even very practical. Twenty years is too long to keep track of several thousand subjects. Many other things might happen to them that would ruin the experiment.

The studies undertaken in the 1950s, first in England and then in the United States, were not experimental studies. The first big U.S. study, sponsored by the American Cancer Society, began by selecting a sample of almost 200,000 men, both smokers and nonsmokers. All subjects were between the ages of 50 and 69. The reason for excluding younger men was that the effects being studied do not normally occur in great numbers until after age 50. And women were excluded because the number of women in the whole population that had already been smoking for 20 years was very small. And the incidence of lung cancer among women at that time was very low. Remember this was the early 1950s.

The subjects, gathered from 394 counties in nine different states, were personally interviewed by a team of 22,000 trained volunteers. They recorded the history of each subject's smoking habits plus other information that was thought might be relevant. All subjects were in good health at the beginning of the study.

The subjects were "followed" for almost 4 years. During that period, nearly 12,000 deaths were reported, more than 2000 from some form of cancer. For each man who died, a death certificate listing the cause of death was obtained. Extra information was sought in cases where cancer was suspected.

As is the case with most studies of this type, the data are usually not reported in actual percentages, but in what are called "mortality ratios." Before looking at the data you should learn what these ratios are and what they mean.

Mortality Ratios

A *mortality ratio* is a ratio of *death rates* from two different samples or populations. A death rate in any group is simply the number of people who died divided by the original total number in the group. (You should remember this from the discussion of population growth in Chapter 7.) So when you see a mortality ratio, you must first determine what two groups are being compared. Also, the deaths may be further classified by specific cause, for example, deaths from lung cancer.

Let us study one example. In the American Cancer Society study, the mortality ratio, including deaths from all causes, for smokers as opposed to nonsmokers, was roughly 2.5. This may be spelled out in more detail as follows:

$$MR = \frac{\text{(number of smokers who died)/(number of smokers in study)}}{\text{(number of nonsmokers who died)/(number of nonsmokers in study)}}$$

$$MR = \frac{\text{death rate of smokers}}{\text{death rate of nonsmokers}} = 2.5$$

What this means, in plain language, is that the percentage of smokers who died from any cause was 2½ *times* the percentage of nonsmokers who died. The smokers in the study were dying at a rate that was 2½ times the rate for nonsmokers.

Mortality ratios are fine for comparing two groups, but they have one drawback. They do not permit you to make any estimate of the *effectiveness* of the suspected causal agent. For example, in this study the mortality ratio for deaths due to lung cancer was about 7. That is, seven times as many smokers died of lung cancer. That was without doubt a statistically significant difference, and, as we shall see, provides good justification for the hypothesis that smoking is indeed a positive causal factor for the incidence of lung cancer. But an MR of 7 can represent very different degrees of effectiveness. If only .1 percent of the nonsmokers died of lung cancer, an MR of 7 means that .7 percent of the smokers died of lung cancer. If these values held for the populations as well, the effectiveness would be only .006 (.007 − .001). But if 2 percent of the nonsmokers died of lung cancer, an MR of 7 means that 14 percent of the smokers died of lung cancer, which would be an effectiveness of .12 (.14 − .02), which is 20 times the previous value. Knowing only that the MR is 7, you cannot tell whether the effectiveness is small or large. This will be very important when we consider the role of causal hypotheses in making decisions. For making decisions, knowing the effectiveness is much more useful than knowing the mortality ratio.

The Basic Data

On the question of lung cancer in particular, the data seem quite conclusive. The mortality ratios for smokers classified by the amount they smoked, in packs per day are shown in Figure 12.5.

Not only is the MR greater for smokers, it increases along with increased consumption of cigarettes. In the study, those who smoked over two packs a day died from lung cancer at a rate over 12 times the rate for nonsmokers. Note that the MR for nonsmokers is exactly one. This is not a matter of the data, but follows logically from the definition of the mortality ratios for smokers as compared with nonsmokers. If this is not obvious to you, go back and look again at the definition of mortality ratios. Nonsmokers are by definition people who smoke no cigarettes per day.

Much of the discussion of the health effects of tobacco focuses on lung cancer simply because the connection between smoking and the lungs is obvious. But the ACS study, like all later studies, showed increased mortality from a number of other diseases. The mortality ratios for various diseases, including lung cancer, are shown in Figure 12.6.

The mortality ratio for death due to heart attacks is particularly interesting. An MR of 2 is relatively low compared, for example, with the MR of 7 for lung cancer. But nonsmokers have a much higher death rate due to heart

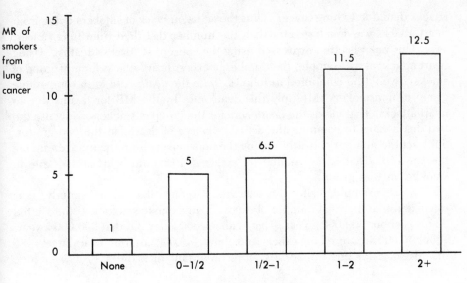

FIGURE 12.5

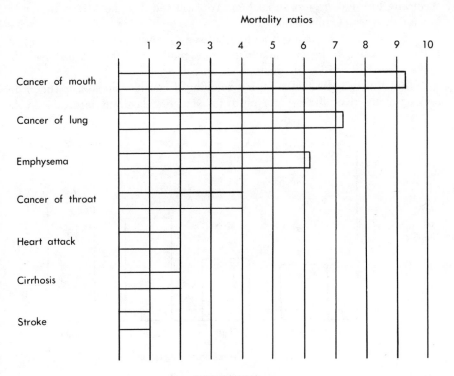

FIGURE 12.6

attacks than due to lung cancer. Thus the actual *number* of smokers dying from heart attacks was much greater than the number that died from lung cancer.

This point is often expressed using the concept of "excess deaths." Using heart attacks as an example, the number of excess heart attacks for any group of smokers would be computed as follows. Take the *death rate* for a comparable group of nonsmokers. Multiply this death rate by the MR for deaths due to heart attacks. That yields the *death rate* for the group of smokers. Now use the two death rates to compute the actual *numbers* of deaths in the two groups. Subtract the number of deaths among the nonsmokers from the number among the smokers. That is the number of *excess deaths* from heart attacks due to smoking in that group.

As an example, using the data from studies like the ACS study, it is calculated that the total number of excess deaths due to smoking in the United States is about 300,000 a year. That is about 800 a day! Of that 300,000 a year, about 40,000 are from lung cancer and about 140,000 are from heart attacks. So the MR of 2 for heart attacks is more important than the MR of 7 for lung cancer.

The impact of smoking on health depends not only on the amount smoked, but also on how one smokes, and on how early one begins to smoke. Figure 12.7 shows the mortality ratio due to all causes for smokers classified according to the degree of inhalation. You will see that the MR is higher for greater degrees of inhalation.

Figure 12.8 shows a connection between mortality ratios and the age at which the subjects *began* smoking. Those that began the earliest had the highest mortality ratios for death from all causes.

Finally, the effect of stopping smoking if one has smoked is shown by looking at the overall mortality ratios for smokers classified according to the

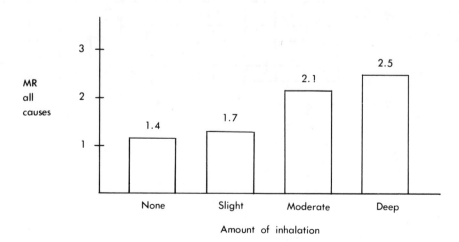

FIGURE 12.7

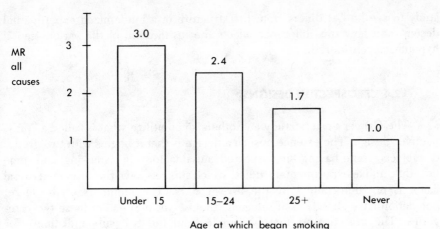

Age at which began smoking

FIGURE 12.8

number of years since they had stopped smoking. The data are given in Figure 12.9.

For the people in this study at any rate, those who stopped smoking, and lived another 10 years, were little different from nonsmokers. The trick is to quit early enough. Note that those who had quit just in the year prior to beginning the study had a *higher* MR than those who never quit. Can you guess why this should be so? (*Hint:* Many people give up smoking only because their doctors insist that they do.)

Now that we have the basic data let us look at the logical structure of this

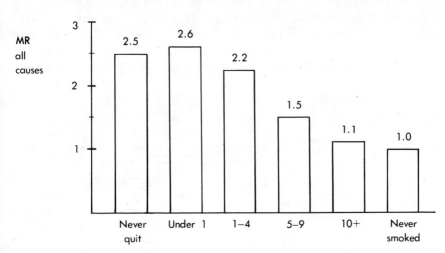

Years since quitting smoking

FIGURE 12.9

study to see how it differs from the structure of a randomized experimental design, and how this difference affects the justification of the various causal hypotheses considered.

12.4   PROSPECTIVE DESIGNS

The Cancer Society study and others since utilize what is called a "prospective" design. The essence of such a design is that it begins with two groups of subjects, some having the suspected causal factor (e.g., smoking) and some not. But, unlike experimental studies, which subjects have the suspected causal factor is not determined by the investigators. The members of the target population have already "self-selected" themselves into one of these two categories. The investigators merely sample from the two subpopulations—for example, smokers and nonsmokers—as they already exist. The "prospective" feature is that the occurrence of the effect, in both groups, is in the future. At the beginning, all subjects in both groups do not yet exhibit the effect under investigation. Thus, people who already had lung cancer were not included as subjects in the Cancer Society study. The strategy is to get two groups that are, on the average, alike in every feature except the suspected causal factor. If there is then a statistically significant difference in the frequency of the effect, that provides the justification for the causal hypothesis. This type of design is pictured schematically in Figure 12.10. The sample of people having the suspected cause is still called the experimental group even though there is no actual experimentation involved. The investigators do nothing to the subjects. They merely collect information about them and see what happens.

Comparing Figure 12.10 with Figure 12.2 makes clear why prospective studies do not provide as "direct" a test of a causal hypothesis as do randomized experimental studies. In a randomized experimental study the two groups can be regarded as samples from the hypothetical populations that define the causal

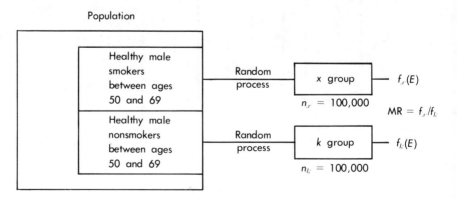

FIGURE 12.10

hypothesis. These samples are created through the process of random selection and direct experimental manipulation. Prospective studies, on the other hand, are more like tests of correlations in that they are based on samples from the real population as it exists. But prospective studies differ from tests of correlations in that you are not just looking at coexisting properties. The two groups are selected from among members of the population who do not yet have the effect. The effect shows up later.

In reading reports based on statistical studies, you must sometimes read very carefully to discover whether the data are based on a one-shot survey or on a genuine prospective study in which subjects initially without the effect are followed through time to see how many in each group later develop the effect. In a one-shot survey, you can ask people whether they had some condition earlier. For example, you can ask people if they smoked marijuana *before* they tried heroin. But the reliability of the information you get about the prior condition is likely to be strongly influenced by whether the "effect" occurred or not. A heroin addict is likely to have a very different perception of earlier involvement with marijuana than a nonaddict, even though they may both have had very similar involvements. Similarly, if you ask someone with lung cancer about their previous smoking habits, you are likely to get an exaggerated report, though the exaggeration might go either way. The subject might feel, "Why me? I didn't smoke that much," or he might feel, "That's what did it. I smoked too much." Either way, one is not getting reliable information. That is the value of *genuine* prospective studies.

Conclusions from Prospective Studies

Let's continue with the example of lung cancer, though we could say roughly the same things about many other diseases. The null hypothesis, then, is that smoking tobacco is causally irrelevant to developing lung cancer.

To derive a probability distribution over possible values of the mortality ratio we need an auxiliary assumption that we have not needed before. This is that the process by which people "self-select" themselves into the category of smoker or nonsmoker is a random process or at least that it not be biased in ways that have any connection with getting lung cancer.

For example, suppose that men are more likely to take up smoking the more insecure they feel in social surroundings. They take up smoking because it makes them feel socially more at ease. This could mean that the subpopulation of smokers has a higher percentage of less socially secure people than the general population. So the process that selects smokers is biased in this way. This need not be a worry, however, unless for some reason the lung cancer rate in this group is higher than it is for other categories of nonsmokers. If the subpopulation of less-secure nonsmokers has a higher incidence of lung cancer than other nonsmoker groups, and if these people tend to become smokers more readily than others, that bias, by itself, could raise the lung cancer MR for smokers even if the null hypothesis were true.

If we explicitly add the auxiliary assumption that the self-selection process is not biased in any way that is relevant to getting lung cancer, then we can proceed in the usual manner. The null hypothesis, together with the sample sizes and auxiliary assumptions, logically implies that the distribution of probabilities for possible values of the mortality ratio has an expected value of 1, with a standard deviation determined by the size of the two groups. Those values of the MR within 2 SD of the mean will not be SSD (.05). (Do you understand that?) And so on.

Thus if we get a statistically significant MR we conclude by the standard form of argument that smoking is a causal factor for getting lung cancer. Indeed, assuming the MR is greater than 1, we conclude smoking is a *positive* causal factor. If we should not get a statistically significant MR, we would conclude ΔH_0,—that is, that any effectiveness smoking may have for producing lung cancer is small, and indeed very small given the large size of the samples involved.

As it is, the MR for lung cancer is clearly statistically significant, so we conclude that smoking is indeed a causal factor for getting lung cancer.

Tobacco and Air Pollution

If one wants to question the justification for concluding that smoking causes lung cancer, the best way is to question the auxiliary assumption that the self-selection process is not biased in any relevant way. This assumption appears explicitly in the second premise of the standard form of argument that is used to justify a simple causal hypothesis. If it is not justified, then neither is the conclusion. You will note, by the way, that the need for this assumption arises just because the design of the study was prospective rather than randomized experimental. In a randomized experimental study, subjects are assigned to the two groups by an explicitly random process, such as flipping a fair coin.

The possibility of a causal connection between air pollution and lung cancer provides a clear example of one relevant way in which the self-selection process could be biased. In 1900, smoking was not common and lung cancer was a rare disease. Today about half of all American men and a third of all American women smoke cigarettes, and lung cancer kills about 40,000 people per year—roughly the number that die each year in automobile accidents. But in 1900, the majority of the population lived on farms or in small towns where there was little air pollution. Today the majority of the population lives in towns and large cities where there is considerable air pollution, much of it from automobiles. This suggests that maybe the increasing incidence of lung cancer since 1900 is due to air pollution and not to smoking. It is known that the percentage of smokers is greater in cities than in rural areas; that is, there is a positive correlation between being a smoker and living in an urban environment. Thus the smokers in a random sample of Americans would tend to be city dwellers, whereas the nonsmokers would tend to be country people. But it is the city dwellers that are subjected to the most air pollution. The prospective study shows a higher lung cancer rate for smokers, but that could be just the result of the bias introduced by the fact that a greater percentage of city dwellers are smokers.

The way to counter this suggestion is to *control* for living environment. The easiest way to do this is to confine one's attention only to subjects in one environment or the other. If the mortality ratio for lung cancer is still statistically significant, then smoking is a positive causal factor independently of what air pollution might do. That is, if all subjects in both the x group and k group are from the *same* living environment, any difference in the lung cancer death rates between these two groups must be due to something else (i.e., smoking).

Table 12.2 shows data not from the ACS study, but from a later prospective study. The data are given not in the form of mortality ratios but in death rates per 100,000 man-years—that is, the number of men studied multiplied by the number of years they were studied. In this study, "urban" means a city of 50,000 or more. "Rural" means farm or very small town.

Figure 12.11 shows the data from Table 12.2, diagramed so as to control for living environment. These differences are all statistically significant. What they show is that smoking is a positive causal factor for lung cancer in either a rural or urban environment. The suggestion that our original conclusion was not justified because of the correlation between urban environments and smoking is refuted.

The data given in Table 12.2 make it possible to test *another* causal hypothesis, namely, that air pollution *itself* is a causal factor for lung cancer. In making this test we would of course control for smoking. The best comparison is that which completely eliminates smoking from consideration—that is, rural nonsmokers compared with urban nonsmokers. The next best comparison is rural smokers with urban smokers, since this at least keeps smoking constant in both groups. These data are diagramed in Figure 12.12. Since these differences are also statistically significant, we can conclude that air pollution is *also* a

TABLE 12.2

DEATH RATES PER 100,000 MAN-YEARS FOR SMOKERS AND NONSMOKERS IN URBAN AND RURAL ENVIRONMENTS

Category	*Death Rates*
Urban smoker	85
Rural smoker	65
Urban nonsmoker	15
Rural nonsmoker	<1

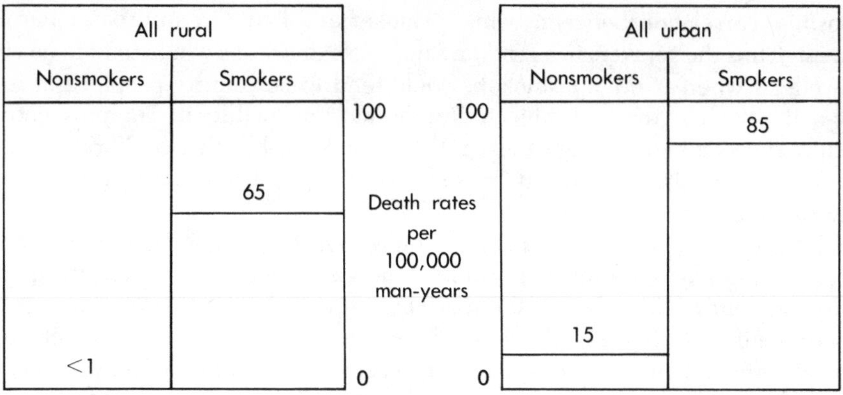

FIGURE 12.11

positive causal factor for lung cancer. That is, both smoking and air pollution are causal factors that produce lung cancer in the overall population.

There is one more thing to be learned from the data in Table 12.2. Since the samples involved are fairly large, we can take the observed frequencies as rough estimates of the effectiveness of the two causal factors in the corresponding populations. What we see is that smoking, with a rate of around 65, is a much more effective cause of lung cancer than air pollution, with a rate of only about 15. Moreover, the lung cancer death rate from the two factors combined (85), is *higher* than the sum of the two rates taken separately (65 + 15 = 80). Acting together they are a more effective causal factor than both acting separately. The two causal factors apparently interact.

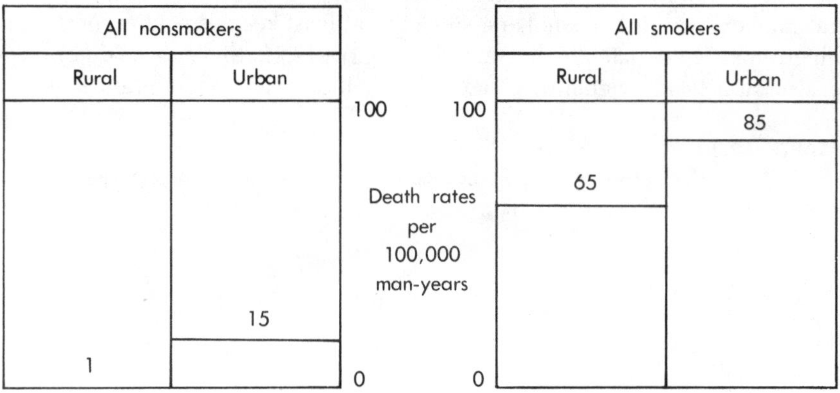

FIGURE 12.12

Controlling for Other Variables

We have just shown that smoking is a causal factor for lung cancer even if air pollution is taken into account. That is, any correlations that exist between living environments and lung cancer have not sufficiently biased the self-selection process to render our conclusion unjustified. We eliminated this possibility by looking at the data when *controlled* for the variable of living environment.

The technique of controlling for other variables is completely general. It can be applied to any variable so long as there are enough subjects in the relevant categories to leave us with a sufficiently sensitive test of the null hypothesis.

In designing a prospective study it is desirable to build controls for other variables into the study right from the beginning. That way one can then make sure that there are enough subjects in each of the resulting categories. In any case, when you read a report of a prospective study, you should note whether any *other* variables have been controlled, and what they are. You might also try to think of other variables that might be operating in the self-selection process in a way that could seriously bias the result. If you can think of some obvious other variables, you should be reluctant to regard the causal hypothesis as being very well justified.

In well-designed studies, other variables are controlled not by considering only subjects who do not have the suspected other factor. Rather, the original control group is carefully selected so as to *match* the experimental group for other variables thought to be relevant. This means that the control group is not a random sample of subjects without the suspected cause. It is better than a random sample in that the subjects are selected to be *more* like the experimental subjects than could be expected from a purely random sample. For example, if one wanted to control for urban versus rural environments right from the start, the investigators would have made sure that the percentage of urban subjects in the control group was the same as that in the experimental group. This eliminates this aspect of environment as a source of any observed statistically significant difference in the death rates. If air pollution is related to lung cancer, which turns out to be true, it should, on the average, have the *same* impact on both the experimental and control groups.

There is still a problem about justifying the auxiliary assumption that the self-selection process was not biased in a strongly relevant way. In any study it is impossible to match for more than a fairly small number of other variables. This means that one cannot eliminate more than a relatively small number of suggested ways that the observed significant difference could have been produced by biased self-selection rather than by the proposed cause. But in principle there are innumerable ways that the self-selection process might be biased. So no matter how many variables are matched, the *possibility* of biased self-selection still remains. Because of this possibility, no prospective study can ever provide as secure a justification for a causal hypothesis as a randomized experimental study. A small question mark always hangs over that auxiliary assumption.

But the question mark can be reduced to a mere possibility. In a later prospective study which involved almost a half-million subjects, a computer was used to match smokers and nonsmokers for a large number of other variables, including age, race, height, living environment, occupation, religion, education, marital state, use of alcohol, amount of sleep, amount of exercise, nervous tension, use of tranquilizers, current general health, and history of cancer, heart attacks, stroke, and high blood pressure. The computer found almost 40,000 "matched pairs," that is, pairs of men who had all these different features in common, except that one was a smoker and one a nonsmoker. The mortality ratios were about the same as for other studies. None of the mentioned factors was responsible for the observed statistically significant difference in death rates.

The data for this matched-pair study would be diagramed as shown in Figure 12.13. This is in general the best way to represent data from a prospective study. It helps to focus attention on the need to consider other variables.

If there is still some "other" factor involved, it is hard to find. Moreover, the studies on tobacco and health do not just show a greater lung cancer rate for smokers. They show differences that vary with amount smoked, degree of inhalation, age of beginning to smoke, etc. To argue that smoking is causally irrelevant to health, one would have to imagine another factor that somehow managed to produce all these differences which make it *look* like smoking is the crucial variable. The tobacco industry may continue to claim that smokers are "a different kind of people" who smoke because they are "different" and also get lung cancer, heart attacks, and so on, because they are "different." But unless they can come up with some *specific* difference that can at least explain the observed differences in mortality, their claim has only the force of a typical "*ad*

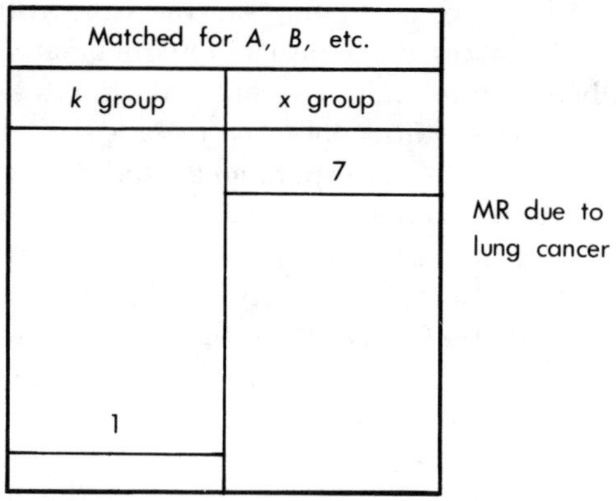

FIGURE 12.13

hoc rescue." (See Section 8.4.) They refuse to grant a needed auxiliary assumption but have no specific suggestions as to what might be wrong.

PRACTICE EXERCISES

12.4 SMOKING AND STROKES

Figure 12.6 shows mortality ratios for smokers from a number of diseases, including stroke. The data are from the Cancer Society study discussed in the text. Assuming that any MR greater than 1.5 is SS(.05), answer the following questions.

A. State the causal hypothesis at issue in standard form with reference to the population of interest—that is, the target population.

B. State the null hypothesis in standard form. Be sure to note the population actually sampled.

C. Following the model of Figure 12.10, diagram the experiment including as much relevant information as you can find in the text. Look for descriptions of the population, experimental and control groups, sample frequencies, and so on.

D. Diagram the data following the model of Figure 12.13. Indicate on your diagram whether the reported difference in sample frequencies (or other measure) is statistically significant.

E. What conclusion is justified by the data? Sketch the "short form" of the appropriate argument.

F. What, if anything, can you conclude about the effectiveness of the causal factor?

G. Summarize your evaluation by noting the main strengths and weaknesses of the study. What is your overall assessment of how well the conclusion has been justified?

12.5 SMOKING AND CIRRHOSIS OF THE LIVER

The Cancer Society study included data on death from cirrhosis of the liver. Use the data in Figure 12.6. Assume that any MR greater than 1.5 is SS(.05).

A. Answer questions A through G of Exercise 12.4 as they apply to the data on cirrhosis.

B. It is a fact about American society that there is a strong positive correlation between smoking cigarettes and drinking alcohol. It is also well established that heavy drinking is a positive causal factor for cirrhosis of the liver. Explain briefly the relevance of these facts for the conclusion you reached in your answers to Part A of this question. In particular, explain how they might be used to question the justification of the conclusion you reached.

C. The text mentions a study based on nearly 40,000 matched pairs of men: half smokers and half nonsmokers. Among the characteristics matched in this study was consumption of alcoholic beverages. The MR for cirrhosis in this study was a statistically significant 1.9. Diagram this data following the model of Figure 12.13. What should you conclude? Does it follow that drinking alcohol does not cause cirrhosis?

12.5 BLOOD CLOTS AND THE PILL

In the 1960s several doctors in Great Britain and the United States discovered a surprising number of fatal blood clots in young women whose medical histories made such a condition extremely unlikely. Even though they had only a handful of such cases, that was many more than they had experienced in the past in otherwise healthy young women. Looking at the medical histories of these few women, they discovered that most of the victims had begun taking oral contraceptives less than a year before the fatal clot appeared. Since the use of such contraceptives had only recently become widespread, the doctors immediately suspected that the pill was responsible for the sudden increase in fatal clots among young women.

The connection between the pill and these few cases of fatal clots could have been a coincidence. So the question arises how better to test the hypothesis that the pill really is a causal factor for fatal clots. Oral contraceptives had already been tested on animals, and this particular problem had not arisen. But then no one was looking for clotting in particular. One could begin new animal studies utilizing randomized experimental designs, but those would take several years to complete.

Similarly with prospective studies. Existing data included no information on blood clots. Given the relatively low incidence of fatal clots in the general population, it would take a year to set up a sufficiently extensive prospective study and 3 to 5 more years for the data to build up. Is there not some faster way to test the hypothesis?

The desire for a "faster" method of testing the hypothesis is medically understandable since, if it is true, you want to know as soon as possible in order to warn potential victims. But even if the effect in question had been less serious, there is still good reason for seeking a simpler test. Experimental and prospective studies are costly and time-consuming, and it is impossible to test for every possible effect. So one wants at least some justification for beginning such studies, that is, some reason for thinking that a particular causal hypothesis might really be true.

In an attempt to test the suggestion, doctors in Britain began a search of records in 19 different hospitals for the previous 2 years (i.e., 1964–1966). They looked for women who had been treated for *nonfatal* clots of the legs or lungs (clots in the heart or brain tend to be fatal). They needed nonfatal cases so that they could locate the women in question and find out if they had taken birth control pills prior to developing the clot. Such information was not generally available in regular hospital records. Their search turned up 58 cases of clots suffered by young, married women with no prior disposition for developing blood clots. Of these 58 women, 26 (i.e., 45 percent) were discovered to have been taking oral contraceptives the month preceding their admission to the hospital for treatment.

By now you should be aware that that 45 percent figure by itself means nothing. If 45 percent of all British women in that age group had been taking oral contraceptives, this is just what one should expect. We need a control group to provide a basis for comparison. In the British study, the control group consisted of 116 married women admitted to one of the participating hospitals for some serious surgical or medical condition *other than blood clotting*. The control group was chosen so as to match the experimental group in age, number of children, and several other possibly relevant factors. Of these 116 control subjects, only 10 (i.e., 9 percent) were found to have been taking oral contraceptives.

This difference in the relative frequency of pill users in the two groups was reported as being "significant," meaning statistically significant. Understandably, the publication of this study produced some panic among women taking the pill. After a brief examination of the data, officials of the U.S. Food and Drug Administration issued statements saying, "Yes, oral contraceptives do seem to cause blood clots, but, no, you should not stop taking them on this account because the risk is very small." Actually evaluating the risk and assessing its relevance for the decision whether or not to take the pill is a topic for Part IV. For the moment let us concentrate on the prior question of how and to what extent this type of study may indeed justify a causal hypothesis.

12.6 RETROSPECTIVE DESIGNS

The British study of blood clotting and the pill utilized what is called a "retrospective" design. The name comes from the fact that the study begins with a sample of subjects that already have the *effect* (e.g., a blood clot) and attempts to look back in time to discover the *cause* (e.g., oral contraceptives). The study is "backward looking"—unlike both experimental and prospective designs, which are "forward looking."

You will note that random sampling plays almost no role in retrospective studies. The experimental group is chosen from among subjects who already have the effect being studied. Women who have been treated in a hospital for a nonfatal clot are obviously not a random sample of women. And the "control"

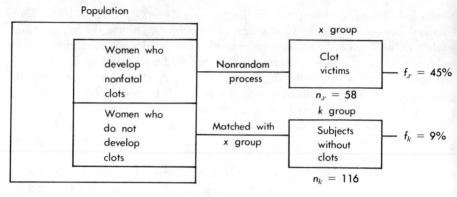

FIGURE 12.14

group is chosen by attempting to *match* the subjects in the experimental group for other variables that might be relevant. You will note that the British study matched each experimental subject, 58 in all, with *two* control subjects, 116 in all. The number of experimental subjects they could find was fairly small. They couldn't increase this number very easily. But they could increase the number of control subjects, so they did. This increased their chances of obtaining a statistically significant difference, if there was indeed any difference in the population to be found. The only situation in which random sampling might be used in a retrospective study is if there are more potential experimental subjects or matched control subjects than are needed. At that point the actual groups might be randomly selected from among the available subjects.

Figure 12.14 pictures the essentials of a retrospective design. You should note in particular that the frequencies observed in the two groups are not frequencies of the *effect*, but frequencies of the *cause*. Similarly, the subpopulations from which the samples come are not those that do or do not have the suspected *cause*, but those that do or do not have the *effect*. In retrospective studies, the location of the cause and effect are, roughly speaking, reversed. Thus you can almost always tell when a reported study is retrospective because the frequencies given will be the frequencies of subjects with the *cause* in groups of subjects that either all have the effect or all do not have the effect. Reports of experimental and prospective studies always report the frequencies of the *effect* in two different groups.

This reversal of the roles of cause and effect in the design also shows up, of course, in the way one would diagram the observed frequencies. The data from the British study are pictured in Figure 12.15. Note that reference to all *other* controlled variables is included in this diagram.

Conclusions from Retrospective Studies

The null hypothesis, of course, is that oral contraceptives are causally irrelevant to the occurrence of a nonfatal blood clot. It is nonfatal clots that were

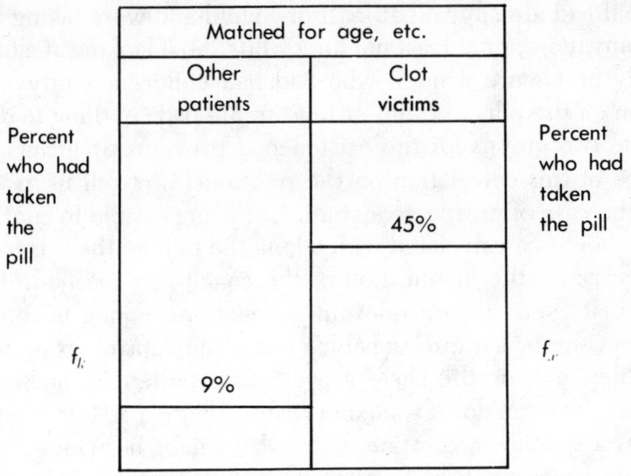

FIGURE 12.15

actually studied, though the inference to fatal clots seems completely unproblematical. In order to deduce the probability distribution for possible differences, d, between f_x and f_k from H_0 and the sample sizes, one has to make some definite assumption about the sampling process and possible correlations in the population. The standard assumptions would be that the sampling was done randomly, at least for the x group, and that there are no correlations among uncontrolled variables that could bias the observed frequencies of pill users in the two groups.

Granting these assumptions, the probability distribution for d may be deduced, and the set of values of d that will count as statistically significant may be determined. The form of the arguments for either H or ΔH_0 would be as in the case of experimental or prospective studies. Given these assumptions, the difference in percentages of women who had been taking the pill was definitely statistically significant at the .05 level. That is not the problem.

The problem with the justification of either of the two possible conclusions is that it is almost impossible to justify the needed auxiliary assumptions. Indeed, they are often known to be false.

Like prospective studies, retrospective studies suffer from the possibility that the self-selection into subpopulations with the effect (e.g., blood clots) and without the effect may be biased in a way relevant to the percentage of subjects taking the pill. It is for this reason that an attempt is made to match the control subjects with the experimental subjects. Matching for a variable (e.g., age) evens out any correlation that might exist between that variable and the cause (e.g., taking the pill).

For example, the subjects in the study were matched according to whether they had any children or not. Women who have been pregnant have a higher risk of getting a blood clot. Also, in England in 1965, most of the women taking the pill had already had at least one child and were taking the pill so as not to have any more, or at least not for a while. So if looking at clot victims you automatically find more women who had had children, you would also find more women on the pill, whether or not the pill had anything to do with clots. Matching the two groups for the existence of previous pregnancies evens out the influence of this correlation on the frequencies of pill users.

As in the case of prospective studies, it is impossible to control for *every* variable that might be correlated with taking the pill. So there is for this reason a question mark on the justification of the conclusion reached. The auxiliary assumption that there are no relevant correlations among uncontrolled variables is always somewhat questionable. But if the control group was matched for all variables that, on the basis of previous experience, might be relevant, that is the most one can do. If someone wishes still to question the justification, that requires a specific suggestion as to what might be wrong.

But retrospective studies suffer from a further possible failure of the auxiliary assumptions, a possibility that is not so serious in the case of prospective studies. This is that the experimental group itself could hardly be said to be a random selection from the population—even from the population of subjects with the effect. Subjects generally get into the experimental group because something special has happened to them—for example, they were hospitalized or saw a doctor. These special features may well be correlated with the suspected cause. For example, it is well known that people who have reason to believe they may develop some conditions are more likely to call a doctor when something suspicious occurs than people who have no reason to be suspicious in the first place. This fact is relevant because women taking oral contraceptives for the first time are generally warned that they may experience various side effects. Thus such women are much more likely to call their doctor about various conditions than women who have not had this warning and are not wondering whether what they are experiencing is due to the pill or not. This means that among all the women who may develop minor clots in their legs or lungs, those taking oral contraceptives are more likely to call a doctor, and therefore more likely to have the pain diagnosed as a clot and to be hospitalized for treatment. Maybe the higher frequency of pill users among the 58 clot patients was simply an artifact of this "reporting factor."

Although it is easy to think of nonrandom selection processes that might produce misleading data, it is generally very difficult to control for such processes. The present example might be dealt with by considering only women whose initial symptoms were serious enough that they would have been hospitalized no matter how weak or strong their inclination to call a doctor for minor aches and pains. It is doubtful, however, that the initial study contained enough subjects in this category to produce statistically significant results. And in any case there are many other selection factors that might have been operative and would be more difficult to control.

As a general rule of thumb, it is probably best always to regard conclusions based on retrospective studies as more or less tentative. A causal hypothesis supported by retrospective data is not to be ignored and a positive result should be taken by scientists and responsible agencies as a reason to undertake other studies, either experimental or prospective. But causal hypotheses supported by retrospective data alone cannot be regarded as being as well justified as conclusions based on prospective or experimental studies.

In general, retrospective data are more like simple survey data than like prospective data. Basing a causal hypothesis on the results of a retrospective study is almost like basing it on a simple correlation exhibited in survey data. And we know that a correlation alone cannot justify a causal hypothesis. The superiority of a retrospective study lies mainly in the reliability of the information and in the care with which the control group is selected. Hospital records and personal interviews are more reliable than questionnaires. And, with care, it is possible to control for most variables that might reasonably be thought to bias the data. But a prospective test of the same hypothesis would almost always be better yet.

Effectiveness and Retrospective Studies

There is an additional disadvantage to retrospective studies. The data they yield allow no estimate of the *effectiveness* of the causal factor. Effectiveness is defined in terms of the percentage of the population that would or would not get the *effect* depending on whether they all had the cause or not. The frequencies of the effect in two samples, as in a prospective study, may be used to estimate the effectiveness of the causal factor in the population. But retrospective studies give you the frequency of the *cause* in groups with and without the effect. There is no way to use these frequencies to estimate the effectiveness of the causal factor. For example, knowing that 45 percent of women who had been hospitalized for clots had been using the pill tells you nothing about the percentage of women using the pill that were hospitalized for clots. But that is the percentage you need to estimate effectiveness.

This feature of retrospective studies is especially serious because the effectiveness of the causal factor is the most useful information when it comes to making decisions. Just knowing that something is a positive causal factor is of limited usefulness in decision making. But that is the very most you can get out of a retrospective study.

On the other hand, it is sometimes possible to *combine* information from a retrospective study with other information to reach an estimate of effectiveness. In the case of the pill studies, British scientists were able to use other information on the occurrence of blood clots and the use of oral contraceptives to determine a rough estimate of the probability of blood clots among women using the pill. This probability turned out to be roughly 15 per million. That was about seven times the probability in the population at large, but still quite small. So the FDA was right. It is a causal factor, but the risk is small.

PRACTICE EXERCISES

12.6 SMOKING AND HEART ATTACKS IN WOMEN

The following is adapted from a newspaper story:

Heavy Smoking Costs Women Years of Life

A team of New York medical researchers said today that a heavy upsurge of cigarette smoking among American women is causing a soaring increase in fatal heart attacks.

These warnings are based on a study of 194 Westchester County, New York, women who died between 1967 and 1971. The researchers autopsied the women and questioned relatives about the women's cigarette smoking habits. The researchers said 41 of the women suffered sudden, fatal heart attacks and the other 153 women died from other causes.

Comparing the two groups, the researchers found 62 percent of the heart attack victims had been heavy cigarette smokers—20 or more cigarettes daily. In contrast, only 28 percent of the women who died from other causes had been heavy smokers.

A. State the causal hypothesis at issue in standard form: "*C* is a positive/negative causal factor for *E* in the population of . . . ."

B. State the null hypothesis. Note any information about the population actually sampled.

C. Following the model of Figure 12.14, diagram the experiment. Identify and fill in as many details as possible from the given report—for example, populations, *x* group, *k* group, sample sizes, sample frequencies.

D. Diagram the data following the model of Figure 12.15. Indicate on your diagram whether the difference in sample frequencies was reported as being statistically significant. If it is not reported, try to make an "educated guess."

E. What conclusion is justified by the data? Sketch the "short form" of the appropriate argument.

F. What, if anything, can you conclude about the effectiveness of the causal factor?

G. Summarize your evaluation by noting the main strengths and weaknesses of the study as it is reported. What is your overall assessment of how well the conclusion has been justified?

SUMMARY AND EXERCISES

The following exercises are adapted from actual reports of scientific studies—mainly from newspapers. In each case your job is to analyze the report as best you can with the information given. Your objective is always to determine whether the conclusion is justified and, if so, how well justified. The following seven steps provide you with general guidelines to follow in your evaluations. Your answer to each problem should thus have seven parts. The last step is the most difficult, but also the most important. It will determine what you have learned, if anything, from the given report.

1. Identify the causal hypothesis *H* at issue. This means picking out both the causal factor and the effect and identifying the population of interest. If you are doing the exercise to hand in, write out the causal hypothesis explicitly in the form: *C* is a positive/negative causal factor for *E* in the population of . . . . In most cases the population of interest (the target population) is fairly broad—such as all humans, all Americans, all women, all men over 50 years old, and so on.

2. State the null hypothesis H_0 that is tested. Note information about the population *actually sampled*. In most cases the population sampled will be only a subpopulation of the population of interest. For example, if the target population is all American women, the population actually sampled might be women belonging to a particular health program. If the population actually sampled looks to be a very specialized subsection of the target population, that is something to note in your overall evaluation in step 7.

3. Identify the design of the study. Is it randomized experimental, prospective, or retrospective? Retrospective studies are the easiest to spot because the *x* group consists of subjects that already all have the *effect*. The data will refer to facts about the percentage of these subjects that have the *cause*. Any mention of random division into two groups—double-blind setup or placebos—tells you immediately that you have a randomized experimental design. If the members of the *x* group already had the suspected causal factor when they were selected, then you have a prospective design. Finally, be on the lookout for cases in which the data merely justify claiming a correlation. Those cases should be handled by the methods of Chapter 11. If you are handing in your work, diagram the study following the model of Figure 12.2, 12.10, or 12.14. Fill in the diagram as completely as you can from the information given. It will help if you put in question marks (?) at places where information is required but not given in the report. Crucial omissions, like sample size, are things to note for your overall evaluation in step 7.

4. Diagram the given data following the model of Figure 12.3, 12.13, or 12.15, depending on the design of the study. In the case of either prospective or retrospective studies, you should note explicitly whether any other variables were controlled or not. Were the subjects "matched" for other possible

relevant characteristics? If there is no mention of such obvious other variables as age, sex, health, occupation, or the like, you might want to note that for your final evaluation.

Finally, you should note whether the difference in sample frequencies, however reported, is said to be "significant," meaning, of course, statistically significant. Note the significance level if that is reported. If there is no mention of significance, you should try to infer from the general context or the actual reported frequencies whether or not the difference is statistically significant. You may just have to go on the general reputation of the source of the report. If the study is said to have been published in the *New England Journal of Medicine* or some such place, and a difference is claimed for the population, you can be reasonably sure that the data were at least statistically significant at the .05 level.

5. What is the conclusion? There are only two possibilities: H and ΔH_0. If the indicated conclusion is H, you should check to see whether this could be due more to a very big sample than to any "interesting" difference in the population. If the conclusion is ΔH_0, you should check to see whether this might be due simply to there being too small a sample to have a high probability of detecting an important effect by getting an SSD. Either circumstance should be noted for your overall evaluation.

6. Check the report for any indication of the actual *effectiveness* of the causal factor in the population. For this you need sample frequencies that you can use to estimate the population ratios. If the sample is small, no estimate of effectiveness is possible unless the sample frequencies are very far apart. If the data are given in the "30 percent more" form, then you cannot tell anything about effectiveness. And of course you can do nothing with retrospective data. It gives the "reverse" ratios. But do read carefully, because sometimes a report will employ supplemental information to give an estimate of effectiveness.

7. The last step is to try bringing everything together to form an overall evaluation of the study. This is what you are most likely to remember long after you have forgotten the specific details. The most important fact for your overall evaluation is the *type of study* involved. No prospective study can provide as good a justification for a causal hypothesis as a good randomized experimental study. And no retrospective study can provide as good a justification as a good prospective study.

The next most important component of your overall evaluation is the relationship between the size of the samples and the existence of an SSD. If there is an SSD with a small to medium-size sample, or if there is no SSD with a large sample, that usually indicates something important about the population. An SSD with a very large sample or no SSD with a small sample is usually less informative.

For both prospective and retrospective studies you should look for indications that *other* variables have been controlled, such as by matching the experimental and control groups in various ways. Often you will have to infer the existence of matching from details of the reports. For example, if all subjects were patients in the same hospital, that indicates that subjects in both groups are pretty much alike. The better the matching, the better the justification for the conclusion. A complete lack of attempts to control for other variables is reason to discount the value of the study. This is especially true if you yourself can think of obvious other variables that could have produced the given data, such as age, sex, race, and so on.

Finally, you might want to give some weight to other aspects of the study. For example, the population actually sampled might be very different from the population to which the causal hypothesis is applied. Skiers in Switzerland may be very different from college students in Kansas.

Now the problem is to put all these things together to reach a single judgment. This judgment can only be fairly rough, but several gradations are clearly possible. (a) The conclusion is very well justified. (*Example:* Saccharin causes cancer in rats.) (b) The conclusion is well justified. (*Example:* Smoking causes heart attacks.) (c) The conclusion is fairly well justified. (*Example:* Oral contraceptives cause fatal blood clots.) (d) The conclusion is somewhat justified. More information is needed. (*Example:* Vitamin C prevents colds.) (e) The conclusion not at all justified by the given data. Forget about it. (*Example:* Smoking cigarettes causes low grades.)

For cases in which you decide that some causal hypothesis is at least fairly well justified, you should ask yourself if this fact is relevant to any decision you might face. Are you a smoker? Do you use oral contraceptives? In this case you want to look for indications of how *effective* the causal factor is. That is what will be most relevant to any personal decision you have to make.

12.7   THE EFFECTS OF MARIJUANA ON HUMANS

In one of the few "scientific" tests of the effects of marijuana on humans, nine marijuana-naive volunteers were tested for various psychomotor and psychological abilities, including hand-eye coordination and ability to concentrate. In this experiment the same subjects served in both the experimental and control groups. That is, they were given either a measured amount of marijuana or a placebo in a series of "sessions." Which subjects got the marijuana and which got the placebo in any particular session was determined randomly by an experimenter other than those that actually conducted the tests. The test scores achieved by the subjects when they in fact had the marijuana were regarded as the experimental data. The scores achieved by the subjects when they in fact had the placebo were regarded as the control data. The average score of the subjects was significantly lower when they received the drug than when they received the placebo.

A similar experiment was carried out with nine experienced marijuana users as subjects. These latter tests, however, could not be done blind because it proved impossible to devise a placebo that could fool the subjects. The average scores achieved by these subjects were not significantly different when given the drug and when not.

One report of these experiments reached the following conclusions:

A.   The performance of experienced drug users is completely unimpaired by the drug.

B.   The performance of neophytes is adversely affected by the drug.

In evaluating this study, you should consider each conclusion separately. One of them is better justified than the other.

12.8   THE XYY SYNDROME

After rumors that mass murderer Richard Speck had an XYY chromosome (normal males have XY), there was considerable speculation that this genetic abnormality was itself responsible for violent behavior. Investigations into the genetic makeup of men imprisoned for violent crimes, such as murder and robbery, yielded an XYY rate of roughly 1 percent for inmates of several large penitentiaries. The incidence of this abnormality among the general noncriminal population was determined to be roughly 1/10 of this (i.e., .1 percent). Given the size of the samples involved, this difference is statistically significant at the .05 level. However, further investigation has shown the XYY males tend to be taller and stronger, but less intelligent, than normal males. It is known that bigger men are more likely to commit violent crimes just because they are more powerful; and less intelligent men are more likely to be caught and imprisoned. This suggests controlling for the variables of size and intelligence—for example, looking at men over 6 feet tall with IQs less than 100. When this is done, the percentage of XYYs in the criminal group remains about 1 percent, but that in the noncriminal group increases to roughly .8 percent. This difference is not statistically significant (.05).

12.9   STUDY SHOWS EXERCISE CUTS HEART-ATTACK RISK

A study of nearly 17,000 Harvard alumni has found that there were fewer heart attacks among those who participated regularly in strenuous exercise than among those who were less active.

The study was begun in 1968 by determining the health and exercise patterns of men who had entered Harvard as freshmen between 1920 and 1950. Of these, 16,936 were found to be free of heart disease. Each man was classified as having a "high" or "low" exercise level depending on whether he expended more or less than 2000 calories a week on exercise. By 1976, 572 of the men had experienced a serious heart attack. But the rate of heart attacks was 64 percent higher in the low-exercise group.

It was suggested that perhaps the high exercisers were simply those people who were more healthy and athletic all along. They exercised more because they were more healthy, not vice versa. But this suggestion was ruled out because a check of men who had participated in varsity sports as undergraduates showed the same pattern of heart attacks depending on their level of activity in later life.

12.10   NEW ROUND OF TESTS SHOW HOW LAETRILE IS INEFFECTIVE

Scientists taking a fresh approach to testing laetrile have found it ineffective against human cancers implanted in mice. Previous tests have used only cancers native to laboratory animals.

Dr. A. Overier, a researcher at the Battelle Memorial Institute in Columbus, Ohio, implanted human cancers in two dozen mice, bred so as not to reject implanted human tissue. Half of the mice were treated with laetrile. There was no noticeable difference between the two groups in the number of cancers that grew to fatal proportions.

12.11 VITAMIN C AND ALCOHOL

Whatever the value of vitamin C in preventing colds, it does seem to counteract the worst effects of alcohol—at least in rats. According to an article in the *New York Times,* two groups of rats were given heavy doses of alcohol two days in a row. One of the groups had been given doses of vitamin C as well. All of the rats fortified with vitamin C survived the ordeal, but for 68 percent of the other group, the experience proved fatal.

12.12 SMOKING, HEART ATTACKS, AND THE PILL

Since the mid-1960s, when the use of birth control pills became widespread, many hospitals and clinics have followed the medical histories of their patients to determine whether women who choose to use the pill experience effects different from those who choose other contraceptives or none at all. In the early 1970s it was discovered that women on the pill had a higher death rate from *heart attacks* than others. Typically one found results like the following (death rate from heart attacks given in deaths per 100,000 women of ages 40–44):

Women not on pill: 7
Women on pill: 24

This difference is statistically significant given the large samples used.

Although the government issued official warnings based on this information, these studies were not controlled for smoking. This is hard to believe given the then well-known connections between smoking and heart attacks (in men). When someone finally got around to controlling for smoking (10 years later), the results were as follows. (Data given in deaths from heart attack per 100,000 women of ages 40–44):

Nonsmokers not on the pill: 7
Nonsmokers on the pill: 7
Smokers not on the pill: 14
Smokers on the pill: 59

Again all reported differences are statistically significant. *Note:* This problem is difficult. There are a number of different hypotheses you can consider, depending on which variable you control. By considering various comparisons you should be able to figure out which is the really important conclusion. You may find it helpful to review the discussion of tobacco and air pollution in the text.

12.13 SCIENTISTS TO STUDY ASPIRIN, HEART ATTACKS

Aspirin, a drug that is taken for everything from arthritis to removing warts, may have still another use: preventing heart attacks.

The most convincing evidence comes from two studies conducted by researchers at Boston University, who since 1966 have been compiling data relating drug exposure to the diagnosis of patients admitted to hospitals.

The first study involved 325 heart attack patients and 3807 other patients. It revealed that only .9 percent of the heart attack patients had been taking aspirin regularly before their attack (for arthritis, for example). Aspirin taking was much higher—4.9 percent—among those patients who had not had heart attacks.

A second follow-up study of 451 heart attack and 10,091 other patients at 24 Boston hospitals showed the same kind of results: 3.5 percent of the heart attack patients said they had been taking aspirin regularly compared with 7 percent of those with other diagnoses.

The studies were reported in the March 9 issue of the *British Medical Journal.*

12.14   LATER STUDIES ON ASPIRIN AND HEART ATTACKS

In recent years it had been noted that people who for some reason regularly ingest moderate quantities of aspirin (e.g., for arthritis) seem to have a lower incidence of heart attacks than others. To test this idea in a reasonably short time, researchers at several New York hospitals obtained a large sample of patients who had already had one heart attack. A randomly selected group of these patients were prescribed regular doses of aspirin. The rest were given ordinary multiple vitamins. Of the 200 patients on aspirin, 23 (or 12 percent) experienced a second heart attack within one year of their first. Of the 160 patients in the other group, 39 (or 23 percent) experienced their second attack within a year. This difference is statistically significant (.05).

12.15   STRESS AND CANCER—IN RATS

Recent studies by the National Cancer Institute have shown that the development of cancer may be linked to emotional stress. In a 1975 study, rats were used to determine the effects of stress. The rats were divided into two groups. Both groups were injected with cancer cells. One group was then placed in a normal environment (normal for rats) and the other placed under adverse conditions that produce stress. The rats in the stress environment died much sooner than those in the normal conditions.

12.16   BENZPYRENE AND LUNG CANCER

It has been known for some time that prolonged contact with the chemical benzpyrene can cause skin cancer. It has also been discovered that small amounts of this chemical are contained in cigarette smoke and in polluted urban air. This suggests that benzpyrene may be the agent that causes lung cancer in smokers and in inhabitants of large cities.

To test this hypothesis, a specialist in environmental medicine studied the health of 5900 roofing workers. These workers breathe large amounts of benzpyrene, which is an ingredient in coal tar pitch and asphalt—both common roofing materials. The amount of exposure to benzpyrene for these workers was estimated to be equivalent to that from smoking 35 packs of cigarettes a day! At the beginning of the study, all of the workers had been members of the Roofers Union for at least 9 years. Six years later the number of lung cancer deaths recorded among these workers (43) was reported to be not significantly different from the number expected in the general population.

12.17 SACCHARIN AND CANCER: HUMAN LINK IS FOUND

The case against saccharin has been strengthened by a new Canadian study linking the artificial sweetener for the first time to human bladder cancer. An earlier study, also Canadian, had found a similar link in laboratory animals.

Researchers for the National Cancer Institute of Canada and three Canadian Universities compared 480 men and 152 women who had developed bladder cancer with equal numbers of both sexes who had not. The scientists found that the frequency of use of saccharin among men with bladder cancer was 60 percent higher than that of the other group of men. While this difference was regarded as statistically significant, the difference in saccharin use in the two groups of women was not.

The U.S. Food and Drug Administration called the latest Canadian research very important, but said that the government would need until fall to evaluate it.

12.18 ARTIFICIAL SWEETENERS DON'T INCREASE RISK OF BLADDER CANCER, STUDY SHOWS

A study published in the July 28 issue of the *Journal of the American Medical Association* concludes that the ingestion of artificial sweeteners, at least at moderate dietary levels, is not associated with an increased risk of bladder cancer. Earlier studies have shown that high doses of saccharin can cause such cancers in rats.

This study involved 519 humans, all patients with bladder cancer at 19 Baltimore area hospitals between 1972 and 1975. These patients were matched with a like number of patients of the same age and sex who did not have bladder cancer.

The researchers questioned the cancer patients and the matched controls about their consumption of artificial sweeteners in carbonated and noncarbonated soft drinks, iced tea, and other liquids, and in salad dressings, candy, ice cream, pastry, gelatin, chewing gum, and other foods. All subjects were also questioned about smoking habits, occupational history, diabetes and other factors that might have been involved.

In light of this data, the researchers questioned the relevance of animal tests to studies concerning the development of cancer in human beings.

12.19 SCREENING FOR BREAST CANCER

Fifteen years ago it was debated whether regular and systematic screening for breast cancer would in fact save lives. Many doctors thought it would; others argued that the methods of detection were not reliable enough and that the primary factor in survival is the individual patient's genetic and general bodily resistance to cancer. To settle this question, the following study was undertaken by the Health Insurance Plan of New York City, a comprehensive medical program operating more than 30 clinics in the New York City area. They selected at random 62,000 women of ages 40–64 from among all women in the program. Half of these (31,000) continued to receive the same services as all other patients. The other 31,000 women were invited to come in for extensive medical interview, a clinical examination, and a mammography (a set of breast X rays). These examinations were repeated yearly for the next 3 years. As expected, some women in both groups developed breast cancer. Although all were treated by the best available methods, some died. Six years after the beginning of the study, the number of deaths from breast cancer was one-third lower in the group that had the regular examinations.

That is, the number of women who died from breast cancer in the group with the special examinations was only two-thirds as great as the number among women not receiving the special screening. This difference is clearly statistically significant (.05). (*Note:* Just for your information, more recent studies indicate that routine breast X rays for women under 40 may cause more cancers than they detect. Many doctors now use X rays only if there is some *other* indication of cancer.)

12.20 THE GREAT SWINE FLU "EPIDEMIC"

In the Fall of 1976 the U.S. Federal Government, fearing a possible epidemic, undertook a program to immunize the whole population against swine flu. By the time about 50 million people (out of a total population of about 200 million) had been inoculated, there seemed to be an outbreak of cases of a rare and usually temporary paralytic condition called the Guillain-Barre syndrome. A quick investigation turned up 94 recent cases of Guillain-Barre. Of these, 51 had received swine flu shots within the past 3 weeks; 31 had not been inoculated, and the inoculation status of the other 12 was unknown. After this discovery the inoculation program was halted and never resumed. Continuing investigation turned up 685 cases of the syndrome that had occurred after the immunization program began. Of these, 354 had been inoculated; 28 of the victims died. (*Hint:* Regard the whole population of 200 million people as the control group.)

12.21

Find a report of a scientific study that involves statistical reasoning. Try newspapers and newsmagazines first. If you can't find anything suitable, try more specialized sources like *Science, Science News, Psychology Today,* etc. Analyze the report following the pattern established in the exercises to Chapters 11 and 12.

Up to now we have been primarily concerned with understanding *scientific* reasoning, that is, reasoning that can justify hypotheses about the various systems that make up our world. Among the hypotheses we have considered, some have little direct relevance to our everyday lives. The hypothesis that the universe is expanding is an obvious case. Other hypotheses are clearly relevant to everyday concerns. The hypotheses that smoking causes heart attacks and that oral contraceptives cause blood clots are clear examples. The hypothesis that the whole world system might "collapse" is another example, although its relevance may seem less immediate.

The reason some scientific conclusions are relevant to everyday concerns is that they bear on *decisions* we have to make. If you smoke cigarettes, for example, the conclusion that smoking causes heart attacks is obviously relevant to your decision either to quit smoking or not to quit. We are now going to examine just how scientific conclusions bear on individual decisions. We know, for example, that riding in automobiles and airplanes causes deaths and injury. Indeed, in the United States more people die each year in auto accidents than from lung cancer. Yet many more people would say that one ought to give up smoking than would argue that we ought to stop driving or riding in airplanes. Is there some "rational" ground for this difference in attitude, or is that just the way people react?

By now it should not surprise you to learn that decisions, like conclusions, can be justified. But there is a major difference. It does not make sense to call a decision true or false. Rather, a good decision is one that leads to the right, or "rational," course of action in the given situation.

Now if by right or correct action you mean the *morally* right action, then we are getting into questions of ethics and moral philosophy. We shall leave such issues to one side and merely consider the "right" action to be the one that has an "appropriate" relation to desired results, whatever they might be.

Even bypassing the issues of moral philosophy, knowing how to justify decisions is not obvious. For example, if you knew that you are one of those people for whom smoking would make the difference between dying at age 40 or dying at age 65, then it

PART FOUR

Values and Decisions

seems pretty clear that not smoking is the "right" decision for you. But of course neither you nor anyone else knows whether you are that type of person or not. Yet you have to make the decision. Is there a right decision in this case, that is, a decision that is rational *now* regardless of what happens in the future? If so, how do you determine what it is?

Learning a little about decision theory can help you to make "better" decisions and also to understand decisions made by others. For most people, just a little understanding of the theory is all that is really useful. We shall concentrate on basic concepts and useful rules of thumb that you can apply every day.

We shall proceed in two stages. Chapter 13 sets out the basic elements that go into any decision. These elements determine the basic "logical" structure of decisions. Of particular importance will be the nature and location of the "value input" in the overall structure of a decision.

In Chapter 14 we shall look at various strategies for making the "best" decision in various situations. As you will see, what strategy is best depends largely on the nature of the available *information*. So in Chapter 14 we shall really be learning how the information from scientific studies feeds into every-day decisions.

13

The Structure of Decisions

Having learned that there is a great similarity in the structure of arguments for different kinds of scientific hypotheses, it should not surprise you now to learn that most decisions can be understood in terms of a single structure. In this chapter we are going to set out the basic elements that go into any decision and see how they fit together in one simple form. We shall pay particular attention to the role of *values* in the general structure of any decision.

13.1 ACTIONS

For our purposes, a decision will always be a decision to *do* something. Thus, making a decision is choosing a course of action. Now there is no decision involved unless there is more than one thing that you might do. That is, you must be facing a choice between two (or more) different possible courses of action. You have the option of choosing one from among a set of possibilities. In the language of decision theory, then, any decision problem involves a set of at least two different possible actions, or, for short, "actions."

Surveying Your Options

Many decisions seem to come upon us with the options already built in. You learn that saccharin causes bladder cancer, so your options seem to be either to avoid saccharin or to go on using it. Similarly with smoking or using oral contraceptives. Either you do or you don't. Those seem to be the only possible actions.

Even in these relatively simple cases, however, your options are rarely restricted to just two. There are almost always intermediate possibilities. For example, you might decide to stop using saccharin in your coffee but continue to drink diet sodas. This cuts down your consumption of saccharin, thus presumably cutting your risks, but is less inconvenient. Similarly, you might decide simply to cut down your smoking to half a pack a day. That lowers your risks, but leaves you with a few crucial cigarettes per day, such as after dinner. Likewise, you might use oral contraceptives every other year, relying on some other method on "off" years.

For more complex decisions, it is usually clear that there are more than two options. At the end of a school year you face the decision of whether to return the following year. You may have the option of getting a job instead. And if a job, then maybe one of several different jobs. Or maybe you could take off a year to travel, picking up work only when you need the money. Even if you don't think about this problem now, you must when you graduate. Should you go on for a higher degree, an MBA perhaps? Should you look for a "permanent" job? If so, where? Near home? In another region of the country? Should you travel first? And so on.

One of the virtues of thinking about decision making is that you become aware of the importance of surveying your options at the outset. Just formulating a decision problem requires that you pick out a set of possible actions from which you will choose. In almost all cases there will be more than just two options open to you, and by thinking about other possibilities you may be led to consider an option that turns out to be the best possibility. If you don't even think of it, you are going to end up choosing some other action that might not be as good.

It is important, then, to survey your options to make sure you are not passing up some possibility that might in fact be your best choice. It is equally important, however, not to get too many options. Once you begin thinking about it, you may find it easy to generate so many possibilities that it becomes impossible for you to sort them all out. It is easy to become overwhelmed with possibilities.

One of the tricks to intelligent decision making, then, is to generate enough options that you have most of the "good" ones, but not so many that you cannot possibly make up your mind. This principle is implicitly recognized by sales people. If you work in a retail store, you are likely to be told always to let the customer look at two or three different possibilities, but no more. Otherwise the customer will not be able to decide and will leave without buying anything.

Organizing Your Options

The principles of decision making require that the possible actions in any decision be organized so that one, and only one, action will be chosen. In technical terms, the actions must be "mutually exclusive" and "exhaustive." We have already dealt with mutually exclusive alternatives when using the addition rule for probability (Section 10.2).

To formulate your actions so that they are *mutually exclusive*, you must state them so that it is impossible that you could do more than one simultaneously. For example, "smoking" and "smoking a half pack a day" are not mutually exclusive. If you smoke half a pack you are also smoking. Mutually exclusive alternatives would be to "smoke a half-pack or less per day" and "smoke more than a half-pack per day." You can't do both.

One further example. Suppose you are graduating and deciding whether to get more education or a job. You are also deciding to move somewhere else or not. You might be tempted to say that you have three options: "school," "work," or "move away." But this is not correct for the purposes of decision making. If by "school" and "work" you mean "full time," then these two are mutually exclusive. But you can both move away and get a job or move away and go to school. So the third option is not exclusive of the other two. Your options must be formulated as follows:

Work	Away	Work away
	Home	Work home
or		
School	Away	School away
	Home	School home

These four possibilities are mutually exclusive. You could not possibly choose more than one.

For your options to be *exhaustive* means that no matter what you do, you are going to do at least one action in the total set. The set of actions in any decision problem must exhaust your possibilities. Thus, if you had already decided that you would either go to school or work, then the above set of four actions would be exhaustive—for the purposes of the decision you have to make. They are not *logically* exhaustive since there are other *logically possible* alternatives. It would be possible, maybe, just to loaf and sponge off your parents for a year. But if you have already ruled out such possibilities, then you may regard the above set as exhaustive.

One of the reasons it is so tempting to formulate decision problems in terms of simple dichotomies like "smoke" and "don't smoke" is that such sets of two options are automatically exclusive and exhaustive. This is because the statement describing the one option is the *negation* of the statement describing the other. (There is no escaping the basic concepts of logic.) But if you always think in such simple terms, you may pass up some really good options.

13.2 STATES OF THE WORLD

No one makes decisions or carries out a course of action in a vacuum. We all operate within a system consisting of other systems, including other humans. The world, actions by others, and even our own actions, all influence the results of the actions we choose to carry out. So in formulating a decision problem we must take account of what the world might be like when we carry out our decision.

Surveying How the World Might Be

When formulating a particular decision problem, you do not, of course, have to survey the state of the whole world. That would be impossible. But you do need to survey that part of the world that might influence what happens as a result of your choosing one course of action or another. You need to consider the states of the world that are relevant to your decision.

Suppose, to take a new example, you are an official of an oil company, and you have to decide whether to commit the company's resources to drilling in a particular new location. In thinking about this decision you must of course have considered the possibility that there is oil in that place. But since even the best geologists cannot be sure that there is any oil there, you would clearly be irrational if you did not also consider at least the possibility that there is no oil to be found in that location. For a small company in financial difficulties, the expense of drilling and not finding anything could mean financial ruin.

As in the case of actions, people tend to think in terms of simple dichotomies—for example, oil or no oil. But if you survey the situation carefully you will almost always find intermediate possibilities or even completely different possibilities to consider. For example, it is not enough that there be oil where you drill, there has to be enough to pay off the investment. So you should consider some of the different amounts of oil that might be found in that spot. In addition, it could happen that a pipeline company is considering putting in a new pipeline that would come very near the proposed site. If they do, that means you can make a better profit on that oil because of reduced transportation costs. If not, your profit will be lower. So that may be another possibility to consider.

Similarly, if you are considering quitting smoking, you must at least consider both the possibility that you are susceptible to heart attacks and that you are not. Either is clearly possible, even though the best doctor cannot tell which is actually the case.

Again, any government or government agency involved in long-term social and economic planning would be irresponsible not to consider the possibility that our world system operates something like a World II system. If it is like a World II system and certain things are or are not done, the result will be a disaster for our children or grandchildren. That is why it is important that such possibilities be explored even if we have no way of determining whether or not they obtain. Even the mere possibility is relevant to some decisions that have to be made.

Describing How the World Might Be

Like possible actions, the various relevant states of the world must be described in terms of exclusive and exhaustive alternatives. It would be very difficult to formulate any simple decision strategies that could apply to overlapping sets of possible states.

In the oil drilling example, "there being oil," "there not being oil," and "the pipeline going in" are not exclusive. The pipeline might go in whether or not you find oil in that place. So once again there are at least four, not three, possible states of the world to consider. These might be set out as follows:

Oil		No oil	
Pipeline	No pipeline	Pipeline	No pipeline

If now there is some other completely different state to consider, that will increase the total set to eight. For example, the major oil-producing countries might raise the world price of oil. That might mean that your company could raise its prices as well. This possibility is clearly relevant to your decision. The easiest way to incorporate this possibility would be to insert the descriptions "increase" and "no increase" under each occurrence of the words "pipeline" and "no pipeline" in the above scheme. That will give you a set of eight mutually exclusive possibilities. It is impossible that more than one of these eight actually occur.

Making sure that the relevant states of the world are described as mutually exclusive alternatives can be done fairly mechanically. It is more difficult to say when your set is *exhaustive*—that is, when one of these alternatives must occur. You might think you have considered everything that is relevant and some new consideration could still emerge; for example, you might learn that some other company is considering offering to buy the rights to that site.

It is clearly important to consider relevant states of the world, but you cannot consider too many because describing them as a mutually exclusive set soon multiplies the number of states beyond what you can keep straight even if you write them all down. That is why large corporations, government agencies, and the military use computers in making decisions. Computers are good at remembering long lists.

13.3 OUTCOMES

When you actually perform some action, your action will take place in some state of the world. This is the "result," or "outcome," of your decision—the performance of some action in some state of the world. Thus, once you have fixed your options and decided which states of the world are relevant, the possible outcomes of your decision are completely determined.

This is a bit abstract. Let us work it out for a simple example. Suppose your decision problem is whether to smoke cigarettes or not. You have ruled out the possibility of cutting down or switching to a pipe. The relevant states, as far as you are concerned, are whether you are susceptible to the effects of smoking or not. These two possible actions and two possible states generate

Possible States of the World

	Susceptible	Not susceptible
Smoke		
Don't smoke		

FIGURE 13.1 Matrix for Simple Smoking Decision.

four different possible results of your decision. The easiest way actually to exhibit these outcomes is in the form of a *matrix*, as in Figure 13.1. Each of the four "boxes" in Figure 13.1 represents a *possible outcome* of your decision problem. The upper left box, for example, represents the outcome of smoking and being susceptible.

The possible outcomes corresponding to a particular decision problem may be described as *action-state pairs*. The fact that each action-state pair determines an outcome means that the *number* of outcomes to consider will always be the *product* of the number of options and the number of states. This is another good reason for trying to keep your decision problems from becoming too involved. It is very easy to generate more outcomes than you can think about.

The reason for making such a fuss about outcomes is that *values* enter decisions via the outcomes. Your values attach not to actions alone nor to states alone, but to action-state pairs, that is, outcomes. It is time we looked at the role of values in decision making.

13.4 VALUE

"Value" is a big general concept, like "truth." Numerous books have been written on the general nature of values. There is even a subject called "value theory," which examines various theories about the nature of values.

To avoid getting lost in generalities let us return to the simple smoking decision. For anyone who smokes, this is not a trivial decision problem, but neither is it so special that it cannot be used as a typical example of decision making. By looking at this example we can get a good idea of the place of values in a decision problem.

The matrix for the smoking decision is shown in Figure 13.1. To simplify the example, let us suppose that being "susceptible" to the effects of smoking means that you are almost certain to suffer some serious consequence (e.g., a heart attack) by the age of 40—if you smoke. "Not being susceptible" means that you will suffer no ill effects at all if you smoke. These assumptions are not completely realistic, of course, but they will help us at the moment. Later we shall reexamine this decision with more "realistic" assumptions.

Once a decision problem has been specified to the point that there is a definite set of outcomes to consider, your values may be used merely to *rank* the outcomes in order of your preference. So your values, whatever they may be, are expressed simply as a "value ranking" over a given set of outcomes. For some decision problems, that is all that is needed to determine the "best" decision.

To get a value ranking you have to be able to answer questions like: Which of the four possible outcomes in Figure 13.1 do you *most* prefer? Which do you prefer *least*? Do you prefer this one to that one, or vice versa? The nature of these questions clearly indicates that the values relevant to a decision are your personal values, if it is your decision.

To proceed with the example, then, imagine yourself in the position of someone who already smokes, and clearly likes smoking. You enjoy it. You may *in fact* be in this position, so no imagination is necessary. But even if you don't smoke, you must know enough people who do that it is easy to imagine how you might respond.

Now, which of the four possible outcomes do you *most prefer?* That is, if it were in your power to put yourself in one of these four situations, which would you pick? If you enjoy smoking, then, given our assumptions, you would certainly pick the outcome in which you have chosen to continue smoking but are not susceptible to the effects of smoking. That, for you, would be "the best of all possible worlds."

Unfortunately the rest of the ranking is not completely obvious, even given our simple assumptions. Most people would probably say that the outcome of being susceptible and having chosen to continue smoking is the worst possibility—that is, the least preferred outcome. It is imaginable, however, that to some people in their early twenties the age of 40 seems so far off, and the current pleasure of smoking cigarettes so great, that smoking and being susceptible might well be preferable to not smoking and being susceptible.

If you do not think this way, you may be inclined to say that anyone who does is "irrational"—or just plain crazy. This raises a really deep question. Can some values themselves be "irrational"? Let us put this question aside for a moment. Decision theory itself does not make such a distinction. It only tries to tell you which *decision* is best, given whatever value ranking you may have. In order to complete the example, let us assume that you like smoking, but not so much that you would prefer a debilitating illness before the age of 40 to giving it up now. That fixes the *least* preferred outcome.

Of the two outcomes involving the action of not smoking, the outcome in which you are susceptible would surely be preferred by most smokers. In that case you are getting something in return for giving up something you like. If you gave up smoking and were not susceptible, you would have done it "for nothing." That surely is less desirable than doing it "for something."

Our value ranking of the four outcomes is now complete. The standard practice is to convert any such ranking into a numerical ordering. Any numbers would do, so long as they reflect the chosen order. The simplest numerical ordering is just to go 1, 2, 3, 4, from least valued to most valued. But 5, 6, 7, 8 would do equally well. So would 2, 13, 21, 37. All that matters is the order. The decision matrix for the smoking decision, complete with value ranking, is shown in Figure 13.2. We will call such a matrix a *value matrix*.

The matrix of Figure 13.2 is complete *except* that it contains no specific scientific information. Presumably some statistics on the relative numbers of smokers and nonsmokers who get heart attacks, and so on, would be relevant to your decision to smoke or not. But so far this sort of information is not represented in the matrix. We shall consider the role of such information in the following chapter (Chapter 14) when we take up specific strategies for actually making the decision. At the moment we are just looking at that part of the structure of decisions that is common to all decision problems regardless of the scientific information that might be relevant.

Returning to values, it is clear that one's value ranking in the smoking problem reflects "deeper" or more basic values such as the desire for health and happiness. If you rank the outcome of smoking and being susceptible the lowest, that is because you value good health in your prime years over the pleasure of smoking the previous 10 or 20 years. But to make rational decisions, it is often unnecessary to measure these more fundamental values directly. Such values need only be considered insofar as they determine the ranking of

	Susceptible	Not susceptible
Smoke	1	4
Don't smoke	3	2

FIGURE 13.2 Value Matrix for Smoking Decision.

the outcomes in some specific decision problem. Ranking such outcomes is much easier than trying to say how much you value good health in general.

Measured Values: Drilling for Oil

Decision theory is widely studied in business schools. One of the reasons is that the business context provides a ready-made measure of value: money. Determining the value of any outcome is just a matter of determining the profit (or loss) that would be associated with that outcome. This is not always easy to do, but there is little doubt about what one is trying to discover. Indeed, one of the reasons that the business world concentrates so much on profit and loss is that it is easily quantifiable, unlike "quality of product" or "customer satisfaction."

Let us set up one decision matrix for a problem in which the values are expressed in dollars. The oil drilling decision is a good case. We shall restrict our attention to just two options: "drill," meaning go ahead and begin exploration now, and "wait," meaning don't go ahead now, but leave open the possibility of doing it later. And let's take account of two different contingencies: there being oil (or not) and there being a pipeline (or not). This means that there will be eight (2 × 4) outcomes to consider.

The financial situation might be as follows. Just sitting on the land and doing nothing costs $10,000 a year in taxes, minimal maintenance, and so forth. That price is fixed whether or not there happens to be oil under the ground. Sinking a well to see whether there is oil will cost $50,000. Thus if there is no oil, the company is out that much, plus the regular $10,000.

On the other hand, if there is oil, you can reliably predict a profit, including maintenance, of $550,000 the first year—assuming you have to transport it yourselves. Taking account of the initial drilling cost, that represents a net gain of $500,000 (i.e., $550,000 − $50,000). However, if the pipeline goes through, you can save $100,000 in transport costs. So in that case you would be ahead $600,000 for the fiscal year.

The completed value matrix for this problem is shown in Figure 13.3. The values are measured in thousands of dollars. Note that if there is no oil, the cost of drilling is the same whether or not the pipeline goes in.

This is as far as we shall carry the problem right now. In Chapter 14 we shall examine several possible strategies for making the decision. Which strategy is best will depend, as in the smoking decision, on the kind of "scientific" information available.

Ranking versus Measuring

It is usually possible just to *rank* a set of outcomes according to their relative desirability. We did this without too much difficulty in the smoking

	Oil		No oil	
	Pipe	No pipe	Pipe	No pipe
Drill	600	500	−60	−60
Wait	−10	−10	−10	−10

FIGURE 13.3   Outcomes for the Oil Drilling Problem.

example. But ranking provides only a very crude measure of relative value—indeed, too crude a measure to be useful in some situations.

The drawbacks of having only a value ranking of outcomes is easily seen in a simple example. Figure 13.4 shows just the value part of the value matrix for the smoking decision (Figure 13.2). The first matrix (1) shows the value ranking given in Figure 13.2. The second matrix (2) shows the same ranking, but with different numbers.

Intuitively one would think that it should make *more* difference to you whether you are susceptible to the effects of smoking if you smoke than if you do not. If you do smoke, being susceptible or not is the difference between good health and serious illness before the age of 40. If you don't smoke, being susceptible or not is just a matter of some missed pleasure.

You might think that this difference is reflected in the first value matrix. The difference between 4 and 1 is 3, which is 3 times the difference between 3 and 2. But this calculation is misleading. All we had was a *ranking* of the outcomes. We chose the numbers 1, 2, 3, 4 to represent this ranking because it is convenient. In both matrices the upper left outcome is the worst, the lower right the next worst, the lower left the next worst, and the upper right the best.

In the second matrix, however, the difference between 9 and 1 (i.e., 8) is only 1⅓ times the difference between 8 and 2. Thus the ratio of the differences has no real meaning. It is merely the result of which numbers we pick arbitrarily to represent the ranking.

If we want such differences in values to be meaningful, we need to do

$$\begin{array}{cc} 1 & 4 \\ 3 & 2 \end{array} \qquad\qquad \begin{array}{cc} 1 & 9 \\ 8 & 2 \end{array}$$

$$(1) \qquad\qquad\qquad (2)$$

FIGURE 13.4   Two Value Matrices for the Smoking Decision.

more than just rank the outcomes. We must attach a more meaningful measure of our preferences to the outcomes.

At this point you might think the whole idea of applying decision theory to cases like the smoking example is misguided. How could one put a dollars-and-cents value on good health or the pleasure of smoking to a smoker? If that is what using decision theory is going to require, it may be fine for business and economic transactions, but not for decisions like whether to smoke or not. But things are not that serious. A real "dollars-and-cents" evaluation is not necessary. We only need the *differences* in values to be meaningful, not the values themselves. Indeed, the only thing required is that the ratios of the differences be meaningful, and that is more easily done.

Measuring Your Values

The fact that decision theory requires only ratios of differences in values to be meaningful means that the *scale* used to measure values is arbitrary. That is, you can pick any number you want for the least-valued outcome and any higher number whatsoever for the most-valued outcome. So all you have to do is make the differences in the numbers you assign to the other outcomes reflect real differences in your preferences for the other outcomes. This can be difficult, as we shall see, but not as difficult as putting a price on good health.

The proof that the scale is indeed arbitrary is not all that complicated, but we shall not bother to explore it here. If you are interested, see the references at the back of the book.

Since the scale is arbitrary, we can pick the one that seems most meaningful. A good choice would be to set the lowest ranked outcome at zero and the highest at 100. This could be thought of as 100 cents, 100 dollars, or even 100 thousand dollars, depending on the problem.

Let's try this out on the smoking matrix. The matrix, with these two values inserted is shown in Figure 13.5. The numbers we have inserted can be understood as follows. You are given $100 to use in buying outcomes in your decision matrix. You have already decided that the outcome of smoking and being susceptible is worth nothing. The outcome of smoking and not being susceptible, however, is worth the whole $100. Now, relatively speaking, what are the other two outcomes worth?

	Susceptible	Not susceptible
Smoke	0	100
Don't smoke		

FIGURE 13.5 Smoking Matrix with Measured Values.

Suppose you were caught in the state of being susceptible. What would you pay to get into the outcome associated with not smoking? That is, what part of your $100 would you pay to be assured of making the "right" decision given that you are in this state? Presumably quite a lot. That was your second highest outcome in the original ranking. And it represents, on our original assumptions, the difference between good health and serious illness before age 40. Would you say $95? Let's try that.

Now suppose that you are in the state of not being susceptible. What would you pay to get from the outcome involving not smoking to the one including smoking. If you value health over the pleasure of smoking, this difference could not be as great as the value difference between the other two outcomes. Try $15. That gives the lower right outcome the value $85. You might wish to write in the numbers 95 and 85, respectively, in the lower left and lower right boxes in Figure 13.5. We will refer back to the completed matrix later.

If you find that you do not agree with my suggestions, assuming, remember, that you are a smoker and like it, that is understandable. Thinking about putting numbers on your preferences for various outcomes is probably not something you have ever tried before. It takes some getting used to. And even so, it is hard to feel very confident of the numbers you do pick. For this reason we shall not rely heavily on the ability to make such comparisons. In many decision problems it is clear what to do given only the roughest idea of the relative value of the various outcomes.

Can the Values Themselves Be Wrong?

Decision theory takes the value matrix as given. It attempts to tell you how to combine your value matrix with scientific information to reach the "best" decision. From the standpoint of decision theory, then, there is no such thing as the "wrong" values. Decision theory may tell you that you made the wrong decision relative to your values, but it will not tell you that you had the wrong values.

This feature of decision theory has prompted the objection that decision theory itself is immoral. Using recommended decision strategies one might "rationally" decide to bomb a city killing thousands of civilians. Is that rational?

But this objection puts the blame in the wrong place. If one is to object to a decision on moral grounds, the finger should be pointed at those who regard bombing civilians as a "reasonable" option and, moreover, place a positive value on outcomes incorporating such actions.

This response, of course, leads to the broader question of whether there is such a thing as the right values, and if so, how one could discover which values are right. That is too big a question to take up here. It is one of the main issues in ethics and moral philosophy.

We shall proceed on the assumption that the value matrix is given. If the decision in question is your personal decision, then the value ranking that matters is your personal ranking of the possible outcomes of the decision. If the decision is a "corporate" decision, the corporation provides the value matrix—presumably in terms of its projected profits or losses. The strategies to be used in actually making the decision apply no matter what particular value ranking or value measure is given.

13.5 SUMMARY: A GENERAL VALUE MATRIX

For any decision to be made, there is a value matrix that ranks the possible outcomes according to some scale of value (preference, desirability, etc.). The value matrix for any decision has the same structure. It consists of a set of possible actions, a set of possible states of the world, and a corresponding set of outcomes (or "results") of the decision—that is, action-state pairs. Let A_1, A_2, and A_3 stand for three (exclusive and exhaustive) possible actions. Similarly, let S_1, S_2, S_3, and S_4 stand for four different (exclusive and exhaustive) possible states of the world in which the chosen action might be performed. These three actions and four states determine 12 (i.e., 3×4) different possible outcomes.

In a value matrix, each outcome is assigned a number that is either its rank in a value ranking or its value measured in some appropriate unit, such as dollars. The values are labeled according to the location of the corresponding outcome in the matrix. Each outcome is located by its row and column. In this matrix there will be three rows and four columns. The outcome in the first row, first column is labeled V_{11} (read "vee-one-one"). The outcome in the first row, second column is labeled V_{12}, and so on. The completed value matrix is shown in Figure 13.6.

In many respects, the hardest part of any decision is setting up the value matrix. Indeed, you have to make a number of prior decisions before you can make the decision in question. You have to decide which actions you regard as possible options, which states of the world are relevant, and which ranking or value measure to put on the possible outcomes. Nor are these decisions independent. The prime consideration in deciding whether some possible state of the world should be included is whether it would make any value difference if that should turn out to be the state of the world when you perform the chosen action.

Thus, even though for purposes of exposition I have presented the components of a value matrix one at a time, you would not in actual practice treat them one at a time. They are all bound up together and you should deal with them all simultaneously when formulating a real decision problem.

Decision theory does not tell you what actions, states, or values to consider. That is up to you. The advantage of knowing something about decision theory at this stage in the decision process is that it provides a framework that can help you sort out the various components of your decision.

	S_1	S_2	S_3	S_4
A_1	V_{11}	V_{12}	V_{13}	V_{14}
A_2	V_{21}	V_{22}	V_{23}	V_{24}
A_3	V_{31}	V_{32}	V_{33}	V_{34}

FIGURE 13.6 A General Value Matrix.

One way it does this is by forcing you to state your options and the relevant states of the world as exclusive and exhaustive sets. Trying to deal with overlapping options or contingencies can be very confusing.

Also, without a clear framework to guide you it is easy to associate your values, or preferences, with actions themselves, or with states themselves. This is bound to lead to some confusion and to some "bad" decisions as well. It is the value associated with the total result—that is, an action in a state—that should guide your eventual decision.

Beyond this, the only general advice would be to survey the scene carefully for additional options and relevant states, but not to include so many that there are more outcomes than you can handle. A three-by-four value matrix is about the most complex that most individuals can keep "in their head."

CHAPTER EXERCISES

Each of the following exercises contains a description of a decision problem. In each case the description contains enough information for you to construct a complete value matrix corresponding to the problem as stated. Your answer to each exercise, then, should be a completed value matrix for that problem.

13.1   BUYING A NEW BATTERY

Imagine you own a car that needs a new battery. The best place in town for batteries sells two varieties of the same make. One carries the designation "three-year" and the other "five-year." The 3-year battery costs $30 and the "five-year" battery costs $40. You are convinced that the 3-year battery will last at least 2 years and the 5-year one no longer than 6 years. So the number of years you will get out of either will be 2 through 6 years. The simplest measure of value is cost per year of use. This means that each outcome is assigned some negative number of dollars and cents representing your expense per year in owning the battery in question. (*Hint:* Your value matrix should have ten outcomes.)

13.2　PREVENTIVE ATTACK

Imagine two countries, *A* and *B*, which have a disputed piece of territory between them. Tension is mounting. The ministers of country *A* meet in emergency session to plot their course of action. They conclude that there are really only two options open to them. One is to offer to negotiate the dispute with country *B*, and the other is to attack swiftly in an attempt to take the territory by force. The main uncertainty is over what country *B* will do. Will country *B* negotiate in good faith or will it try to attack first? The ministers of *A* agree that the best thing would be to manage to attack while *B* is preparing to negotiate. That way they can get most of the territory with minimal losses. The worst possibility would be to have *B* attack while they are trying to negotiate. In that case, *B* would get most of the territory. However, if both countries attack simultaneously, being fairly equally matched they will probably each end up with about half the territory—after fighting a short war. If *B* would negotiate, they could probably end up agreeing to split the territory without a fight. (Use the numbers 1, 2, 3, 4 to rank the outcomes. If you think the best decision is obvious just from the value matrix, you are right. In Chapter 14 you will see why.)

13.3　THE WORLD II PROBLEM

Imagine thinking about the future of the world from the standpoint of the ministers of the world's most powerful countries. There are two possibilities. Either the world system is enough like a World II model (see Chapter 7) that collapse is a possibility, or it is like some economic models that postulate mechanisms that would prevent a collapse. The world powers have two options. Either they can take measures to cut the growth rate or they can continue to promote growth. If the world is a World II system, cutting the growth rate is the best action. If the world is like the standard economic models, continuing to grow is the best action. Overall, it would be better if the world were not a World II system, since trying to curb the growth rate is more difficult than letting growth continue "naturally." (Use the numbers 1, 2, 3, 4 to rank the outcomes.)

13.4　BUYING DURING INFLATION

Imagine you have decided to buy a new camera. Your only hesitation is whether to buy it now or next year. As it turns out, all the cameras you like are made in either Japan or Germany. The price of high-quality imported goods has been inflating much faster than your income. You wonder whether this rate of inflation will continue or not. If it continues, and you delay a year, you will lose "value" in that you will get the same camera but pay more for it. Of course, if the inflation in such goods declines to about the level of everything else, including your income, it will not make any monetary difference whether you wait or not. (Assume your decision to buy now or later is to be based purely on monetary considerations. Give the outcome in which you buy the camera now and there is no excess inflation the value zero. You get just what you have paid for. Call gains or losses from this amount $+x$ or $-x$, respectively.)

13.5 GENERATING ELECTRICITY

A. Imagine that the officials of your state's electric company are trying to decide whether a needed new generator should be powered by coal or by a nuclear reactor. Suppose they have reached the point where the only unknown is the nature of new federal regulations on air pollution. If the regulations stay as they are now, a coal generator will involve a net cost of $50 million. On the other hand, if the regulations are tightened, equipment to "clean" the smoke will cost another $10 million. The cost of the nuclear plant is $55 million. (Express the values in negative millions of dollars.)

B. Just as the decision is about to be made, it is learned that another federal agency is considering raising the standards for containment of nuclear materials. This changes the nature of the decision since the regulations being discussed would raise the cost of the nuclear plant another $10 million.

13.6 VACATION PLANS

A. Vacation is coming and you are trying to decide whether to go to Florida with your friends or to go home. Your main worry is about the weather. If the weather in Florida is good, it would be worth the money to go. But if the weather is bad, you would rather be home with your family and old friends. And besides, it's free. On the other hand, if you decide to go home, you prefer for the weather in Florida to be bad so you won't feel that you are missing something. But you don't tell that to your friends. (Use the numbers 1, 2, 3, 4 to rank the four outcomes.)

B. While you are deliberating, you learn that your roommate's parents are getting divorced and that your roommate has decided not to go home for the vacation. You wonder whether you should invite your roommate to come along with you, either to Florida or home. This new option introduces a new contingency as well. Your roommate may or may not accept your invitation. (Of the 16 outcomes you might consider, only 12 are genuine possibilities. It is *not* part of this exercise that you should give a value ranking for this set of outcomes. You might, however, try doing it using your own feelings about which outcomes are best. This will give you some idea how difficult it is to handle even a relatively small value matrix. That may explain why most people don't want even to think about more than two or three options.)

14

Decision Strategies

A value matrix defines a decision problem, but only rarely does the value matrix by itself determine which action you should choose. A higher-valued outcome is of course better than a lower-valued one. But in a typical value matrix, which action leads to the best outcome depends on which state the world happens to be in.

If you knew which state was going to occur when you acted, making the decision would be easy. You would just choose the action that gave you the highest-valued outcome for that state. We shall state this as an explicit "decision strategy" shortly. The more general point, however, is that determining the best action depends on what you know about the states.

Decision strategies are generally divided into three categories, depending on the nature of the information available. The case in which you definitely know which state will occur is called "decision making with certainty." The opposite extreme, in which you are assumed to know absolutely nothing about which state will occur, is called "decision making with uncertainty." The intermediate case is knowing the *probability* of occurrence for each state. This type of decision making, which uses statistical information, is called "decision making with risk." We shall treat each of these cases in turn, stopping along the way to examine interesting examples of each type.

14.1 DECISION MAKING WITH CERTAINTY

One of the main lessons of this text has been that we can never be absolutely certain of the truth of any scientific conclusion. No matter how well justified our conclusion, the fact that we have used an inductive form of argument means that it is always *possible* that we reached a false conclusion from true premises. So the idea that we might be absolutely certain which state of nature will occur is an idealization. What is called decision making with certainty is simply a situation in which the evidence for the occurrence of one

305

particular state is so good that we ignore the small chance that we are wrong. If you are unwilling to ignore such small chances, then all your decision problems will involve uncertainty or risk.

The knowledge of which state will occur may take one of two forms depending on how the actions and states are related. In some problems, the actions and states are *independent;* that is, which state will occur in no way depends on which actions you choose. In the oil drilling problem, for example (Figure 13.3), whether there is oil in that place or not is quite independent of whether the company decides to drill. Deciding to drill is not going to make the oil disappear. Let us look at an example of this type and then consider the case in which the actions and states are not independent.

Actions and States Independent

Exercise 13.1 provides a simple and convenient example. The problem was deciding whether to buy a 3-year battery or a 5-year battery for your car. The 3-year battery costs $30 and the 5-year battery costs $40. It was assumed that neither battery would last less than 2 nor more than 6 years. Our ranking of the outcomes was based on "cost per year of operation." Performing a few simple calculations, the completed value matrix is as shown in Figure 14.1.

Now suppose that you pick up an issue of *Consumer Reports* that provides you with the following very interesting information: There is no real difference in the two batteries; only the outside looks different. Actually, either will last from 3 to 5 years, depending on the use, and particularly on the climate. Long, cold winters are hard on batteries. It turns out, however, that the useful life of any of these batteries is quite predictable, given the climate. For your particular climate, you can be almost certain that either of the two batteries will last you around 4 years.

The impact of this particular information on the decision problem can be represented quite graphically. Simply cross out all the outcomes associated with the states that you know will not occur. This is done in Figure 14.2.

Thus, given the information that either battery will last 4 years, you have

	2	3	4	5	6
3-year	−$15	−$10	−$7.50	−$6	−$5
5-year	−$20	−$13.33	−$10	−$8	−$6.66

FIGURE 14.1   Value Matrix for Exercise 13.1.

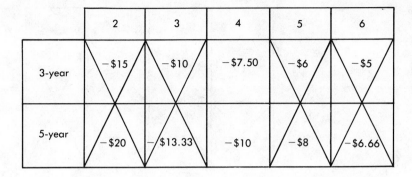

FIGURE 14.2 Value Matrix with Information Added.

only two outcomes left to consider, one associated with each possible action. Now your decision is easy. Putting a higher value on one of several outcomes means, by definition, that you would choose that outcome if you could. But now you *can* choose which of the two remaining outcomes you get because each of these outcomes is associated with one and only one of the two possible actions. To get the highest-valued outcome you simply do the action associated with that outcome. In the example, a value of −$7.50 is higher (less negative) than −$10. So you choose the action that leads to the −$7.50 outcome; that is, you buy the 3-year battery.

Before formulating this obvious strategy as a rule, let us look at the situation if the actions and the states are not independent.

Actions and States Not Independent

We can use the same example if we assume that the information we get is different. Suppose, then, that both the 3-year and the 5-year batteries come with a guarantee that the company will refund a portion of your purchase price if the battery gives out before the guaranteed time. This insures you that the 3-year battery will cost you no more than $10 a year and that the 5-year battery will cost you no more than $8 per year. But if you are hoping maybe to do better than that, you are out of luck. *Consumer Reports* tells you that there is virtually no chance that either battery will go a year beyond its guaranteed life. In fact, the company would rather refund a small portion of your purchase price since the refund is in the form of credit against the purchase of a new battery. They want to keep selling you new batteries.

This information, too, can be incorporated into the original value matrix by crossing out those outcomes that you know will not happen. The result is Figure 14.3. This time the state that occurs depends on which action you take. Buying the 3-year battery gets you the state in which your battery (or replacement) lasts 3 years at a cost of $10 a year. Buying the 5-year battery guarantees

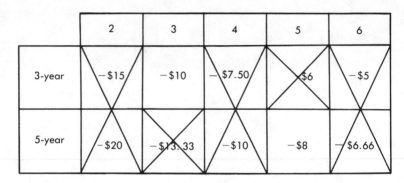

	2	3	4	5	6
3-year	−$15	−$10	−$7.50	−$6	−$5
5-year	−$20	−$13.33	−$10	−$8	−$6.66

FIGURE 14.3 Value Matrix with Information Added.

you the state in which your battery (or replacement) lasts 5 years at a cost of $8 per year.

The basic nature of the problem, however, is the same as when the actions and states are independent. You know that for each action there is only one outcome compatible with your knowledge of the situation. So you can literally choose which outcome you get. You of course take the action that yields the highest-valued outcome. That's what placing a higher value on an outcome means. In this case, the highest-valued (least-negative) outcome corresponds to buying the 5-year battery, which costs you $8 a year rather than $10.

The Highest-Value Rule

Decision making with certainty, then, is defined as a decision problem in which your knowledge of the whole situation tells you exactly what outcome will occur for each possible action you might perform. In such a context, the "rule" for making the decision is obvious:

> *Rule for Decision Making with Certainty* (Highest-Value Rule): Choose the action associated with the highest-valued outcome compatible with your knowledge of the states.

If all decisions were made in a context of "certain" knowledge of the states, there would be little need ever to consider a "full" value matrix. We did in this example because we assumed we had the decision problem and *then* got the knowledge that made it a decision problem with "certainty." This sort of thing might happen in real life, but usually you get the information when you consider the options. So you need never consider more than one outcome per action. Apart from surveying the available options, most of the work of making the decision goes into ranking the outcomes. If you can manage to rank the outcomes, you know immediately what you should do.

Unfortunately, not all decisions can be made in such fortunate circumstances. You often have to act in situations in which all the available information still leaves you with several possible outcomes for each action. After disposing of one small difficulty, we shall turn to these more interesting cases.

What If the Available Outcomes All Have the Same Value?

Continuing with the battery example, suppose you learned that the 3-year battery would last 3-years, but that the five year battery would last only 4 years, and no guarantee is offered. These two outcomes have the *same* value. Your cost in both cases would be $10 per year of use. There is no one highest-valued outcome. What should you do?

In such a situation you have two options. You can decide that the decision problem as it stands represents everything that you might regard as relevant to the decision. In that case, you really value the two outcomes equally and it makes no difference to you which action you choose. The rational thing would be to flip a coin, any coin. It doesn't even have to be a fair coin. The idea is just to make a choice and get out of there. You could even let the salesperson make the choice for you––in which case you would probably end up with the more expensive battery. But if all you really care about is cost per year of use, that should not matter to you.

Most people do not like the option of making an "arbitrary" decision. So what most people would do in such a situation is *change the decision problem* by finding some *additional* relevant state of the world that yields differently valued outcomes for the two (or more) actions.

In the present case, for example, one might consider the fact that buying the 3-year battery means coming in after only 3 years for a new battery. That is a chore most people like to avoid. On the other hand, buying the 5-year battery means handing over $40 now rather than just $30. You might well just as soon have the extra $10 in your pocket right now. Or you might consider the possibility of your car being stolen or wrecked before the 4 years are up. In that case you wouldn't get all the use of the battery that you had paid for. Better to lose a $30 battery than a $40 one.

In any case the principle is clear. If applying the highest-value rule does not yield a value difference between two options, either make the decision arbitrarily (e.g., by flipping a coin) or change the decision problem so that you do get a value difference between the options. Which you do depends on your determining whether there are any other relevant states, or whether it is even worth thinking about other relevant states.

14.2 DECISION MAKING WITH COMPLETE UNCERTAINTY

In real life one is hardly ever completely uncertain which of the states will occur. One can almost always come up with at least a rough estimate of the relative probabilities of the various states. To take account of such estimates, however, would take us into the category of *known risk*, which is the last of the three categories of decision problems. So the category of decision making with complete uncertainty, like that of certainty, is somewhat of an idealization. But it can be a useful idealization in cases in which there really is very little relevant information and perhaps no clear way to utilize what little there is.

To make a decision assuming complete uncertainty about the states is really to make the decision on the basis of the value matrix alone. Thus, if there is a "best" action, it must be determined which is best solely on the basis of the value matrix. That does not seem like an easy thing to do. And indeed it is not. But there are some systematic ways to go about it, and these are worth knowing.

Independence of Actions and States

There is one clear respect in which the phrase "complete uncertainty" is quite misleading. There may be one important thing you do know about the states, namely, that they are *independent* of the actions. For example, we may not have any idea at all whether the whole world socioeconomic system is like a World II system or not. But we do know that whether it is or not may have nothing to do with the actions we are contemplating. What happens, of course, depends on what we do, but the nature of the system itself may not. Similarly, you may have no idea whether you are susceptible to the effects of smoking, but you know that whether you are or not will remain unchanged if you decide to quit. Your decision will not change your susceptibility, although it may change what happens to you.

As a general rule, any decision problem in which the actions and states are dependent may be *reformulated* as a decision problem in which the actions and states are independent. For example, the battery problem could be reformulated by describing the states in the form: The 3-year battery lasts X years and the 5-year battery lasts Y years. To cover all combinations when each may last from 2 to 6 years will require 25 states with this form of description. Thus the problem will look more complicated, but you can be sure that the actions and states are independent. How long the batteries last is clearly independent of which you happen to choose.

Thus, unless there is some explicit indication to the contrary, our discussion of decision making with uncertainty will assume that the actions and states are *independent*.

Better (and Worse) Actions

There is one way in which one action can be clearly better (or worse) than another quite apart from any information about the states. Moreover, this way of being better (or worse) can exist even if the value matrix represents only a ranking of the outcomes.

The inflation problem of Exercise 13.4 provides a good example in which one action is clearly better than another. The problem involved buying an imported camera, but it would be the same for any item if prices are inflating faster than income. Let us think in more general terms and simply state our options as "buy now" and "buy later," where what is being bought is left open. The relevant possible states of the world are that there is rapid price inflation and that there is not. The matrix, with the values inserted, is shown in Figure 14.4.

The values in Figure 14.4 were determined as follows. If you buy now, you get the object for what it is worth now. It is convenient to identify the current price with "what the object is really worth." So you can think of yourself as losing nothing if you buy it now. You pay just for what you get, and thus suffer no loss nor make any gain. Thus the value "zero" for V_{11}. (See Section 13.5 for this notation.)

Now if there is no inflation, it makes no difference whether you buy now or later. You pay the same. But there is a difference if you wait. If there is inflation, but you buy now, you still don't lose anything. You got it for what it was worth. But if you wait, and there is inflation, then you lose the amount of the inflation. You get the same goods, but pay more. Thus the entry $-\$X$ for V_{22}.

If you think a moment about the matrix in Figure 14.4, you should be able to see that buying now is obviously the better action. If you buy now, you can't lose. If you buy later, you might lose.

There is a better way of thinking about this "can't lose" feature of "buying now." There are only two possible states of the world considered. If the first state (no inflation) occurs, then both actions are equally good. If the second state (inflation) occurs, then the first action is better. But, by assumption, one of these two states must occur. So "buying now" is as good as or better than

	No inflation	Inflation
Buy now	0	0
Buy later	0	$-\$X$

FIGURE 14.4 Value Matrix for Buying During Inflation.

"buying later" no matter what happens. It doesn't matter that you may be completely uncertain which state will occur. If you choose to buy now you cannot possibly end up worse off than you would by buying later.

Let us formulate this idea as a general definition.

> For any two actions, A_1 and A_2, action A_1 *is better than* action A_2 if and only if, there is at least one state for which A_1 has a higher valued outcome than A_2 and there are no states for which A_2 has a higher valued outcome than A_1.

If A_1 is better than A_2, we will say that A_2 *is worse than* A_1. Note that this definition of better and worse actions can be used even if your value matrix is only a ranking of the outcomes.

Eliminating "Worse" Actions

It would be silly to take a "worse" action if a "better" one is available. Thus if a value matrix for any problem exhibits an action that is worse than some other action, you can simply forget about the worse of the two. Under no circumstances would you decide to do the worse action.

As we saw, the requirement that your actions be formulated as exclusive and exhaustive alternatives tends to multiply the number of options you have to consider. Now you see that some of these extra options might be automatically eliminated once you get your values in place. Some of them may be worse than others.

It may be helpful to state a specific rule for the elimination of "worse" options.

> For any two actions, A_1 and A_2, if A_2 is worse than A_1, A_2 may be eliminated from further consideration.

Note that this rule applies only when the actions and states are *independent*. The definition of "better" and "worse" actions assumes independence.

Best Actions

In the inflation example, buying now is not only a better option than buying later, it is the *best* action available. When there are only two possibilities, the better of the two is automatically "best." But, in general, an action can be best no matter how many other options there are. It merely needs to be better than *all* the others. Let's state this as an explicit definition.

> Action A is the *best* action (BA) in a set of possible actions if and only if it is better than every other action.

Quite clearly, if your value matrix for a decision problem exhibits a best action, you should take it. We can state this simple principle as our first rule for decision making with uncertainty, the "best-action rule":

> *First Decision Rule for Decision Making with Uncertainty:* If action A is the best action in the value matrix, do it.

Thus, in the inflation example, you should buy now. That is clearly the best action.

As a sidelight, the inflation example of Figure 14.4 shows how decision theory can be used to "explain" a very interesting, and dangerous, phenomenon. If enough people in a society *believe* that there is going to be price inflation, the existence of that belief itself can cause the phenomenon that makes the belief true (i.e., inflation). It does not take a course in decision theory to realize intuitively that buying now is the right thing to do in an inflationary situation. If enough people act on their belief that there will be inflation, that will increase the demand for products. In normal periods, increased demand tends to drive up prices. So, given the usual connection between demand and price and given the obvious principle of taking a best action, the belief that there will be inflation tends to be self-fulfilling. This is one case where enough people believing that something will be true can make it true.

The same principle of decision making shows why one of the usual ways of dealing with inflation doesn't work. This is to try to get people *voluntarily* to hold off some of their spending "for the good of everyone." This leaves everyone with the decision to hold off or not depending on what everyone else does. The matrix for this decision looks just like the matrix for the original example, except that now the two relevant states of the world are "Enough other people hold off spending" and "Not enough other people hold off spending." If they do, there will be no inflation; if they don't, there will. It remains true that, *for each individual,* "buying now" is still the best action.

The same sort of reasoning applies to corporations wishing to raise prices and unions seeking to raise wages. Doing it now is a best action. Exhortations to "hold off" do not change this situation. If people who make policy are serious about bringing down inflation, they have to do something to change the value matrix of individuals and other decision-making units in the system—like putting a flexible sales tax on goods purchased *now.* If what you had to pay extra in tax now were as much as you stood to lose by inflation later, buying now would no longer be so clearly the best action. Presumably fewer people would do it, there would be less demand, and prices would not rise.

But one must be careful not to try to get too much out of such a simple principle as the best-action rule. There may be other powerful mechanisms working as well. But thinking in terms of how these other mechanisms might influence the value matrices of decision makers can provide a good way of understanding what is happening in a complex social system.

Satisfactory Actions

One might well wonder whether there is not some way to specify actions which, while not "best," are nevertheless "good enough." What would it mean for an action to be "good enough"?

The general idea seems to be something like this. In any decision problem the person making the decision has some idea of the lowest value that would be regarded as a satisfactory outcome of the problem. Thus, if there were an action that would *guarantee* at least this minimum value and no action guaranteeing any higher value, one would choose the action with the guaranteed minimum satisfactory value.

Now let us try to make this general idea a little more precise. First we need to specify the minimum value level that would satisfy the decision maker. This we do in the following definition.

> The *satisfaction level* of a decision maker for a given decision problem is the minimum value (or value rank) that the decision maker regards as a satisfactory result of the decision.

A satisfactory action, then, would be defined as follows:

> Action A is a *satisfactory action (SA)* in a given decision problem if and only if every outcome associated with action A has a value at least as great as the decision maker's satisfaction level for that problem.

We can now give our second rule for decision making with complete uncertainty, the "SA rule."

> *Second Rule for Decision Making with Uncertainty:* If action A is the only satisfactory action in the value matrix, choose it.

As an example, let us take another version of the battery problem. Suppose the batteries carry no guarantee, so that you are stuck with whatever happens. But suppose you have reliable information that the 5-year battery will definitely last 5 years, no more and no fewer. But the 3-year battery is less predictable. It may last 3 or 4 years, and there is no way of telling which. Using our earlier price information (Figure 14.1), and reformulating the states so that they are independent of the actions, we get the value matrix shown in Figure 14.5.

The matrix in Figure 14.5 has no best action. If the 3-year battery only lasts 3 years, you are better off buying the 5-year battery. But if the three-year battery will go four years, you are better off buying it. Make sure you understand this.

Even though the matrix in Figure 14.5 shows no best action, there could be a satisfactory action. If you are the buyer, and if you think that paying $8 per year's use of a battery is acceptable to you but paying $10 is not, then you should buy the 5-year battery. That choice is, for you, a satisfactory action. You

	3-year lasts 3 and 5-year lasts 5	3-year lasts 4 and 5-year lasts 5
Buy 3-year	$-\$10$	$-\$7.50$
Buy 5-year	$-\$8$	$-\$8$

FIGURE 14.5

might do better with the 3-year battery, but you also might not. Buying the five-year battery guarantees a satisfactory result in either case.

Our rule for satisfactory actions was restricted to those cases in which there is just one satisfactory action. If there is more than one, you have a further decision problem. Which of the satisfactory actions should you choose? The answer is that you should "gamble," but we do not yet know what that means when making a decision with uncertainty.

Gambling

Every value matrix has at least one highest-valued outcome. Most have only one. Let us assume only one for the moment. Then it is natural to define a gambler as a person who "goes for all the marbles" regardless of the risk. This means, in the present context, choosing the action that is associated with the highest-valued outcome. If you do this, you may not get that outcome, of course, but you couldn't get it at all if you chose some other action.

Formulating this idea as a rule for decision making (the gambler's rule), we get:

Gambler's Rule for Decision Making with Uncertainty: Choose the action associated with the highest-valued outcome.

Remember that this rule applies to decision making with uncertainty when the actions and states are independent and there is only one highest-valued outcome.

The smoking decision provides a good example of the gambler's rule in action. For convenience, this matrix is repeated as Figure 14.6.

The ranking in Figure 14.6 was determined assuming the decision maker enjoys smoking (see Section 13.4). Thus, a person with this problem who follows the gambler's rule would smoke and hope not to be susceptible to the

	Susceptible	Not susceptible
Smoke	1	4
Don't smoke	3	2

FIGURE 14.6   Ranked Value Matrix for Smoking Decision.

effects of smoking. There is just one highest-ranked outcome, and smoking is the only action that might lead to that outcome.

The matrix for the smoking decision exhibits a common pattern. The action with highest-valued outcome is also the action with the lowest-valued outcome. So if you gamble and "lose," you lose badly. You get the worst possible result. The old saying "Nothing ventured, nothing gained" seems to describe this type of value matrix.

In case there is more than one outcome with the same "highest" value, the consistent strategy would seem to be looking at the second-highest values, and taking the action corresponding to the highest second-high value as well as the highest value overall. If there is more than one of these actions, go to the highest of the third values, and so on. This strategy must in the end get you a unique decision.

The overriding question, of course, is "Should you gamble?" Let us postpone this question until we have looked at one other strategy.

Playing It Safe

Having given a clear account of what it means to "gamble" in a decision-making context, let us next try to give an equally clear account of what it would be like to be cautious. In general, a cautious person is one who is more concerned with avoiding bad consequences than getting a shot at good ones. A cautious person will pass up a chance at a very good result in favor of the guarantee of a less good result—if the action leading to the very good result might also lead to a very bad one.

The strategy we shall define is in fact a very cautious strategy, more like "playing it safe" than just being cautious. In the context of decision making, playing it safe is defined in terms of the "security level" of an action.

The *security* level of any action is the value (or rank) of the *lowest*-valued (or lowest-ranked) outcome associated with that action.

Informally, the security level of an action is the *worst* outcome you might get if you choose that course of action. But this is looking at the dark side of things. Looking at the bright side, if you choose the action in question, you are guaranteed of getting an outcome at least as valued as the security level. You can't get lower, since the security level is by definition the lowest possible.

Granted that it is nice to know where the bottom is, one would always prefer the bottom to be as high as possible. This idea is embodied in the following "play-it-safe rule" for decision making with uncertainty.

> *Play-It-Safe Rule for Decision Making with Uncertainty:* Choose the action with the greatest security level.

Among decision theorists, this rule is known as the "maximin rule"; that is, it tells you to maximize your minimum possible value.

In the smoking decision (Figure 14.6), the action of not smoking has the highest security level, namely, 2. If you are playing it safe, you will choose to quit smoking.

If there happen to be two actions with the same high value for their security level, you would presumably see which had the highest next-lower value. If these are the same, look for the highest next-lower value, and so on.

Gambling versus Playing It Safe

Figure 14.7 again shows the smoking matrix of Figure 14.6. Off to the right are the highest value and lowest value, respectively, for each of the two possible actions.

The gambler's strategy looks at the high values for each action and takes the action with the highest high value ("maximax"). The play-it-safe strategy looks at the low value for each action and takes the action with the highest low values (maximin). Is there some reason to think that one of these strategies is inherently "more rational" than the other in cases of uncertainty?

	Susceptible	Not susceptible	High value	Low value
Smoke	1	4	4	1
Don't smoke	3	2	3	2

FIGURE 14.7 Smoking Matrix with High and Low Values.

The accepted answer among philosophers and decision theorists alike is no. Taking a best action or a satisfactory action does seem to be the "correct" thing to do, but if neither of those two rules applies, then no uniquely rational choice is possible.

The main reason for taking the time to formulate the strategies for gambling and playing it safe is to have a clear definition of these two attitudes toward a decision problem. If you want to gamble or to play it safe, you know clearly what to do. And if you want to understand decisions already made, by others and even by yourself, one of these two rules may correctly describe what has been done. You may be able at least to understand what is happening, even if you cannot justify it.

The underlying reason for this unhappy situation is easy to spot. We are dealing with decision problems in the absence of any specific *information* about the states of the world other than that they are possible and that they are independent of our actions. It should not be expected that it is always possible to make a "rational" decision in the absence of specific "scientific" information. The situation is quite different if there is more information, as we shall soon see.

14.3   THE WORLD II DECISION

Exercise 13.3 contains a version of the "World II decision," that is, deciding what, in general, to do, given the possibility that our world is like a World II system. In the exercise, the decision was viewed "from the top," that is, from the standpoint of world leaders who have at least some influence on what is or is not done. In this section we shall look at the decision both "from the top" and "from the bottom"—which is the view that most of us have.

The View from the Top

In its simplest form, the decision problem has two possible actions and two possible states. The states are either that the real world is like a World II system or that it is like some economic model in which there are natural mechanisms to prevent growth from destroying the system. The two options are taking deliberate steps to limit growth and not taking such steps. The matrix for the problem is shown in Figure 14.8.

The value ranking has the same structure as the smoking decision. The worst outcome is being in a World II system in which no actions are taken to limit growth. The "best possible outcome" is being in another type of system and not taking any special measures to limit growth. And having taken measures to limit growth in a World II system is better than having unnecessarily taken such measures in a different system.

As we saw in Chapter 7, there is no really good evidence that could tell us

	Real world a World II system	Real world a different system
Take no actions to limit growth	1	4
Take actions to limit growth	3	2

FIGURE 14.8 Value Matrix for World II Decision, viewed "From the Top."

what type of system we are in. No "global" model is formulated precisely enough to test over a relatively short period of time. So the situation is pretty much one of decision making with uncertainty. No one's guess as to what our world is really like is much better than anyone else's.

Clearly neither of the actions is a "best" action. So the best-action rule is of no help. Is there, then, a satisfactory action? That is an interesting question. If there is, it has to be the action of taking steps to limit growth. No one could regard a total collapse of the world's socioeconomic system as a "satisfactory" outcome. Clearly the third ranked outcome is satisfactory. So the only question is whether taking growth-limiting measures when we need not have done so is a "satisfactory" outcome.

At this point one has to ask "satisfactory to whom"? The people who hold positions of political and economic power. They are the ones making the decisions. As a matter of fact, it seems that "they" don't regard limiting growth, if it would turn out to be unnecessary, as a satisfactory outcome. It is not even completely clear that "they" regard limiting growth if it *would* turn out to be necessary as satisfactory.

If no action is a satisfactory action, the only question left is whether to gamble or play it safe. It looks like most people in power are gambling. The few serious efforts to limit growth on a large scale have met strong opposition. Maybe a few people in powerful positions want to "play it safe," but not many.

The View from the Bottom

The "view from the bottom," that is, the view of the ordinary citizen, regardless of country, is quite different. More specifically, the possible states of the world are the same for everyone. But the actions are different. No individual citizen can take actions to limit growth on a world scale. Individuals can at most take actions to limit their own consumption of resources—including the consumption that results from having children.

	Real world a World II system	Real world a different system
Take no steps to limit consumption	3	4
Take steps to limit consumption	2	1

FIGURE 14.9   Matrix for World II Decision, viewed "From the Bottom."

The trouble with the view from the bottom is that limiting one's own consumption cannot make any difference on a global scale. Even if we do live in a World II system, whatever any individual did *now* could not make any noticeable difference to what happens in a hundred years. But it makes a lot of difference to how one lives now. And for most people, that difference is pretty strongly negative. Limiting consumption means giving up lots of things that most people do not want to give up. Thus, while everyone would prefer we not be in a World II system, there is no reason to prefer limiting consumption and lots of reasons not to. So the full decision matrix for any individual looks like that shown in Figure 14.9.

The matrix in Figure 14.9., you should realize immediately, has a "best" action. It is not to limit one's personal consumption. The situation is just like that of buying goods during inflation. For any individual, not limiting consumption is a best action, even though the result in a hundred years might be a disaster.

Looking at the whole problem through the windows of decision theory, the only "solution" would seem to be for the major societies of the world to reorganize their social and economic systems so that it pays most individuals to limit consumption. Just what in particular should be done is a difficult question. How it could be done is a more difficult question. And how it could be done "fairly" is an even more difficult question. Unless answers to these questions are found, and acted upon, we are all caught up in a great gamble that our world is not too much like a World II system. If the gamble does not succeed, it will be our children and grandchildren who will pay.

14.4   DECISION MAKING WITH RISK

Making decisions in cases of near complete uncertainty about the states is obviously risky business. When decision theorists talk about risk, however, they have in mind "known" or "controlled" risk. The risk is known or controlled by *probabilities*. So decision making with risk is defined as making a decision knowing the probabilities of all the possible states. Knowing the probabilities is a type of knowledge that is clearly intermediate between "certainty" and "complete uncertainty."

Up to now, all the decision rules we have considered can be applied even if one's values are expressed merely in terms of a ranking of the possible outcomes. Rules for decision making with risk are not so inclusive. They can be used only if the values are measured to the extent that ratios of differences in value are meaningful. (See Section 13.4.) This makes sense because probability can be thought of as a kind of "measured knowledge." If you are to combine probabilities with values, the values have to be "measured" to roughly the same extent. If they were not, the combination would not be meaningful.

We shall begin by examining one special strategy for making decisions with *uncertainty*. This is a convenient place to begin because this strategy involves converting uncertainty into probability. It thus treats decision making with uncertainty as a special case of decision making with risk.

The Rationalist Strategy

The "rationalist" strategy, like all strategies for dealing with risk, requires measured intervals of value. The example of drilling for oil provides a good case of this type. Let's reduce the problem to one with just two actions and two states. We shall use the values given in Figure 13.3, in which no pipeline is involved. The resulting decision matrix is shown in Figure 14.10. The values are expressed in thousands of dollars.

Suppose we regard this decision as being made in uncertainty. The decision maker has no idea whether there is oil there or not. This is not a realistic assumption since oil companies in fact have some idea where there is oil and where not, but it makes a good example.

	Oil	No oil
Drill	$500	− $60
Don't drill	− $10	− $10

FIGURE 14.10 Simple Oil Drilling Matrix.

Following the general method for dealing with decisions in uncertainty, you should look first for a "best" action. In this case there is none. Make sure you know why.

The next thing is to inquire about a "satisfactory" action. This requires knowing the "satisfaction level" of the decision maker. Most companies, however, would not regard any loss as "satisfactory" even if they cannot avoid it. So let us assume that neither action is a satisfactory action.

The gambler's rule says "drill," whereas the play-it-safe rule says "don't." The trouble with both of these rules in the present case is that the differences in value are fairly extreme. Losing $60,000 is a lot worse than losing $10,000. But gaining $500,000 is much better still. How about "averaging" the values of the possible outcomes?

If you average the value of the outcomes for not drilling, you of course get −$10,000. But if you average the values of the outcomes for drilling, you get ($500,000 − $60,000)/2 = $220,000. So the average value of the outcomes if you drill is much greater than the average value if you don't. This suggests the rule of choosing the action whose outcomes have the greatest average value. In the present example, following that rule would lead you to "drill."

Let's write out this suggestion as an explicit rule.

Rationalist Strategy for Decision Making with Uncertainty: Choose the action whose outcomes have the greatest average value.

For the moment we shall assume that only one action has the "highest" average.

The rationalist strategy is deceptively simple. There is a lot more to it than meets the eye. The rationalist is saying that since we are uncertain which of the two states will occur, we should treat them equally. That is what taking the average does. By adding the values of the outcomes and dividing by the number of states, one is weighing each outcome equally.

The rationalist strategy is usually justified by saying that if you are completely uncertain which state will occur, you have no reason to treat them differently. The lack of any reason to treat the states differently justifies treating them all the same. This is why it is called the rationalist strategy.

In our overall scheme for organizing decision rules, we should count the rationalist strategy on roughly equal footing with the gambler's rule and the play-it-safe rule. It is a rule one can use when facing uncertainty, if there is no "best" and no "satisfactory" action to choose. We shall not try to determine whether the rationalist rule is itself more justifiable than, say, the play-it-safe rule. It does have the disadvantage that it requires a value measure, and not simply a value ranking, over the outcomes.

Expected Value

The rationalist strategy treats the states *as if* they all had the same probability. Taking the average of the values of two outcomes, for example, is just like assuming that each state has probability of ½ and then calculating the "expected value" of each action. The rule is then to choose the action with the greatest expected value. Now let's see what the expected value of an action looks like if the probabilities are *not* equal. This will provide the basis for a rule that can be applied whenever the states can be assigned probabilities, whether equal or not.

Let us begin with a simple example. Suppose you are at a carnival and come across a game of chance. It consists of a balanced wheel divided into four equal parts. Three are blue and one is red. In order to play the game you have to pay $1. Then the wheel is spun. If it stops with the pointer in the red, you win $5. If it comes up blue, you get nothing. Your problem is to decide whether to play the game or not. The game and your decision matrix for the problem are shown in Figure 14.11.

The main difference between this matrix and others we have used is that it contains the probabilities of the two possible states. These probabilities are an added bit of information that must be included if we are to treat the problem as one of known risk.

The values used in Figure 14.11 represent your net monetary gain or loss for each possible outcome. If you play and win, you get $5. But you paid $1 to play, so your net gain is only $4. If you lose, you have lost $1. If you don't play, then you neither gain nor lose no matter which color comes up. Note that these values are all strictly monetary. It would be possible to include nonmonetary values as well, such as the excitement of watching the wheel spin and knowing that you might win $5. But such values are hard to measure, even for oneself. So we shall stick to the easily measured monetary values, at least for the moment.

Before defining the expected value of an action, let's look at several ways of approaching it informally. One way is to imagine playing the game a large number of times. On the average you could expect to win once for every three times you lost. So imagine just four games. If things went just as expected, you would gain $4 on the one win and lose $3 on the three losses, for a net gain of $1 in four games, which is 25 cents a game.

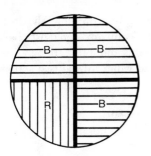

	Red Pr = 1/4	Blue Pr = 3/4
Play	$4	−$1
Don't play	$0	$0

FIGURE 14.11   **Simple game of chance and corresponding matrix.**

Another way of looking at it uses the probabilities more directly. If you play, you have a ¼ chance of gaining $4 and a ¾ chance of losing $1. But these are exclusive alternatives, so you can add the results. One fourth of $4 is $1 and ¾ of −$1 is −$.75, which adds up to $.25 (i.e., 25 cents).

Now let's look at the explicit definition of the *expected value* of an action.

> The *expected value (EV)* of an action is the *weighted sum* of the value of its possible outcomes, the weights being the probabilities of the corresponding states.

One computes a weighted sum by first multiplying each value by its corresponding weight and then adding the resulting products. For example, the expected value of playing the game represented by the matrix in Figure 14.11 is:

$$EV(\text{Play}) = (1/4 \times \$4) + (3/4 \times -\$1) = \$1 - \$.75 = 25 \text{ cents}$$

This, of course, is just what we calculated earlier.

We can now explicitly state the "expected-value rule" for decision making in cases of known risk.

> *Expected-Value Rule for Decision Making with Risk:* Choose the action with the greatest expected value.

If there happens to be more than one action with the greatest expected value, then you can simply eliminate all the other possible actions and treat the remaining matrix as a case of decision making with uncertainty. Having used your information about the probabilities to determine the expected values, you have no information left that could be used to decide which of several remaining actions to choose, if the remaining actions all have the same maximum expected value. In practice this does not happen very often, so you need not worry much about this eventuality.

In our problem with the carnival game, the rule tells us to play the game since the expected value of not playing is obviously 0, and .25 is greater than 0. That is, if you play the game your expectation is to win 25 cents a game. If you don't play, your expectation is 0.

As a further example, let us return to the oil drilling problem shown in Figure 14.10. This time, however, suppose that the company geologists have determined that hitting oil in sites like this one happens about one time in five. So $\Pr(\text{Oil}) = 1/5$ and $\Pr(\text{No oil}) = 4/5$. The expected values of the two actions are then easily calculated.

$$EV(\text{Drill}) = (1/5 \times \$500) + (4/5 \times -\$60)$$
$$= \$100 - \$48 = \$52 \text{ (in thousands)}$$

$$Ev(\text{Don't}) = (1/5 \times -\$10) + (4/5 \times -\$10)$$
$$= -\$2 + -\$8 = -\$10 \text{ (in thousands)}$$

The second calculation is really a waste of time since the value is the same for both outcomes, but it is nice to see that the "formula" does work.

Applying the "Expected Value Rule" leads to the decision to drill. That action has a substantially greater expected value. Thus, even though the probability of hitting oil is only 1/5, the value of the oil if you get it is great enough to "offset" the low probability. It is the *product* of the probability and the value that matters, not just the magnitude of either separately.

Expected Value with Actions and States Not Independent

Having partial knowledge, like having complete knowledge, allows consideration of problems in which the actions and states are not independent. In this case, the probabilities for the states will be different for different actions. So the probabilities, like the values, attach directly to outcomes, not to states. We shall set up a simple example.

Suppose you are at the carnival and are about to play the game represented in Figure 14.11, when you spot another, similar game across the fairway. Perhaps this other game has a higher expected value. You decide to check it out, thus enlarging your options. Suppose you have already decided to play the first game if the new one has a lower expectation, so we can forget about the option of not playing at all.

The new game has a wheel with 10 equally spaced numbers rather than colors. It costs $2 to play, but the payoff is $20 if the number 10 comes up.

So now your problem has two possible actions: game one or game two. There are two possible states, which we can call "win" and "lose." The net monetary gains and losses for each outcome are easy to calculate. The only major difference from the previous matrix is that the probabilities attach not to the states but to the *outcomes*. Winning has a different probability for each of the two games. The completed matrix is shown in Figure 14.12. Determining the "correct" decision—that is, the one given by the expected-value rule—is left as an exercise for the reader.

	Win	Lose
Game one	$4 Pr = 1/4	−$1 Pr = 3/4
Game two	$18 Pr = 1/10	−$2 Pr = 9/10

FIGURE 14.12

Is the Expected-Value Rule Really Best?

It is worth pausing for just a moment to consider whether the expected-value rule is really the best possible rule in cases of known risk. What if someone confronted with the carnival game of Figure 14.11 were to insist on ignoring the probabilities and playing it safe—which means not playing the game. After all, this person reasons, you cannot actually get the expected value of 25 cents on any particular play. You are either going to win $5 or lose your $1. There is no intermediate outcome. Moreover, the person might say, losing a dollar is not a satisfactory outcome, but losing nothing is. So not playing is the better action.

The usual answer to this sort of worry is that the true value of money for any person is not necessarily given by the actual dollar amounts. A child at the carnival with only $1 to spend would be foolish to risk it all on one game. You are no different in principle. Would you play that game if the price was $1000 for a chance at winning $5000? What if the price were $100,000? Or $10,000,000? If you had $1,000,000, you most likely would not risk it for a 1/4 chance of increasing it to $4,000,000. You'd take the million and run. So the value of $1,000,000 to you is much more than just 1/4 of $4,000,000. For most people, in fact, there is little difference in *real value* between one and four million dollars.

Thus, if the outcomes were measured in your "true" value units, not necessarily dollars, you would be willing to use the expected-value rule. You would not always ignore the information and play it safe.

14.5 TO SMOKE OR NOT TO SMOKE

We earlier (Section 14.2) examined the smoking decision in the context of decision making with uncertainty. In that case, smoking has to be regarded as a "gamble" for most people. But the real situation is not one of uncertainty. It is a case of known risk. There is a fair amount of information available on the health effects of smoking.

In this section we shall examine this decision in more detail from several points of view. The decision on using oral contraceptives in light of data on fatal blood clots has the same structure as the decision to smoke or not. So we can apply what we learn in examining the smoking decision to that case as well.

Smoking and Lung Cancer

You should by now know enough about both smoking and about decision making to realize that considering only the possibility of getting lung cancer is taking too narrow a view of the real decision problem. There are many other possible states that are relevant to the decision, the possibility of heart attacks being the most obvious. But there is a nice lesson to be learned by beginning with lung cancer as the only contingency considered.

To be specific, let us view the problem from the standpoint of someone between the ages of 20 and 25 who started smoking before the age of 20, say 16 or 18. Most people do not explicitly decide to smoke. It is something one just sort of falls into. Like using marijuana, the main factor at work in determining who smokes tobacco is whether one's friends (and parents) smoke.

Let's consider only two options. One is to continue smoking at a "normal" rate, say one pack to one and a half packs per day. The other option is to quit now. In general there are other options one could consider, but let us not complicate the example. If you can see what is going on in this simple example, you can work out more complicated versions for yourself.

If one's concern about the dangers of smoking is restricted to the possibility of getting lung cancer, the simplest set of states to consider would be just getting lung cancer sometime and not getting it ever. In principle it might be better to consider more specific states—for example, getting lung cancer before the age of 40, after 40, or not at all. But to keep the example simple, we shall stick to just two states.

Since there is good evidence that smoking is a positive causal factor for getting lung cancer, we have good scientific reasons for thinking that these actions and states are *not* independent. But that is fine. When dealing with known risk, it is possible to formulate the decision problem even if the states depend on the actions.

Figure 14.13 shows the matrix for this decision. The values, but not the probabilities, are indicated in the matrix. Now let me explain the value measure shown in this matrix.

For a smoker, the best outcome would be to smoke and not get lung cancer. In keeping with our general scheme for measuring values (Section

	Get lung cancer	Don't get lung cancer
Continue smoking	V = 1	V = 100
Quit smoking	V = 0	V = 99

FIGURE 14.13 Smoking Decision with Minimal Values.

13.3), that outcome gets the value 100. The worst outcome would be to quit smoking and get lung cancer anyway. So that outcome gets value zero.

The one unit difference in the other two values represents how much a smoker values smoking. This value matrix represents the smoker as being willing to pay only $1 out of $100 to be sure of making the "right" decision in either state. For example, it represents a smoker being willing to pay only $1 out of $100 not to quit smoking given that lung cancer will not be the result. This does not seem like much value to place on smoking. In fact, it is quite a lot, as we shall see.

If you are tempted to say that "continuing to smoke" is a "best action," you have made a mistake. That rule applies only when the actions and states are independent, which in this problem they are not. The fact that this mistake is easy to make may help to explain why people who don't want to quit smoking are so anxious to deny that smoking causes lung cancer. If you regard getting lung cancer as being independent of smoking, and regard it as completely unknown whether you will get lung cancer or not, continuing to smoke looks like a best action. "If you get it, you get it. If you don't you don't. Who knows? But I know that I like to smoke."

Turning, finally, to the probabilities, what we need to know is the percentage of smokers age 20 to 25 who eventually get lung cancer and the percentage of nonsmokers age 20 to 25 who get lung cancer. These probabilities measure the true risk of smoking for someone 20 to 25 years old.

It may at first surprise you to learn that estimates of these probabilities are difficult to find. For example, no such figures are to be found in the complete version of the famous 1964 "Report of the Advisory Committee to the Surgeon General of the Public Health Service." That was the report that led to warning labels being printed on all cigarette packs and advertisements.

When you think about it, however, it is easy to see why these probabilities are not generally available. Most of the detailed data on smoking and health comes from prospective studies on people (mostly men) over 40 years old. These studies have usually been carried out over a period of 3 to 5 years. So what is known is how many people *in the study* died of lung cancer during a 3- to 5-year period. Most published reports of the data do not even give you these numbers. They tell you the *mortality ratio* of smokers from lung cancer. (Recall Section 12.3.) Among smokers, the death rate from lung cancer is roughly *ten times* the rate for nonsmokers. This fact, however, is of little help in decision making. It only tells you the *relative risk* in a group of the type studied. It does not tell you your risk.

The other figure you are likely to see is the annual *number* of deaths due to lung cancer. In the United States this is currently about 40,000 *per year*. We know that about ten times as many smokers as nonsmokers get lung cancer. So that works out to about 36,000 smokers dying from lung cancer each year. The total number of smokers is estimated to be about 70 million. Thus the probability of a smoker dying from lung cancer in any given year is about 36,000/

70,000,000, or about 1/2000. In deaths per 100,000 population, which is a standard measure, that is 50/100,000. (Remember what we said in Chapter 7 about learning to think in round numbers.)

Just for comparison, you might recall that the number of deaths occurring in auto accidents in the United States is roughly 50,000 per year. Assuming all 200 million Americans ride in automobiles, that works out to be 1/4000 or 25/100,000. So, if you smoke, your chances of dying from lung cancer in any particular year are about double your chances of being killed in an auto accident.

That ratio of 1/2000 for smokers dying from lung cancer in the United States is just for one year and for all ages. Your probability of dying from lung cancer in your whole lifetime, beginning from age 20, must be much greater than that.

The best estimate I have seen is from the large matched-pair study that involved about 37,000 each of smokers and nonsmokers ages 40 to 75. In roughly 3 years, there were 110 deaths from lung cancer among the smokers, which is roughly 1/300. There were 12 cases of lung cancer among the 37,000 nonsmokers, or roughly 1/3000. But that was only over a 3-year period. The probabilities for a group beginning at age 20 to 25 must be higher than that. Let's settle on 1/200 for smokers and 1/2000 for nonsmokers. That may be a little high, but, as you will see in a moment, it doesn't matter. Also, we are assuming that quitting before age 25 is as good as never smoking. This is not strictly true, but it is close.

Going back to the matrix in Figure 14.13, the probability for the outcome "continue smoking and get lung cancer" is 1/200. For "continue and not get lung cancer" it is, of course, 199/200 (total sum rule.) For quitting, the probabilities are 1/2000 and 1999/2000 respectively. All we need do now is calculate the *expected value*.

$$EV(\text{Smoke}) = (1/200 \times 1) + (199/200 \times 100) = 99.5$$
$$EV(\text{Quit}) = (1/2000 \times 0) + (1999/2000 \times 99) = 98.9$$

So the expected value of continuing is greater than the expected value of quitting. According to the expected-value rule, the rational decision is to continue smoking!

Has something gone wrong? No. It is just that the difference in probability between 199/200 and 1999/2000 is not great enough to offset even the one unit of value between 100 and 99. For the expected value of quitting to be greater, either the probability of lung cancer would have to be *greater*, say 1/100 rather than 1/200, or the value difference would have to be *less*, say 1 out of 500 or 1 out of 1000.

But 1/200 is already a fairly high estimate of the probability for lung cancer for smokers. And hardly anyone could claim to measure their values in units smaller than 1 in 100. In short, the danger posed by lung cancer seems an acceptable risk.

For most people, the decision to smoke is comparable to the decision to drive, though perhaps not quite as "good." Most people would put a higher relative value on driving than on smoking. And the risk of death in an auto accident is only half that of lung cancer for a smoker.

If you smoke, however, it is too soon to rejoice. Smoking and quitting have about the same expected value if you consider only the risk of death from lung cancer. But that is not the only risk. We did not get the "wrong" answer, we have been considering the wrong problem!

One final point. Our analysis of the smoking decision based on the contingency of getting lung cancer again points up the difference in viewpoint between individuals and the society as a whole. Just looking at lung cancer, the individual may see smoking as an acceptable risk. But from the viewpoint of public health officials, 36,000 extra cases of lung cancer a year is a big worry. It places a large burden on public medical facilities and health insurance, including Medicare. Such considerations can justify public officials in discouraging smoking through taxation and other means. The problem of reconciling individual freedom with social needs is ever present.

Smoking and Health

A wise decision on smoking requires that you look at all the "relevant" consequences. It would be impossible to take account of the various consequences singly, but they can be dealt with all together. There is information available on the *overall* life expectancy of smokers and nonsmokers, if you look for it. Here again the tendency is to report data that shows that smoking *reduces* life expectancy. But that data is not directly relevant to decision making. You need to know your risks directly, not the relative risk in some different population.

Figure 14.14 shows estimates of life expectancy published by the American Cancer Society. These are obviously relevant to our decision problem.

Sometimes it is easier to formulate a decision problem in terms of the information available rather than try to find the information required by a problem already formulated. In this case, the information given allows us to consider only the states "die before age 65" and "live past age 65." If you are under 25, it may be difficult to think about being 65. Data for age 30, 40, or even 50 might be more relevant. But the source at hand does not have this information.

As before, the options under consideration are continuing to smoke about a pack and a half a day and quitting altogether. And again we assume that quitting before age 25 is about as good as never smoking. As for values, let's use the same "minimal" values we used before. So the matrix for our new problem, complete with values and probabilities, is as shown in Figure 14.15. The expected values are shown to the right of the matrix. You may want to check to see that these are indeed the correct figures.

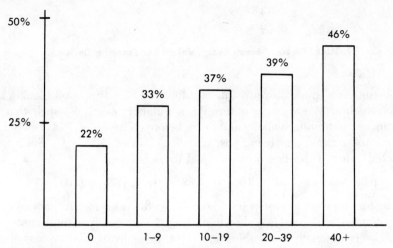

FIGURE 14.14   Percentage of 25-year-old Males Expected To Die before 65.

In our new problem, then, quitting smoking has the higher expected value by a wide margin. Applying the expected-value rule, the correct choice is to quit.

In the matrix of Figure 14.15, the difference between a probability of .62 and a probability of .78 is more than enough to offset the one-point difference in values between the two corresponding outcomes. The only reply that a smoker might make is that the value difference is just too small. A smoker in the state of living past 65 would clearly pay more than $1 out of $100 to get from the outcome involving quitting smoking to the outcome of continuing to smoke. How much more?

Let's turn the question around. By how much would you have to decrease the value difference between dying before 65 and living past 65 in order for the two actions to have the *same* expected value? That value difference would mark

	Die before age 65	Live past age 65	Expected Value
Continue smoking	V = 1 Pr = .38	V = 100 Pr = .62	62.4
Quit smoking	V = 0 Pr = .22	V = 99 Pr = .78	77.2

FIGURE 14.15

14	100
0	86

FIGURE 14.16 "Break Even" Values for Smoking Decision.

the "break-even" point on your scale of values. This is a simple problem in high school algebra. We want to determine a number, call it V, such that the outcome of continuing smoking and dying before 65 has value V. Likewise, the outcome of quitting and living past 65 will get value $(100 - V)$. Setting the expected value of the two options equal to each other,

$$[.38 \times V] + [.62 \times 100] = [.22 \times 0] + [.78 \times (100 - V)]$$

Solving for V, we get a value of just about 14. So the value matrix that makes the two options equally good is as shown in Figure 14.16. So if you think continuing to smoke is rationally justifiable, you must value living past 65 less than a difference of 86 points on a scale of 100. That seems a high relative value to place on smoking over living into retirement, but we are not in the business of questioning a person's values. We are merely pointing out what relative values the decision to continue smoking represents.

Most people don't have much feeling for rating their values. Let us put these numbers in a form that is more easily understood. Suppose that you were giving grades to outcomes. At 100, the outcome of smoking and living past 65 gets an "A+". But living past 65 and quitting smoking gets only a "B". If you continue to smoke, you are saying that for you smoking makes the difference between an "A+" life and a "B" life. Can smoking really be that important?

Sometimes it can. The story is told of a famous scholar who tried hard to give up smoking. Once he stopped smoking, he found he couldn't write. He decided that being able to do his life's work was worth the risks of smoking. But this is an extreme case. Few people can claim such an "excuse."

Confronting Your Values

Decision making with risk is usually presented as a process of using known probabilities together with the values of outcomes to determine which action one should choose. This works fine when the values are given in dollars and cents, as in the case of a lottery. But for more qualitative values, like the value of smoking to a smoker, this approach works less well. In such cases it is sometimes helpful to put the decision problem in a form that does not require an *explicit* assessment of values. Rather, your values are confronted indirectly through the contemplation of the choice to be made.

A simple way to do this is to imagine each possible action replaced by a jar containing 100 marbles. You need one jar for each possible action. (You don't have to imagine. You could actually get jars full of marbles.) The marbles must come in different colors. Each possible state of the world is assigned a color. So you need as many different colors as the problem has states. With dependent actions and states, the probabilities of the states are different for each action. You represent the probabilities of the states by the relative number of balls of the various colors. A state with a probability of 1/2 for a given action would be represented by having 50 of the 100 marbles being the color assigned to that state.

Once all the jars are set up correctly, your decision to choose one of the actions is equivalent to deciding which jar to pick. Having picked a jar, you must select one marble at random from that jar. The color of the marble selected tells you which state, and thus which outcome, you have achieved. Since you know the relative numbers of colored marbles in each jar, in picking a jar you are implicitly weighing your desire for the corresponding action against the probabilities of the various states. Imagining reaching in and picking a marble is a graphic way of confronting your true feelings about the decision.

Now we shall apply this device to the smoking decision shown in Figure 14.15. Let a red marble represent the state of dying before the age of 65. A blue marble will represent the state of living past 65. The jar that represents the option of continuing to smoke will have 38 red marbles and 62 blue ones. The jar representing the option of quitting has 22 red marbles and 78 blue ones. You have to select a single marble at random from one of these two jars. Selecting a red marble means that you will die before the age of 65. Which jar do you choose? Your choice is pictured in Figure 14.17.

This way of presenting the problem brings you face to face with your values because you have to ask yourself: Is continuing to smoke *worth* facing those additional 16 red marbles in the "smoking" jar? If you smoke, only you can answer that question.

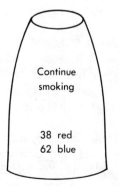

Continue smoking

38 red
62 blue

Quit smoking

22 red
78 blue

FIGURE 14.17

Blood Clots and the Pill

When data linking oral contraceptives with fatal blood clots first became known (see Section 12.5), the headlines, of course, read "Pill Causes Fatal Blood Clots." Many people were, quite naturally, rather alarmed. Some decided immediately to stop taking the pill. Meanwhile, those who read the fine print learned that the U.S. Food and Drug Administration (FDA) regarded the evidence as being only preliminary results of retrospective studies. Moreover, they were recommending that women not go off the pill in mid-cycle, and later only after consultation with a doctor and after considering other methods of contraception. Were those who immediately went off the pill acting too hastily? Was the FDA giving good advice? These are questions we can now answer for ourselves and could have at the time using information that was readily available.

In the first place, we now know enough not to jump to conclusions just because something has been shown to *cause* some undesirable effect. What matters for decisions is how *effective* a cause it is. And once that is known, one may still find that the risk is outweighed by the benefits of the causal agent.

We also know that effectiveness cannot be estimated from retrospective data. So even knowing the details of the relevant studies is not much help in making a rational decision. But it did not take demographers and epidemiologists long to come up with fairly reliable estimates of the relevant probabilities. What they came up with, and what the FDA used in making its recommendations, was this: Among women of childbearing age, the frequency of fatal blood clots is about 2 per million. Among women in that age group who take oral contraceptives, the frequency is roughly 7.5 times greater, or 15 per million. The newspapers, of course, latched onto the figure of 7.5 times greater. That is indeed a good indication that the pill is a causal factor for fatal blood clots. But the relevant information for making a decision about taking oral contraceptives involves a figure of 2 in a million. That is indeed a very small risk. And multiplying it by 7.5 still leaves a very small number. That is, 1.5 per 100,000, which is low compared, for example, to one's yearly chance of being killed in an auto accident (25 per 100,000).

By now you should have noted that the structure of this case is very similar to that involving the connection between cigarette smoking and lung cancer. The causal factor multiplies the death rate by 5 or 10, but the rate is so low to begin with that even multiplying it by 10 leaves a risk low enough to be outweighed by even a fairly small difference in values. But let us go through the details of the decision.

Contrary to what many people seem to think, including some doctors and public health officials, the choice is not between oral contraceptives and none at all. Just formulating the options this way already biases the decision in favor of taking the pill. The realistic options are between oral contraceptives and other types, of which there are several.

Since the data concern fatal blood clots, the possible states of interest are whether one develops a fatal clot or not. You should now be aware, of course, that ultimately this is much too narrow a set of relevant states on which to base the decision. But the question we are considering at the moment is whether it was rational to stop taking the pill on the basis of the data on fatal clots. These are the right states to consider in answering that question.

Values, of course, are a personal matter. But as in the case of smoking, we can learn a lot about the decision by considering what would for most people be the minimal positive value in favor of oral contraceptives. Other things being equal, most women prefer the pill to other available contraceptives. The question is just whether other things are equal. This time let us measure the values on a scale of 0 to 1000—a very sensitive scale indeed. And we shall assume a value difference in favor of the pill of only 1 unit. With these assumptions, the decision matrix, complete with probabilities, is as shown in Figure 14.18.

As you can see from Figure 14.18, the expected value of taking the pill is still greater than that of using other contraceptives, even though the positive value put on using it is minimal. The reason for this should be intuitively clear to you by now. With a probability of only 15/1,000,000 of experiencing a clot when taking the pill, the expected value for this action will be almost the same as the value associated with not getting a clot. In short, even though the probability of dying from clots is 7.5 times greater if you take the pill than if not, it is still much too small to offset even a very small positive value placed on using the pill rather than other contraceptives.

At the time this data were first released, the FDA argued that even with the risk of fatal clots, taking oral contraceptives is a lot safer than pregnancy and childbirth. Among women not using contraceptives, this risk is roughly 210/1,000,000, which is about equal to the yearly risk of death in an auto accident. Now this figure is misleading in that the alternative to using oral contraceptives is not using nothing at all. There are other effective methods. But women who

	Get fatal clot	Do not get fatal clot	Expected Value
Use pill	1 15/1,000,000	1,000 999,985/1,000,000	999.985
Use other contraceptive	0 2/1,000,000	999 999,998/1,000,000	998.998

FIGURE 14.18

just stopped taking the pill without first securing an effective alternative put themselves, at least temporarily, in this category. Remember too that when this all happened, abortion was illegal. It is thus a reasonable guess, though one hard to support with solid data, that among women who decided to go off the pill because it causes fatal clots, more died from the effects of a badly done illegal abortion than would have died from blood clots. The results of making a bad decision can sometimes be quite tragic.

It is fairly clear now that deciding not to use oral contraceptives because of the risk of fatal clots was not a good decision. But just as one should not decide to smoke or not based solely on the risk of lung cancer, so one's decision to use oral contraceptives should not hinge on just this one possible effect. One wants a broader measure of the overall health consequences of oral contraceptives. As in the case of smoking, life expectancy would be a good measure. Unfortunately, such information is also not as widely disseminated as it should be. There have been several prospective studies in progress for 10 years or more. So far there have been no clear differences in life expectancy between women using the pill and women not. But many effects, like cancer, take 20 years or more before they begin to show up. Thus one should be on the lookout for reports on these studies and be prepared to reconsider the decision in light of any new information.

14.6   SUMMARY OF DECISION STRATEGIES

All the decision strategies discussed in this chapter are summarized in the "tree" on page 337. Keeping this diagram in mind will permit you to categorize any decision problem and proceed to an appropriate decision rule. The only complicated part of the tree is the branch for decision making with uncertainty. This branch has several subbranches depending on whether the matrix exhibits a best action or a satisfactory action. If neither, you are left with a personal choice of which rule to follow.

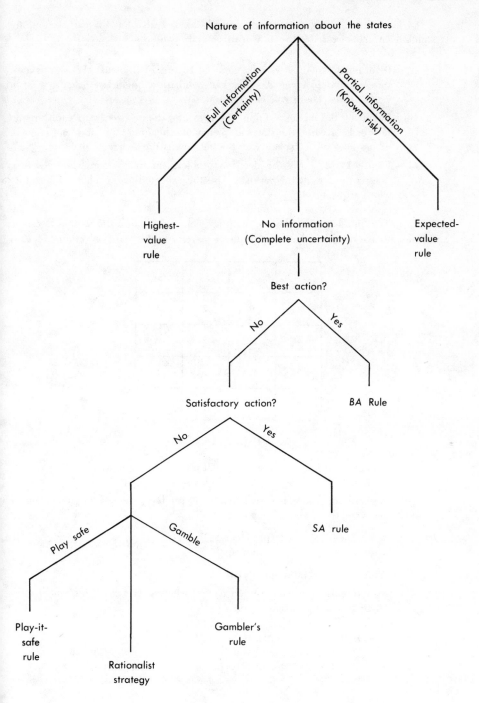

FIGURE 14.19

CHAPTER EXERCISES

14.1 In this chapter there were several "loose ends" left for the student to straighten out. This exercise directs you to several of these loose ends.

A. Reformulate the "battery decision" of Figure 14.1 so that the actions and states are *independent*. Assume that each battery might last either 3, 4, or 5 years. (*Hint:* There will be nine states.)

B. Following the model of Figure 14.4, exhibit the decision matrix for buying now or later when the states are "enough others hold off" and "not enough others hold off." Explain why buying now is the "correct" decision.

C. Figure 14.12 shows the decision problem for the person who sees two different games at the carnival. Use the expected-value rule to determine which game to play.

14.2 This problem provides you with a quick review of all the various decision strategies discussed in this chapter. All questions refer to the matrix of Figure 14.20.

	S_1	S_2	S_3
A_1	8	2	7
A_2	10	3	9
A_3	12	6	1
A_4	4	5	11

FIGURE 14.20

A. Suppose you know that state S_3 will occur no matter which action is performed. Which action should you choose?

B. Suppose you know that each action will produce one and only one state as follows:

A_2 produces S_2. A_1 produces S_3.
A_4 produces S_1. A_3 produces S_2.

Which action should you choose?

C. Now suppose you have no information whatsoever concerning which state might follow any action.

(i) Is there a best action?
(ii) Which action should you choose if you are playing it safe?
(iii) Which action should you choose if you are gambling?
(iv) Which action would you choose if you follow the "rationalist strategy"?

D. Which action, if any, would you choose if your satisfaction level were: (i) 4, (ii) 6?

E. Consider now the "reduced" decision problem which involves only actions A_1 and A_2 and states S_1 and S_2. For this question ignore all other outcomes in the original matrix.

    (i)   Suppose you still have no information concerning the occurrence of any state. Which action would you choose?

   (ii)   What rule did you follow in making this choice?

F. Still considering the reduced problem with only four outcomes, suppose you are now given additional information concerning the probabilities of states relative to the possible actions.

    (i)   The information is that no matter which action you choose, $Pr(S_1) = 1/4$ and $Pr(S_2) = 3/4$. Which action should you choose? Exhibit any necessary calculations.

   (ii)   Which action is best if the information is as follows?

        If A_1, then $Pr(S_1) = 3/4$ and $Pr(S_2) = 1/4$.
        If A_2, then $Pr(S_1) = 1/4$ and $Pr(S_2) = 3/4$.

        Exhibit any necessary calculations.

14.3 A Raffle A friend stops you on the street saying that his fraternity is having a raffle and you now have the opportunity of buying the very last ticket for only $5. The prize, which has been donated by a rich alumnus, is worth $1000 cash. Being somewhat suspicious, you inquire as to the number of tickets being sold and are assured there are only 250. Indeed, the ticket in question has the number 250 on it. Having a spare $5 on you, you pause to consider whether to buy the ticket or not.

Set up the decision matrix for your situation, including all the relevant information given above. Assume that your interest in buying a ticket is based solely on consideration of monetary gain or loss (and not on friendship, love of gambling, etc.). If you apply the standard decision rule for this type of problem, which action should you choose? Be sure to exhibit any necessary calculations.

14.4 Life Expectancy and the Pill Suppose that reliable studies were to show that if a woman takes the pill continuously from age 20 to age 45, her chances of living to the age of 65 are 70 percent. If she does not take the pill, her chances of living to age 65 are 80 percent. (These figures are fictional; no such reliable data exist.)

A. Set up the decision matrix for the problem of whether to take the pill or not. Use any value scale you wish.

B. Represent the decision as a choice between drawing colored marbles from one of two jars.

14.5 Generating Electricity Go back to Exercise 13.5. If you have not already done so, construct the value matrices for both parts A and B. Assume that the utility company is operating under complete uncertainty as to what the government agencies will decide.

    A.   For the matrix of Exercise 13.5 part A, which action is best if the directors of the utility want to gamble? Which if they want to play it safe? Which if they try the rationalist strategy?

    B.   For the matrix of Exercise 13.5, part B, which action is best if the directors of the utility want to gamble? Which if they want to play it safe? Which if they try the rationalist strategy?

14.6 The Prisoners' Dilemma The following is a classic problem in decision theory. Read the "plot" and then answer the questions that follow.

> Suppose that you and a friend are engaged in some illegal activity and are "busted." The district attorney puts you in different cells and then comes to you with the following "bargain." The evidence against you both is pretty good, but the case would be stronger if one of you would provide more information. So the DA promises that if you talk you will get off with at most 6 months—provided your partner doesn't also talk. On the other hand, if you don't talk and your partner does, you are likely to get 3 years in the state prison. Knowing a bit about decision theory, you realize that there are two other outcomes to consider. So you ask the DA what happens if neither of you talks, and if both of you do. Being an honest person, the DA admits that without some additional evidence, you will both get off with at most a year. If you both talk, on the other hand, you will both likely get 2 years.

    A.   Set up the decision matrix for your situation. Use negative years in jail as a measure of the value of the outcomes.

    B.   If you treat your situation as being one of complete uncertainty and follow recommended decision strategies, what should you do? Which decision strategy have you used in reaching this decision?

    C.   The DA goes to your partner with the same "deal." Suppose your partner also treats the situation as one of uncertainty and applies the standard rules. What is going to happen to both of you?

    D.   Do you think your overall situation could be improved if you and your partner were allowed to discuss the problem with one another? How might that change the decision problem? How might it not change the problem?

14.7 The Tragedy of the Commons The same logic that operates in the prisoners' dilemma can be illustrated in more realistic contexts. One well-known example is called "The Tragedy of the Commons."

> In the nineteenth century it often happened that a number of families shared a small field in common. This field was used for raising supplementary animals. To keep the example simple, let us suppose that the families all raise cows, and that at the moment each family has nine. They all sell milk and butter to nearby townspeople, each earning roughly $95 annually. Things have been prosperous,

and each family has been adding a new cow every year or so for a number of years. But the Commons has about reached its limit. It can't take much more grazing.

Suppose you are one of these people, and you are wondering whether to add a new cow this year. You realize that if everyone adds a new cow, the yield per cow will go down due to lack of good grass, so everyone's net gain will drop to $90 a year. On the other hand, if you add a new cow and no one else does, that won't hurt the field so much and you can expect your income to go up to $100. The other side of this coin, however, is that if everyone else adds a cow and you don't, the yield per cow will go down and you will net only about $85.

A. Set up the decision matrix for your situation using the values given.

B. If you treat your situation as one of uncertainty regarding the actions of your neighbors, what decision rule applies? What should you do if you follow this rule?

C. What happens if each of the other families analyzes the situation in the same way?

D. Can you suggest a way that everyone could achieve a better outcome?

14.8 Imagine several steel companies located along the same river. They all both take in water from the river and dump wastes into it. Suppose each of the companies is considering expanding operations in spite of the fact that the river is getting dangerously polluted. By making reasonable monetary and economic assumptions, show in detail how the situation of the steel companies could be exactly like that of a group of villagers in Exercise 14.7. The river is their Commons.

14.9 Analyze the example of Exercise 13.2 (Preventive Attack) following the pattern of "the prisoners' dilemma." Using the appropriate decision matrix, explain what happens if both sides act "rationally" given their situation of mutual uncertainty. What is the "moral" of the story?

14.10 Imagine that the Disease Control Center in Atlanta has just issued a bulletin that there is an impending flu epidemic. The flu is serious but only rarely fatal—typically leaving one unable to do much for a solid month. Fortunately there is a very effective vaccine ready to be used, and it is being made available free to anyone who requests it. But unfortunately this flu vaccine has the side effect that it may cause temporary paralysis, which, though again rarely fatal, also has the effect of leaving one incapacitated for a month. The officials are wary of making recommendations and are leaving the decision whether to take the vaccine or not up to individuals.

A. Set up the decision problem giving your possible actions and the relevant possible states of the world, but leaving out everything else. (*Hint:* There are more than two relevant states of the world to consider.)

B. The way things are, the disvalue of getting either flu or paralysis is roughly equal. If you were so unlucky as to get both, you would be incapacitated the same length of time—you would just be more miserable. Also, assume that you have no great aversion to the process of getting a shot; it is only the possible effects that worry you. In this situation it would be reasonable to measure the relevant values in terms of days of illness, for example, getting the flu is worth −30 days. Now fill in the values in your matrix. (The relative measure of all relevant values is completely determined by the statement of the problem thus far.)

C. If you have done everything right so far, you will see that treating the problem as one of complete uncertainty leaves you with no way to make the choice according to standard decision rules. In any case, you know that such decisions are best made on the basis of the relative *risks* involved, and these have not yet been given. Suppose you investigate further and read that the vaccine is indeed very effective against the flu, which is very contagious. In the whole population, the estimated risk of getting flu *without* a shot is 1/10; but *with* a shot only 1/10,000 (1000 times less). That is good. But reading further you see that the probability of getting the paralysis if you do *not* take the shot is only 1/100,000; but if you *do* have a shot it jumps to 1/100 (1000 times more). That is bad. You are going to have to make your decision on the basis of the relative expected values involved. You cannot refuse to play. Only one further assumption is needed to complete your matrix. This is that the probabilities of flu and paralysis are independent. Getting or not getting flu does not make it more or less likely that you will get the paralysis or vice versa. Now complete your matrix by inserting the relevant probabilities. (You need not figure these exactly; small differences of 1/100 or less won't matter.)

D. Finally, which action has the greatest expected value? Give the approximate relative expected values of the two actions according to the standard method of calculation. (You may be able to answer this question intuitively even if you are not sure of all the probabilities in your matrix. If so, explain your answer.)

REPRESENTATIVE
ANSWERS TO EXERCISES

Chapter 1

1.1 A. (a) Your only grounds for disagreement would be that either the researcher or the reporter simply did not know what they were doing. This is not likely.

B. (a or b) In this context, "significant" does not mean "important" or "interesting." It has a purely statistical interpretation which is roughly that the difference is so great that there is little chance it is just a random fluctuation. Most likely something is producing a real difference.

C. (b or c) The worry here is that maybe there are not more such crimes during full moons, only that for some unknown reason more of them get *reported* when the moon is full.

D. (b or c) Maybe there was something special about the Louisville area during those two years that led to the increase in the number of sexual crimes reported.

E. (c) The headline is catchy, but unjustified. The data consists merely of a statistical correlation which by itself provides no evidence of causation.

F–I. (c) The article contains no support for any of these assertions other than the authority of the "experts" (i.e., the sociologist and the astrologer)—and they disagree.

J. (e) For an average person living in that area during those two years, the chances of being a victim in one of those reported crimes could not be more than about 400/N during the full moon period and about 300/N during each of the other periods—where N is the total population of the area. So your chance of being a victim was actually ⅓ (33%) greater during the full moon period! In either case, however, the chances are small. Assuming a population of roughly 400,000 (a reasonable guess for a city like Louisville), the full moon period for the whole two years gives a chance of 1/1000, or .1%. The other periods give a chance of 3/4000, or .075%. Between .1% and .075% there is not much to choose.

K. (b or c) See comments on J.

L. (e) See comments on J.

Chapter 2

2.1.   A. A contingent statement.

B. Not a statement.

C. A tautology.

D. A definition.

E. A contradiction if taken literally. Our linguistic training leads us to try to reinterpret such statements so that they are not contradictory, as in (F).

2.2.   A. (c)

C. (c)

E. (a)

Chapter 3

3.1.   A. (d) If you answered (e) you are making it impossible to say that an argument ever fails to justify its conclusion.

C. (d)

E. (c)

3.2.   A. (c)

C. (b)

F. (a)

Chapter 4

4.1.   A. Taxes being cut is sufficient for the rate of inflation increasing. (Or, an increasing rate of inflation is necessary for taxes being cut.)

C. Being sixteen is necessary for being issued a driver's license. (Or, not being at least sixteen is sufficient not to be issued a driver's license. Or, not being issued a driver's license is necessary for not being at least sixteen. Or, being issued a driver's license is sufficient for being at least sixteen.)

E. Being ratified by two-thirds of the states is necessary for a proposed amendment to the US Constitution to become law.

4.2. C. If you are issued a driver's license, then you are at least sixteen years old. (Or, if you are not at least sixteen years old, then you will not be issued a driver's license.)

E. If a proposed amendment to the US Constitution becomes law, then it has been ratified by two thirds of the States. (Or, if a proposed amendment to the US Constitution is not ratified by two thirds of the states, then it does not become law.)

4.3. B. (a)–(b) If Smith does not satisfy the language requirement, then he can't graduate.
He is not satisfying the requirement.
Thus, he is not going to graduate.
(c) If P, then Q.   "P" is "Smith does not satisfy
P.   the language requirement."
Thus, Q.   "Q" is "Smith does not graduate."
(d) Affirming the antecedent. Valid.
Or, (a)–(b) If Smith graduates, then he has satisfied the language requirement.
He is not satisfying the requirement.
Thus, he will not graduate.
(c) If P, then Q.   "P" is "Smith graduates."
Not Q.   "Q" is "Smith fulfills the
Thus, Not P.   language requirement."
(d) Denying the consequent. Valid.
Note: Either way you get a valid argument with the same conclusion.

C. (a)–(b) If there is a tax cut this year, then economic conditions will improve.
There will not be a tax cut this year.
Thus, economic conditions will not improve.
(c) If P, then Q.   "P" is "There is a tax cut this
Not P.   year."
Thus, Not Q.   "Q" is "Economic conditions will
                        improve.
(d) Denying the antecedent. Invalid.

4.4. B. Not P and Not Q.

Chapter 5

5.1. A. (H) This follows immediately from the theoretical hypothesis that a shooting star and the earth form a Newtonian system.

B. (T) This is just a statement of the "law of dominance" for Mendelian systems.

E. (N) This is tricky. That the apple will accelerate follows from the hypothesis that it is part of a Newtonian system. But that it will fall with a specified value of the acceleration does not. That requires additional, factual information about the system, e.g., the mass of the earth. The hypothesis that the earth and the apple are a Newtonian system does not contain any information about the exact mass of the earth.

H. (N)

I. (N) See comments on E.

5.3. The reference is most likely to Keynes' *theory* of economics. The other two possibilities don't make much sense.

5.4. They were saying that a nucleus is like a drop of liquid, e.g., a drop of water, and using analogies based on the known properties of drops of liquid to make conjectures about the structure of the nucleus. This model seems to have been about as useful as the raisin pudding model of the atom.

5.5. You might want to say that the system is partly deterministic and partly stochastic, but you were not given that option. The best choice is stochastic. Mendelian inheritance systems are deterministic for some states, e.g., when both parents are true-breeding.

5.9. If two hybrid, brown-eyed parents have a blue-eyed child (this result has a probability of only one quarter), would you say that the choice of the child's eye color was an act of "free will"? The physicists' claims make about as much sense as this. The moral of the story is that even brilliant scientists may spout nonsense when they get outside their specialty. Being an authority on "quantum mechanics" does not make you a philosopher.

Chapter 6

6.1. A. The initial conditions are the observed positions of Uranus. The prediction is that there would be another planet at the specified place and time. The only auxiliary assumption you can infer from the presentation is that nothing *else* interferes.

B. Condition 1 is true because the prediction is calculated (deductively) from H together with IC and AA.

C. Condition 2 is true because it was very unlikely that there should just have happened to be a heretofore unknown planet at that time and place—unless the hypothesis was very close to being right.

D. Prediction successful.

E. H is (to a very close approximation) true.

6.2. The straightforward conclusion is that H is not true, within the limits of approximation that we can measure. Its deviations from the truth are detectable. But given the many previous successes of Newton's hypotheses, most scientists continued to believe that "something else" is happening. It was not until Einstein that the deviations of Mercury were explained. The motions of Mercury provided one of the original tests of Einstein's "general" theory of relativity. It is now agreed that Newton's theory is only "approximately" true, where the degree of approximation, however, is not closer than we can measure, but merely "close enough" for most practical problems. So you must in general distinguish when "approximately true" means "so close we can't tell the difference" and when it means "close enough that we don't, for this particular purpose, care about the difference." Newton's hypotheses are "not approximately true" on the first reading, as the Vulcan example shows, but may be "approximately true" on the second—it all depends on the purpose. In fact, Newton's theory yields hypotheses that are "close enough" for most purposes, e.g., launching artificial satellites and space probes.

6.3. A. Memory in rats is a chemical system.

B. The prediction in this experiment is that the injected rats will learn faster than those not injected. The initial conditions include such things as the fact that the one group was trained, the other not, the average time of learning, etc. There must be lots of auxiliary assumptions not stated. For example, that they got enough of the right chemicals and that the chemicals went to the right places in the brains of the injected rats. And so on.

C. The prediction follows deductively, given enough auxiliary assumptions. So Condition 1 is true. The prediction is not very precise. It only says that the injected mice will learn noticeably faster. It does not say by how much. However, the prediction is highly unusual. It is difficult even to imagine how the training time might be reduced if not by the transfer of "memory chemicals." So Condition 2 is pretty well justified too.

D. Prediction very successful. Better than they could have hoped. Memory is a chemical system.

6.6 A. PH is that light in a Newtonian particle system. WH is that light is a wave system.

B. The particle prediction, PP, is that light should travel at a precisely specified amount faster in water than in air.

C. The wave prediction, WP, is that light should travel a precisely specified amount slower in water than in air.

D. Yes. Both Conditions 1 and 2 are satisfied for the particle hypothesis since (1) PP was deduced from PH together with IC and AA and (2) the prediction states a precise value for the relative speed of light in water.

E. Yes. Same reasons as in D.

F. PP was false and WP was true.

G. The particle hypothesis is not approximately true relative to the precision of the experiment.

> If [PH and IC and AA], then PP.
> Not PP and IC and AA.
> Thus, Not PH.

H. The wave hypothesis is true (at least to within a close approximation).
> If [Not WH and IC and AA], then very probably Not WP.
> WP and IC and AA.

Thus (inductively), WH.

Chapter 7

7.1. A. (c)

B. (b)

C. (b)

D. (c) Remember that R = B − D.

7.4. (B).

7.7. Inflation is an exponential process. At 7 percent a year, the price will double in 10 years. The house would cost you $100,000.

7.9. The birth rate has remained constant or declined. The main factor is a dramatic *decrease* in the *death rate* due to drugs, chemicals, and better public health. DDT eliminated the mosquitos that caused many deaths from malaria. Vaccinations eliminated smallpox. Minor infections that used to cause death are easily treated with modern drugs. Water supplies are safer. So are food supplies. And so on.

Chapter 8

8.1. An ad hoc rescue.

8.3. The Jeane Dixon fallacy. If you predict it often enough, you are bound to be right eventually, no matter what your theory.

8.5. There is one main alternative hypothesis based on the "theory" that Easter Island is a completely terrestrial system. According to this hypothesis, the statues were carved and placed by the natives on the island. Von Daniken tries to refute this hypothesis by saying that the statues are too big to have been moved and set up by only a small number of people—less than 2000, he says. And there are no trees on the Island. And there was no active trade with other places. And so on.

If you view things this way, then the two hypotheses pretty well cover all the possibilities—the statues were the work of extraterrestrial agents or terrestrial agents. But there are many different specific versions of each hypothesis, and, in particular, many versions of the second. Von Daniken does not come close to refuting all of these. For example, there are no trees now. But what reason does he give for saying that there were no trees at the time. So the statues could have been rolled on logs after all. What about stone rollers? Or even smooth pebbles from the seashore? Fifty tons sounds like a lot. But how many men would it take to move fifty tons on rollers or loose gravel. A typical car weighs about two tons, and one man can push a car on a flat surface. So it could not take more than a few hundred men to move fifty tons, once they got it rolling. And setting it up? Build an incline of rocks and stones. Slide the statue up and over the top, feet first. The hats? Just make the incline bigger. The hats were at most ten tons. And these aren't the only ways one could imagine it being done by ordinary mortals.

In sum, the argument by elimination fails to justify the ETV hypothesis because not all the terrestrial alternatives have been justifiably refuted. Indeed, refuting them all would be almost impossible.

8.7. Just ask yourself the following question: If the hypothesis about aquarians is false, is it unlikely that one could find a number of famous scientists who had been born under that sign? Of course not. There are lots of famous scientists, and their birthdays are spread around the calendar. So Condition 2 is false and cannot be used to justify the hypothesis. The problem is that one tends to think that what is explained is that Galileo, for example, was an aquarian. On a random basis, the chances are only about 1 in 12 that any particular person is an aquarian. But the "prediction" is not that Galileo in particular is an aquarian, just that some famous scientists are. If Galileo had not been an aquarian, some other equally famous scientist would be cited.

As an alternative test, you might try looking up the birthdays of everyone who has won a Nobel Prize for science. The hypothesis predicts that "substantially" more will have been born under aquarius than under any other sign.

8.8 You need only know that heavy use of makeup can cause complexion problems. One would not need the pyramid to predict that her skin would clear.

This problem illustrates a difficult point. Your judgment as to whether Condition 2 is satisfied does depend on what you know. If you did not know that staying away from cosmetics for two weeks would tend to improve almost anyone's complexion, you might have judged that the result was unlikely if the hypothesis is false. Or worse, you might not even have been told that the woman used cosmetics when the case was presented. As a general rule, when you are presented with what seems like a good test of a highly suspicious hypothesis, hold off unless there is some way you can check to see if you have been given all the relevant information.

8.9. This example also illustrates the dangers of judging that Condition 2 is true on too little information. As a matter of fact, it has long been known that psychosomatic conditions, like some headaches, can be cured by hypnotic suggestion. For some people, apparently, the suggestion that their illness was caused by something in an earlier life is particularly effective in curing their condition. But this only proves the effectiveness of this sort of hypnotic suggestion. It proves nothing about the actual existence of "life after death." That is, there being no "life after death" does not make it any less likely that a treatment involving such hypnotic suggestions will work. From the passages quoted, it seems that even some therapists have been taken in by this fallacy.

Chapter 9

9.2.   A. A simple statistical hypothesis stating that the percentage of the islanders having lice is pretty close to 100 percent.

B. Having lice is positively correlated with being healthy.

C. Having lice is a positive causal factor for good health. Note that the everyday word "promotes" roughly means "is a positive causal factor for."

D. Being healthy is a positive causal factor for having lice. The expression "leads to" also generally indicates a positive causal factor.

9.3.   A. The poll indicates that among "Americans" there is a positive correlation between belonging to a church and claiming to have had a "religious experience."

C. The claim is that using DES is a positive causal factor for breast cancer (among women, of course).

9.5.   The information is that there is a positive correlation between being widowed and dying from a heart attack among women who are either married or widows. The conclusion is that being widowed is a positive causal factor for death from heart attack. These are two very different statements, as your diagrams should show. Correlations are symmetrical and causation is not. In addition, nothing is said about the ages of the women. The average age of widows is much greater than that of all married women, and the rate of heart attacks is greater for older people. So the correlation may be due simply to an age difference in the two groups.

Chapter 10

10.1.   A. Since Freshman and Sophomore are disjoint properties, $Pr(F \text{ or } J) = Pr(F) + Pr(J) = .50 + .15 = .65$.

D. Since S and N represent disjoint properties, $Pr(S \text{ or } N) = Pr(S) + Pr(N) = .25 + .10 = .35$.

10.2.   A. Since class and sex are not correlated, $Pr(F \text{ and } W) = Pr(F) \times Pr(W) = .50 \times .50 = .25$.

10.3. A. Check first to see that the property (F or J) is not correlated with the property M. Since they are not, Pr((F or J) and M) = Pr(F or J) × Pr(M). From problem 10.1 (A) you have Pr(F or J) = .65. So the answer is .65 × .50 = .325.

C. The property (F and W) is disjoint from the property S. From 10.2 (A) you know that Pr(F and W) = .25. Thus, Pr((F and W) or S) = Pr(F and W) + Pr(S) = .25 + .25 = .50.

10.5. You should get: Pr(3 S) = .016; Pr(2 S) = .141; Pr(1 S) = .422; and Pr(0 S) = .422.

10.6. B. .022.

10.7. B. .386.

Chapter 11

11.1. A sample of nearly 85,000 is huge. Estimates should be accurate to within a percentage point and differences of one or two percentage points should be statistically significant. The only reservation is that the sampling was done only for drivers in cities. That leaves out people in rural areas and people driving on turnpikes. It would be best to restrict all conclusions to the population of city drivers.

From the fact that 18.5 percent of motorists in cars with belts had them fastened, you could confidently (95%) estimate that between 18 percent and 19 percent of urban motorists use their seat belts. Note that the sample here must be somewhat less than 84,000 since not all cars observed would have been equipped with belts. But a high percentage of cars on the road now do have belts.

You can estimate that roughly 45 percent of Volvo drivers use their seat belts. You can't be sure of the interval around 45 percent since you are not told how many Volvo drivers were observed. A reasonable guess would be around 500, so you could be fairly confident of a percentage within three or four percentage points of 45 percent. A similar analysis applies to the reported percentage (14%) of Cadillac drivers using their belts.

You might also consider the correlation between using belts and driving a Volvo—where the comparison is with drivers of all other cars (average around 18%) or drivers of Cadillacs (14%). These differences should be statistically significant.

There is a positive correlation between being a woman and using seat belts. The difference between 20.6 percent and 17.3 percent should be SS (.05), although the correlation is not very strong (roughly .033). Note that there must be 94 drivers whose sex was not recorded.

There is apparently a positive, though not very strong, correlation between being young and using seat belts. You are not told the numbers in each group, but a 5 percentage point difference would be SS (.05) with roughly 500 of each age group.

There is a fairly strong positive correlation (roughly .153) between living on the West Coast and using seat belts. There surely were enough subjects in each group to make the difference between 12% and 27.3% an SSD (.05).

The positive correlation between driving a 1974 model car and using seat belts is taken as evidence for the causal claim that having an interlock system made, i.e., caused, people to use their belts. But it is also possible that people bought 74 models because they approved of and used seat belts. Some people put off buying cars that year in the hope that the regulation would be lifted. Which causal factor was stronger would be impossible to judge from the data given.

Finally, there apparently was not a statistically significant difference in seat belt use by those with buzzers or warning lights and those without these systems. It is possible to have causation without there being a corresponding positive correlation, but the lack of any correlation is generally a good indicator of lack of causation as well. Note, however, that the relevant sample sizes are not given.

11.4.   In everyday language, "depends on" expresses a causal relationship. So the hypothesis appears to be that liking yourself is a positive causal factor for deriving pleasure from work. Imagine the scores on the personality tests divided into two categories, "high" self-assurance and self-esteem, and "low." Similarly, imagine dividing job rankings into "positive" and "negative." Then the reported result is that there is a positive correlation between "high" self-esteem and "positive" job ranking. The two "tended to go hand in hand." You have to take it on faith that the difference in the frequencies in the two groups was SS (.05). With a sample of over 100, this is surely possible. But the correlation could just as well be taken as evidence that liking your work is a positive causal factor for liking yourself. Which is to say, neither causal hypothesis is justified. The "dependence" is merely a correlation.

11.7.   This is a typical "professional" poll. A large sample (1,603) has been "stratified" so as to represent the correct proportions of different races, religions, ages, etc., in the real population. The only obvious bias is that caused by using telephone interviews. People who don't have phones, or are not home much, will not be correctly represented in the sample.

The basic finding is the statistical hypothesis that 53 percent of adult Americans say they prefer "shared" over "traditional" marriage roles. (This is the same as saying that 47 percent prefer "traditional" over "shared.") Given the large sample, you can be confident that the percentage would be over 50 percent in the overall population.

In addition, there is a positive correlation between being under forty years old and preferring "shared" roles. With roughly 800 in each age group, the difference between 61 percent and 41 percent $(100 - 59)$ is obviously SS (.05).

There is also a positive correlation between being a working mother and believing that working women make better mothers than women who do not work. Again, though the sizes of these two groups were not given, the difference between 43 percent and 14 percent is striking, and the source of the information is reliable. They would be unlikely to report differences as being interesting if they were not statistically significant.

Finally, you might wonder whether the information is completely reliable. *Saying* one prefers the shared roles is not the same as *really preferring* them. Some people, especially younger people, may feel social pressure to give "lip service" to the ideal of "shared" roles when in fact they would prefer a more "traditional" role structure. One must guard against simply assuming that people will reveal their real preferences in a five-minute telephone interview.

11.9.   This case involves a test of a statistical hypothesis which is then used to argue against a causal hypothesis. The causal hypothesis is that being at a "critical" point in one's biocycle is a positive causal factor for mishaps such as auto accidents. The test focuses on the "null" causal hypothesis that the critical time is irrelevant. It is assumed that if critical times are irrelevant to mishaps, then accidents should occur randomly with respect to people's biocycles. Assuming just chance happenings, the probability of being at a critical time when an accident occurred was determined. It was the corresponding statistical hypothesis that was tested by examining a sample of 205 accidents. The report says that the frequency of these accidents occurring during critical periods was not statistically significant relative to the null hypothesis. If there is any difference, it is too small to be reliably detected with a sample of 205.

11.14. In this problem you have *three* categories of response to consider: (a) unqualifiedly against death penalty in any circumstances; (b) favor only for "heinous" crimes; (c) favor more extended use of death penalty. Of the roughly 300 people interviewed, 25 percent were unqualifiedly against. That yields an estimate of roughly 20 percent to 30 percent (with 95 percent confidence) for Pasadena as a whole.

In the sample, half as many men as women were unqualifiedly against. If roughly half the respondents were men, this difference was most likely SS (.05). So you could conclude there is a positive correlation (in Pasadena) between being a woman and being unqualifiedly against the death penalty. Your only worry here is that maybe only a very small number of men were interviewed. In this case, "half as many" might not be statistically significant.

For women admitting to having had an abortion, the percentages were: (a) 6%; (b) 83%; (c) 11%. For women reporting not having had an abortion, the percentages were: (a) 26%; (b) 54%; (c) 20%. Here there are two worries about the meaningfulness of these numbers. First, we are not told how many of the women interviewed admitted having had an abortion. If that number were very small, e.g., 16 (1/16 = 6%), these differences might not be SS (.05). Second, this is the kind of question that leads people to give false responses. So the reported figures could in fact be way off. One should not put too much confidence in this data. The implied negative correlation between having had an abortion and being completely against the death penalty is not well justified.

As for causal hypotheses, it is admitted that the causal relationship could go either way, or both at once. But of course there might not be a causal relationship at all.

Finally, even if the data is reliable and statistically significant, you might question the conclusion that women who have had an abortion had "a lower regard for human life" because they are less likely to be unqualifiedly against the death penalty. Indeed, the data given shows that women in the sample who had an abortion were only *half* as likely to favor *extended use* of the death penalty. So you could say that these women had *greater* regard for human life. The truth might be that their regard for human life is tempered by the realization that there are always exceptions. Thus the high percentage (83% vs. 54%) favoring capital punishment for "heinous" crimes only. You must always be wary of reading too much into a correlation—even one that is well-established.

Chapter 12

12.1 A. H: Taking 1 gm. of vitamin C a day is a negative causal factor for upper respiratory infections in humans. In everyday language: "Vitamin C prevents colds."

B. H_o: Vitamin C is causally irrelevant to the occurrence of colds. The population sampled consisted of skiers in Switzerland.

C.

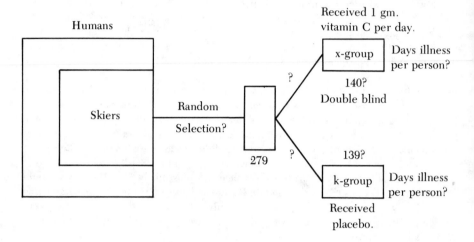

The question marks at the points where random selection should have occurred indicate that there was no *explicit* statement that the sampling was random. But given the context it is a safe assumption. Also, we don't know just how many were in each group since 279 is an odd number. Nor are we told the actual frequencies of illness in the two groups.

D.–E.

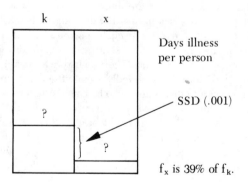

F. The conclusion is the causal hypothesis, H, as stated in A.
　　If [H_o and IC and AA], then, with Pr $\geq$ 95%,
　　there will be no SSD (.05).
　　There was an SSD and IC and AA.
　　Thus (inductively), H (i.e., not H_o)

Don't let the fact that the reported significance level was .001 confuse you. We have taken 95% as a sufficiently high probability for a good inductive argument. If the actual probability is 99.9%, that is all the better. All we need is that the significance level be 5% or less, so that the probability of reaching a true conclusion from true premises is at least 95%.

G. Since all we are told is the *relative* difference in the incidence of colds in the two groups (f_x was 61% less than f_k), we can make *no* estimate of the *effectiveness* of vitamin C in preventing colds.

H. From the report, this appears to have been a careful study with a moderately large sample. For these subjects, at least, a gm. of vitamin C a day does seem to have lowered the incidence of colds. On the other hand, the population sampled (skiers in Switzerland) is not typical, so you are left wondering if the same would be true, for example, of college students in the United States. And the fact that we are only told the relative difference in the incidence of colds diminishes the value of the report as a source of information to use in deciding whether taking vitamin C is worthwhile. You can't tell from this report whether the difference is great enough to make it worth the cost and trouble.

12.2.   A. H: Same as in (A) of Exercise 12.1.

B. H_o: Taking vitamin C is irrelevant to the incidence of colds. The population sampled consisted of potential volunteers in Salisbury.

C. Your diagram should show the following: target population of humans and subpopulation of potential volunteers; question marks at the places where random sampling should occur; sample sizes of 91, 47, and 44; indication that the x-group received 3 gm. of vitamin C a day and the k-group none; the frequency of colds in the two groups, i.e., 18/47 and 18/44 respectively.

D–E. Your diagram should show the two reported frequencies with an indication that the difference was not SSD. Note that the significance level was not stated. Assume .05 since the study was done by reputable scientists.

F. The conclusion is ΔH_o. The argument is given in Section 12.2.

G. The difference between the two reported frequencies was .41 − .38 = .03. That is pretty close to zero. On the other hand, the sample was only about 50 in each group. So the effectiveness in the population could be anything from roughly −.10 to +.10.

H. This was again a careful study. You might worry a bit because there was no mention of random selection or double-blind procedures. Maybe the experimenters were biased. Also, we don't know anything about the "volunteers." Are they vagrants? If so, their general health might be much less good than that of the population at large, and this could make a difference. Finally, inoculation might give a much stronger exposure to cold viruses than one normally gets. It could be that vitamin C is not effective against high levels of exposure, but is against lower levels.

12.3. A. In terms of the concepts we have learned, Pauling is complaining that the Δ in ΔH_o is too large. The sample size used did not provide a high probability of detecting a moderate degree of effectiveness (e.g., .10). Indeed, the effectiveness would have to be around .20 for there to have been a high probability of getting a statistically significant difference in the sample frequencies.

B. Pauling is perfectly correct. On the other hand, we can be very confident that the effectiveness of vitamin C against direct exposure to cold viruses is less than .20.

12.4. A. H: Smoking is a positive causal factor for strokes in humans.

B. H_o: Smoking is causally irrelevant to strokes. The population sampled consisted of American men over 50.

C.

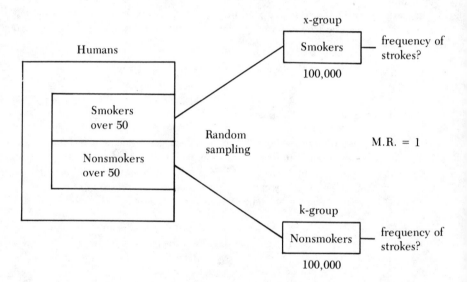

The question marks are attached to the frequency of strokes because these frequencies are not reported—only their ratio is given.

D.

No Controls	
k group	x group
1	1

MR   (table)   MR

No SSD (.05)

The expression "No Controls" appears in the diagram because the data are for all subjects. They have not been broken down into further subcategories. Neither have the two groups been explicitly matched for other characteristics. Even though you cannot record the actual frequencies, you can record mortality ratios. There is no SSD.

E. ΔH_0. Smoking is not a causal factor for strokes, that is, any effect it might have is not detectable by this study. In this case there is no obvious reason to be suspicious of the extension from men over 50 to all humans.

F. Failure to find an SSD (.05) with a very large sample indicates that the effectiveness of smoking in producing strokes is at most very low. With such a large sample, even low degrees of effectiveness have a high probability of yielding an SSD (.05).

G. This was a careful prospective study with a very large sample. The conclusion is well justified. The fact that the data does not include controls for other variables is not a major drawback because the result was no SSD. The main reason for adding extra controls is to guard against misleading SSDs, not a misleading lack of an SSD. The latter case is a possibility, of course, but in general very unlikely.

12.5. A. (a) H: Smoking is a positive causal factor for cirrhosis of the liver in humans.
(b) H_0: Smoking is irrelevant to cirrhosis.
(c) Diagram the same as for Exercise 12.4 except that the mortality ratio equals two and applies to cirrhosis rather than stroke.
(d) Diagram as in Exercise 12.4 except that the M.R. for the x-group is two.
(e) Reject H_0. You can fairly safely conclude that smoking causes cirrhosis in all

humans as well as in men over 50.

> If [H_0 and AA and IC], then with Pr = 95%,
>> there will be no SSD (.05)
> There was an SSD (.05) and AA and IC
>> Thus (inductively): Not H_0.

(f) Since you are given only mortality ratios, you can't say much of anything about the effectiveness except that it is not zero. With large samples, even a low degree of effectiveness has a good chance of producing an SSD.

(g) It is a good prospective test with a large sample. The only worry about the justifiability of the conclusion is the lack of controls for other variables.

B. Knowing that there is a positive correlation between smoking and drinking alcohol suggests that maybe the higher frequency of cirrhosis in the x-group was due to alcohol rather than smoking. Due to the positive correlation, a random sample of smokers would contain more drinkers than a random sample of nonsmokers.

C. The top portion of your diagram should say something like: "Matched for consumption of alcohol and other factors." It should show an M.R. of 1.9 for the x-group (smokers) with an indication that this is SSD (.05). You should conclude that smoking is a positive causal factor for cirrhosis. If there had been *no* SSD (.05) in the matched study, you would have concluded ΔH_0, i.e., that the effectiveness of smoking in causing cirrhosis, if any, is too low to show up with samples of 40,000. The only other possible hypothesis is that smoking *by itself* does not cause cirrhosis, but in conjunction with alcohol causes noticeably more cases than alcohol alone. To test this hypothesis you would have to compare smokers and nonsmokers with both groups containing only nondrinkers. Lack of an SSD in this case would provide justification for concluding that smoking itself is more or less irrelevant to cirrhosis.

12.6. A. H: Smoking is a positive causal factor for heart attacks among American women.

B. H_0: Smoking is causally irrelevant to heart attacks. The population sampled consisted of women in Westchester.

C.

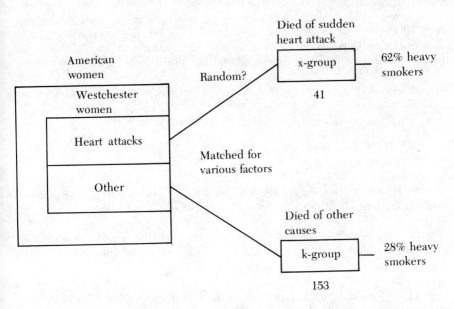

The question mark on "Random" is because it is not stated whether these 197 women were all the women in Westchester who died between 1967 and 1971 or only a selection. If a selection, how were they selected? If all the cases, are the deaths in these years a random sample of deaths in a wider time span? We don't know. There was no explicit mention of matching between "experimental" and "control" subjects, but from the fact that they were all from a single county, you can infer that they were a fairly homogeneous group. If you know anything about Westchester county, you know that they were very homogeneous indeed. Probably most of them even drove similar late model station wagons.

D.

Well-matched	
Other causes	Heart attack
	62%
28%	

Percent heavy smokers

You are not explicitly told that the observed difference is statistically significant, though the article surely presumes that it is. Recalling some of the numbers in Table 11.1, you know that a sample of 150 is very unlikely to be more than 10 percentage points off the true ratio in the population. Similarly, a sample of 50 is unlikely to be much more than 15 points off. These don't even add up to the 34 points spread in the above percentages. It is a safe conclusion that the difference is SS (.05).

E. Reject H_0. Generalizing H to all American women is less well justified, but not obviously mistaken.

> If [H_0 and IC and AA], then with Pr = 95%,
>> there will be no SSD (.05).
> There is an SSD (.05) and IC and AA.
> Thus (inductively), Not H_0.

F. Remember that this is a retrospective study. From the data given you cannot conclude anything about the effectiveness, except that it is not zero.

G. As retrospective tests go, this one is pretty good. The samples are moderately large and the subject should have been fairly well matched. And the observed difference was very large. The conclusion must be taken seriously.

12.7. This is a careful experimental study. The part dealing with "marihuana naive" subjects was double blind. You are left wondering what population was sampled to get the "volunteers" and whether the sampling could have been random. The main source of concern, however, is the small sample—only nine in each experiment. The conclusion that the performance of experienced marihuana users was "completely unimpaired" by the drug could not possibly be justified. With only nine subjects, the scores with the drug and without could have been quite different without that difference being "statistically significant." All you can say is that their performance was not drastically impaired. For the "neophytes," on the other hand, the "adverse" effects must have been quite substantial to produce a statistically significant difference with only nine subjects. This conclusion is well justified.

12.8. This study is retrospective. The conclusion before controlling for height and IQ would be that an XYY chromosome is a causal factor for violent behavior. But this conclusion could not be taken as much more than an "indication" that further research should be done. Men in prisons are hardly a random sample of men prone to violent behavior. Indeed, the further research showed ways in which the sample of men in prisons is biased. Prisons tend to be populated with men who are larger than average and have lower IQs than average—both of these factors being associated with the XYY chromosome. When these factors are controlled, the conclusion is that having an XYY chromosome is not a detectable causal factor for violent behavior. That is, its influence is not detectable with the sample size employed. You are not told what this was, but given the small differences that were reported to be statistically significant, it seems moderately large samples were used.

12.12. This is clearly a large-scale prospective study. The conclusion indicated by the original data is that the pill is a causal factor for heart attacks in women over 40.

However, this conclusion should have been regarded with some skepticism due to the failure to control for obvious other factors like cigarette smoking. The controlled data are very interesting. Comparing all nonsmokers, some on the pill and some not, the conclusion is that the pill is not an effective causal factor for heart attacks. The death rates in these two groups are the same. Looking only at people not using the pill, smokers have twice the death rate from heart attacks as nonsmokers. Smoking is a causal factor for heart attacks. The puzzling comparison is smokers with and without the pill. This difference is statistically significant. The explanation seems to be that the pill interacts with smoking to produce a much higher rate of heart attacks than smoking alone. This is possible even if the pill itself does nothing. (In fact there is a small but statistically significant difference between nonsmokers with and without the pill, but this was left out of the problem to keep it from being more difficult than it is.)

12.13. Both studies are retrospective. Given the large samples and the fact that these studies were reported in a respected medical journal, you may safely infer that the reported differences are at least SS (.05). The conclusion is that taking aspirin is a negative causal factor for heart attacks. We are not told whether any other factors were controlled, though perhaps some were. Otherwise, these are about as good a pair of retrospective studies as you will ever see.

12.14. This is an example of an interesting special type of randomized experimental design known as "high risk" design. All the subjects in both groups are of a type known to have a higher rate for the effect than the population at large. The rate of second heart attacks among people who have already had one attack is much greater than the rate for first attacks in the general population. Because a much higher rate of the effect can be expected for such groups, it is possible to get statistically significant differences with much smaller samples (and much sooner) than with other subjects. The assumption, of course, is that anything that works to prevent a second attack would also work to prevent a first. The well justified conclusion is that aspirin does prevent heart attacks, though its effectiveness is not large—something around .10 for people with one previous attack.

12.16. This is a prospective study with a large sample. There is no explicit mention of other variables that were controlled, but there is no obvious reason why roofers should not be a random sample of the general working population. The conclusion, well justified, is that Benzpyrene has very little, if any, effectiveness in producing lung cancer.

Chapter 13

13.1. See Figure 14.1, p. 306.

13.2. See the answer to Exercise 14.9.

13.3. See Figure 14.8, p. 319.

13.4. See Figure 14.4, p. 311.

13.5. A. See the answer to Exercise 14.5.

Chapter 14

14.2. A. A_4.

B. A_1.

C. (i) No. (ii) A_4. (iii) A_3. (iv) A_2.

D. (i) A_4. (ii) None.

E. (i) A_2. (ii) Best Action Rule.

F. (i) $EV(A_1) = 14/4$. $EV(A_2) = 19/4$. Choose A_2.
(ii) $EV(A_1)$ increases to $26/4$. Choose A_1.

Note that a best action will always have the highest expected value if the actions and states are independent.

14.5. A.

	New Regulations	Current Regulations
Coal	-60	-50
Nuclear	-55	-55

If they gamble, they should choose coal. To play it safe, they would choose nuclear. Both actions have the same average value so the rationalist strategy is no help. They should change the problem, perhaps by considering things other than cost (e.g., the difficulty of storing nuclear wastes).

14.6. A.

	Partner talks	Partner does not talk
You talk	-2	$-.5$
You do not talk	-3	-1

B. Talking is a best action for you.

C. Talking is also a best action for your partner. Acting rationally will land you both in jail for two years.

D. If you could confer, you might agree not to talk. On the other hand, if you begin to wonder whether you should trust your partner to keep the agreement, you are back where you started—assuming your partner also wonders whether you can be trusted.

14.9.

Country B

		Negotiate	Attack
Country A	Negotiate	3	1
	Attack	4	2

Following the best action rule, both countries will attack immediately and end up fighting a short war to achieve no more than they could by negotiation. The moral is that countries, like people, should not let relationships with their neighbors deteriorate to the point where neither can be confident that the other will not attack first. If both sides had this knowledge, they would no longer face the decision in a state of uncertainty and would not be caught in the "prisoner's dilemma."

Suggested Readings

PART I.   BASIC CONCEPTS OF SCIENTIFIC REASONING

The quickest way to learn something useful about a new concept in logic or philosophy is to look it up in the *Encyclopaedia of Philosophy*, edited by Paul Edwards. These articles will provide you with additional references if you want to read still further.

There are numerous textbooks in logic now published. Simply by browsing through those available at your library you should be able to find some that suit your needs and interests. Wesley Salmon's *Logic* covers about the same material as this text, but in somewhat greater detail. It is relatively short and available in paperback. For many years the most widely used text has been Irving Copi's *Introduction to Logic*. For a somewhat more advanced and technical treatment geared to scientific applications I would recommend Patrick Suppes' *Introduction to Logic*. For a less technical book that emphasizes practical reasoning skills see Michael Scriven's *Reasoning*. If you are interested in more theoretical issues, for example, What makes a valid argument valid? you might look at W. V. Quine's *Philosophy of Logic* (a short paperback) or Peter Strawson's *Introduction to Logical Theory*.

Modern discussions of the justification of conclusions by inductive arguments go back to Section VII of David Hume's *Inquiry Concerning Human Understanding* (1748). For more recent versions of Hume's arguments see Chapter VI of Bertrand Russell's *Problems of Philosophy* or Wesley Salmon's *Foundations of Scientific Inference*. Richard Swinburne has collected a number of recent essays in a book titled *The Justification of Induction*. My own views on inductive justification are developed in a paper entitled "The Epistemological Roots of Scientific Knowledge," which appears in *Induction, Probability and Confirmation: Minnesota Studies in the Philosophy of Science*, Vol. 6, edited by Grover Maxwell and Robert Anderson.

PART II.   REASONING ABOUT THEORIES

The account of theories used in this text has been developed and elaborated by a number of recent authors. For a brief and relatively nontechnical account see Patrick Suppes' essay, "What Is a Scientific Theory?" in *Philosophy of Science Today*, edited by Sidney Morgenbesser. Another version of the approach, presented at a more advanced level by Bas van Fraassen, may be found in the book *Paradigms and Paradoxes*, edited by Robert Colodny. The approach is developed at length in Wolfgang Stegmüller's recent book *The Structure and Dynamics of Theories*. Stegmüller draws heavily on Joseph Sneed's *The Logical Structure of Mathematical Physics*. Except for the first two chapters, the latter book is not for beginners.

The development of philosophical thought about the nature of theories is nicely reviewed by Frederick Suppe in the introduction and postscript to his book *The Structure of Scientific Theories*. For an elementary presentation of the older, but still more widely accepted view, see Rudolf Carnap's *Philosophical Foundations of Physics* or Carl Hempel's *Philosophy of Natural Science*. The Carnap book is both more general and more elementary than the title suggests. Hempel's book is shorter, more elementary, and in paperback. Ernest Nagel's *The Structure of Science* contains a classic presentation of the standard view, though at a more advanced level.

For a brief look at the historical development of Newton's physics, see Richard Westfall's *The Construction of Modern Science*. Gerald Holton and Duane Roller have written an elementary and historically sensitive text, *Foundations of Modern Physical Science*, which covers Newtonian theory. The centuries of debate over the nature of forces are reviewed in Max Jammer's *Concept of Force*. Chapter 12 of Patrick Suppes' *Introduction to Logic* contains an up-to-date definition of a classical mechanical system, but you have to know some mathematics to read it.

Mendel's own paper contains one of the easiest and best introductions to Mendelian genetics. It is reprinted, for example, in *Classic Papers in Genetics*, edited by James Peters. For a modern treatment consult any recent text such as Monroe Strictberger's *Genetics*. For good philosophical discussions of genetics and modern biology see David Hull's *Philosophy of Biological Science* or Michael Ruse's *The Philosophy of Biology*.

Mary Hesse's *Models and Analogies in Science* defends the view that analogies play a major role in the justification of scientific hypotheses.

For other accounts of the testing and justification of scientific hypotheses, see the books by Carnap, Hempel, Nagel, and Salmon mentioned above. A somewhat different approach which denies a sharp distinction between contingent and noncontingent statements is presented by W. V. Quine and Joseph Ullian in *The Web of Belief*. I would especially recommend the opening essay in Karl Popper's *Conjectures and Refutations*. Brian Skyrms' *Choice and Chance* provides a nice compact introduction to the problems of justifying scientific hypotheses.

A number of good examples, including the phlogiston theory, are explored in the *Harvard Case Histories in Experimental Science*, edited by James Conant. A good source of biological examples is *Hypothesis, Prediction, and Implication* by Jeffrey Baker and Garland Allen. Examples in psychology may be found in any good text such as James McConnell's *Understanding*

Human Behavior. More examples from the recent history of geology may be found in *Continents Adrift*, edited by J. Tuzo Wilson.

There is a serious and responsible critique of *The Limits to Growth* entitled *Models of Doom*, edited by H. S. D. Cole and others.

Books arguing the case for extraterrestrial visitation either now or in the ancient past appear regularly. Serious critiques are hard to find. The official report, *Scientific Study of Unidentified Flying Objects*, edited by Edward Condon, contains much interesting information, but is very poorly organized and difficult to read. A much more readable, though less extensive, source is *UFO's: A Scientific Debate*, edited by Carl Sagan and Thornton Page.

A somewhat dated but still fascinating source of popular scientific theories is Martin Gardner's *Fads and Fallacies in the Name of Science.*

PART III. CAUSES, CORRELATIONS, AND STATISTICAL REASONING

The analysis of correlation and simple causal hypotheses presented in Chapter 9 is relatively new. I have elaborated this approach in a paper entitled "Causal Systems and Statistical Hypotheses," which appears in *Applications of Inductive Logic*, edited by L. J. Cohen and Mary Hesse. This paper includes an analysis of simple causal hypotheses that applies if the individuals in the population are regarded as stochastic rather than deterministic systems.

The philosophical literature on statistical inference tends to be quite general and abstract. I have reviewed this literature in "Foundations of Probability and Statistical Inference," which appears in *Current Research in the Philosophy of Science*, edited by Peter Asquith and Henry Kyburg. This review includes nearly 200 references. A representative sample of philosophical works would include *Logic of Statistical Inference* by Ian Hacking, *The Logical Foundations of Statistical Inference* by Henry Kyburg, and *Inference, Method, and Decision* by Roger Rosenkrantz.

Texts on probability and statistics are even more numerous than texts on logic. If you wish to dig more deeply into these subjects, you should have little difficulty finding a text that meets your needs. However, most texts are designed to teach you how to do statistics, not merely how to understand statistical conclusions. One book that emphasizes concepts over technique is *The Nature of Statistics*, by W. Allen Wallis and Harry Roberts. This is the first part of a longer book, *Statistics: A New Approach*, by the same authors. An interesting book that teaches basic ideas through examples is *Statistics: A Guide to the Unknown*, edited by Judith Tanur. The best systematic text I know that uses only elementary arithmetic is *Basic Concepts of Probability and Statistics* by J. L. Hodges and E. L. Lehman. A good text at a more advanced level is *Introduction to Statistical Theory*, by P. G. Hoel, S. C. Port, and C. J. Stone. In addition, there are texts in statistics written for many special fields, for example, biology, psychology, sociology—even history.

For more information about the use of marijuana and other drugs, see Erich Goode's *Drugs in American Society*. A good source of data about cigarette smoking is *Tobacco and Your Health* by Harold Diehl.

Finally, there are several short paperbacks dealing with statistical fallacies. These provide a source of often quite amusing examples. One is Darrell Huff's *How To Lie with Statistics*. Huff and Irving Geis have recently written a new book along similar lines entitled *How To Take a Chance*. Another somewhat more systematic book is Stephen Campbell's *Flaws and Fallacies in Statistical Thinking*.

PART IV.   VALUES AND DECISIONS

Decision theory, like statistical inference, has not been a part of the standard philosophy curriculum. Most philosophical work on decision making is quite advanced. A good example is Richard Jeffrey's *The Logic of Decision*.

Decision theory is much studied in business schools and often combined with other subjects such as operations research. For a short, elementary text covering roughly the same material as Part IV of this text, see *The Structure of Human Decisions* by David Miller and Martin Starr. Miller and Starr emphasize business examples. The classic elementary text written from a mathematicians' point of view is *Elementary Decision Theory* by Herman Chernoff and Lincoln Moses. For a somewhat different point of view, though still elementary and emphasizing business examples, see *Decision Analysis* by Howard Raiffa.

INDEX

---

* Asterisk indicates material in an exercise.
[1] Italics indicates page on which a concept is defined.